STP 1255

Airflow Performance of Building Envelopes, Components, and Systems

Mark P. Modera and Andrew K. Persily, Editors

ASTM Publication Code Number (PCN):
04-012550-10

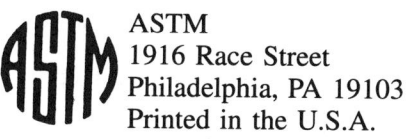

ASTM
1916 Race Street
Philadelphia, PA 19103
Printed in the U.S.A.

Library of Congress Cataloging-in-Publication Data

Airflow performance of building envelopes, components and systems /
Mark P. Modera and Andrew K. Persily, editors
(STP ; 1255)
Papers presented at a symposium held in Dallas/Fort Worth Airport,
TX on 10–11 October 1993; sponsored by ASTM Committee E-6 on
Performance of Building Constructions and its Subcommittee E06.41 on
Infiltration Performances.
Includes bibliographical references and index.
ISBN 0-8031-2023-0
1. Buildings—Airtightness—Congresses.   2. Air flow—Congresses.
3. Ventilation—Congresses.   4. Indoor air pollution—Congresses.
I. Modera, Mark P., 1956–   .   II. Persily, Andrew K.   III. ASTM
Committee E-6 on Performance of Building Constructions.
Subcommittee E-6.41 on Infiltration Performances.   IV. Series: STP
special technical publication ; 1255.
TH6025.A37   1995
693'.83—dc20

95-23430
CIP

Photocopy Rights

Peer Review Policy

Each paper published in this volume was evaluated by three peer reviewers. The authors
addressed all of the reviewers' comments to the satisfaction of both the technical editor(s) and the
ASTM Committee on Publications.

The quality of the papers in this publication reflects not only the obvious efforts of the authors
and the technical editor(s), but also the work of these peer reviewers. The ASTM Committee on
Publications acknowledges with appreciation their dedication and contribution to time and effort on
behalf of ASTM.

Printed in Fredericksburg, Virginia
September 1995

Foreword

This publication, *Airflow Performance of Building Envelopes, Components, and Systems,* contains papers presented at the symposium of the same name, held in Dallas/Fort Worth Airport, TX on 10–11 October 1993. The symposium was sponsored by ASTM Committee E–6 on Performance of Buildings and its Subcommittee E06.41 on Infiltration Performance. Mark P. Modera, Lawrence Berkeley Laboratory, University of California, Berkeley, and Andrew K. Persily, National Institute of Standards and Technology (NIST), U.S. Department of Commerce, served as co-chairmen of the symposium and are editors of the resulting publication.

Contents

Overview

Airflow in buildings has been a topic of interest since people began to build shelters. One of the earliest issues was orientation with respect to the wind, including both augmentation and reduction of indoor air motion. The issues of interest expanded to include thermal buoyancy effects, as well as the provision of air for ventilation and combustion as building shells became more airtight. The issues of current interest encompass natural and mechanical ventilation systems, pressure control, control and dilution of air pollutants, control of the thermal environment (including distribution of ventilation air, heat and cooling), and control of airflow through leaks in the building envelope. As the numerous types of airflow in buildings have been studied over the years, the level of detail of the knowledge available and required relative to those types has continued to increase.

ASTM Subcommittee E06.41 (Infiltration Performances) began writing consensus standards on measurement techniques for air infiltration in 1975, focusing on measuring the airflow through building envelopes with tracer gases (current version: E 741-93, Test Method for Determining Air Change in a Single Zone by Means of Tracer Gas Dilution), and the air leakage characteristics of building envelopes (current version: E 779-87, Test Method for Determining Air Leakage Rate by Fan Pressurization). Since 1975, ASTM Subcommittee E06.41 has written a number of additional standards addressing more of the details associated with understanding airflow and air leakage in buildings (E 1186-87, Practice for Air Leakage Site Detection in Building Envelopes; E 1258-88, Test Method for Airflow Calibration of Fan Pressurization Devices; E 1465-92, Guide for Radon Control Options for the Design and Construction of New Low-Rise Residential Buildings; and E 1554-94, Test Method for Determining the External Air Leakage of Air Distribution Systems by Fan Pressurization).

In parallel with the efforts of E06.41, ASTM Subcommittee E06.51 (Component Performance of Windows, Curtain Walls, and Doors), has been addressing the airflow performance of various building components, starting with the performance of various building shell components in the laboratory (E 283-91, Test Method for Determining the Rate of Air Leakage through Exterior Windows, Curtain Walls, and Doors under Specified Pressure Differences Across the Specimen; and E 1424-91, Test Method for Determining the Rate of Air Leakage through Exterior Windows, Curtain Walls, and Doors under Specified Pressure and Temperature Differences Across the Specimen), and extending to the installed performance of those components (E 783-93, Test Method for Field Measurement of Air Leakage through Installed Exterior Windows and Doors).

To aid in the consensus standards process, as well as to provide concentrated documentation of progress in the area of building airflow and air leakage, ASTM Subcommittee E06.41 has sponsored a technical symposium (and subsequent Special Technical Publication) approximately once every five years. The first of these, Building Air Change Rate and Infiltration Measurements (*ASTM STP 719*), held in 1978, was focused principally on measurement techniques, with some limited data taken by researchers. The 1984 symposium, Measured Air Leakage of Buildings (*ASTM STP 904*) was focused on relatively large sets of field data. The third symposium, Air Change Rate and Airtightness in Buildings (*ASTM STP 1067*) was held in 1989 and was divided fairly evenly between analyses of measurement techniques and compilations of field measurement results.

The symposium that forms the basis of this book attempted to document current research efforts for and knowledge of airflow through the building envelope as a whole, a topic that has been the focus of the earlier symposia put together by ASTM Subcommittee E06.41, as well as to address the growing interest and concerns associated with airrflow through the various components that make up that building envelope (e.g., windows), and the systems that serve to distribute air (and heat or cooling) within a building. Like the 1989 symposium, this symposium was split between analyses of measurement techniques and summaries/analyses of field data, with somewhat more emphasis on the latter. This symposium served to bring together the expertise within at least two ASTM subcommittees: E06.41 and E06.51.

The symposium was divided into four sessions, each session focusing on a particular aspect of building airflow. These four sessions included, in order of presentation: (1) Air Movement, Ventilation, and Indoor Air Quality, (2) Window Air Leakage, (3) Envelope Heat and Mass Transfer, and (4) Envelope and Distribution System Leakage.

The first symposium session, Air Movement, Ventilation, and Indoor Air Quality, was chaired by Andy Persily, and included five papers. The major focus of this session was building airflows under more complex circumstances, in particular larger buildings with multiple zones, and pollutant entry from soil gases. The Palmiter et al. paper provided one of the few published sets of simultaneous measurements of airflow through all pathways in a multifamily building. The Dols and Persily paper compared and contrasted ventilation measurement results obtained for a large office building with several different techniques, whereas the Shaw and Reardon paper reports on changes in the airtightness of six office buildings over a period of about 20 years. The Kozik et al. and Williamson et al. papers explored two aspects of soil gas entry and dilution in the field, the Kozik paper focusing on the impacts of duct leakage and resulting house pressurization/depressurization on radon entry and dilution, and the Williamson paper focusing on whether or not slab flooring provides an effective barrier to soil-gas entry.

The Window Air Leakage session, chaired by Mark Modera, was comprised of four papers addressing the airtightness of window systems. The papers in this session focused on the various issues associated with understanding how windows will leak in the field as opposed to under ideal conditions in a laboratory. The first of the Kehrli papers described the relatively recently-approved ASTM standard for window air leakage, E 1424, which includes the impacts of temperature differentials in laboratory measurements of window air leakage. The second Kehrli paper was a simulation-based analysis of the energy implications of window air leakage for several window products placed in four different building designs in various U.S. climates. The Louis and Nelson paper outlined an approach for quantifying window perimeter leakage, which is not accounted for in current ASTM standards. The Proskiw paper addressed the issue of perimeter leakage for the particular case of residential construction, including cost data for various types of rough opening seals.

The third session, Envelope Heat and Mass Transfer, was chaired by William Brown, and consisted of four papers. In particular, two of the papers addressed airflow and moisture in the envelopes of manufactured housing, and two papers addressed the interactions between conduction heat flows and airflows in the building envelope. The Tenwolde et al. paper addressed the impacts of airflow on moisture accumulation in the walls of manufactured homes, and the Burch paper addressed the same problem for the roof cavities of those structures by means of computer simulation. The Claridge et al. and Jones et al. papers dealt with the important issue of how airflow through building walls interacts with fourier conduction within those walls.

The final session, Envelope and Distribution System Leakage, was chaired by David Saum. This session contained three papers analyzing measurement techniques, and one paper focusing on field measurement results. The Levin et al. paper presented comparisons of various air leakage standards, focusing on the impacts of different test pressures and variations in flow exponents. The Proskiw paper presented field data on variations in whole building airtightness over a three-year period. The Sherman and Palmiter paper presented an error analysis of various single-zone fan pressurization measurement protocols and analysis procedures. The Modera paper compared and contrasted the performance of two alternative techniques for measuring duct system leakage that were recently incorporated into a new ASTM test method (E 1554).

As was the case with earlier ASTM symposia on building airflow, the papers presented and published herein serve both to enlighten us on the progress that has been made in our understanding of building airflows, as well as to point out where that understanding is lacking. This is true both for our ability to predict those flows and their impacts by means of computer modeling, as well as for our ability to standardize measurement techniques that can provide accurate airflow characterization data in a minimum amount of time. This symposium represents a small step forward in our continuing quests to improve our understanding of building airflow, and to use that understanding to provide the building community with standards that can provide accurate data as cost-effectively as possible.

The publication of this book was made possible by the efforts of a large number of individuals. We, as the editors and organizers of this symposium, would like to thank all of those individuals, starting with the authors, who provided the substance upon which this book is based. We would also like to thank the ASTM editorial and conference organization staff who took on the thankless task of assuring that the symposium and the publication of this book actually came to pass. Finally, we would also like to thank the session chairmen for their efforts.

Mark P. Modera

Lawrence Berkeley Laboratory
    University of California, Berkeley, CA
    symposium co-chairman and co-editor

Andrew K. Persily

National Institute of Standards and Technology
    U.S. Department of Commerce
    Gaithersburg, MD
    symposium co-chairman and co-editor

Air Movement, Ventilation, and Indoor Air Quality

Larry Palmiter,[1] Jonathan Heller,[1] and Max Sherman[2]

Measured Airflows in a Multifamily Building

REFERENCE: Palmiter, L., Heller, J., and Sherman, M., **"Measured Airflows in a Multi-family Building,"** *Airflow Performance of Building Envelopes, Components, and Systems, ASTM STP 1255,* Mark P. Modera and Andrew K. Persily, Eds., American Society for Testing and Materials, Philadelphia, 1995, pp. 7–22.

ABSTRACT: A method has been developed to measure real-time airflow in multifamily buildings. This method uses a multi-tracer measurement system (MTMS) and simultaneous measurements of pressure, temperature, and environmental conditions. These measurements are evaluated along with the results of blower door and flowhood fan tests to develop a complete picture of the airflow patterns in multifamily buildings due to temperature differences, wind, and mechanical ventilation.

Six units in an unoccupied three-story multifamily building in Portland, Oregon were tested for a period of eight days during February and March 1992 using this method. The apartments were equipped with timer-controlled ventilation fans that were set to come on at regular two-hour intervals throughout the test period, enabling a thorough evaluation of the effect of various fan conditions on the flow of air from the outside as well as between units.

The average ventilation rate from natural driving forces alone (wind and stack effects) was about 40 m^3/h per unit, or 0.22 air changes per hour (ACH). The average ventilation rate during the periods with all ventilation fans running was 75 m^3/h (0.41 ACH). There was also a significant amount of inter-apartment airflow, which was dominated by flow from lower units to the unit directly above from temperature-driven stack effects. Operating all exhaust fans together had little effect on the interzone flows; however, operation of a single exhaust fan significantly increased the interzone flow to that apartment from all adjacent units. The percentage of total airflow that was outdoor air was found to be lower for higher units than for lower units.

KEYWORDS: infiltration, tracer gas techniques, multifamily buildings, indoor air quality, ventilation, multizone airflow

Measurement of airflow in residential buildings is important for obtaining information about heating and cooling requirements, humidity levels, and indoor air quality. With increasing emphasis on energy efficiency, new building standards are mandating tighter buildings with lower infiltration levels. This raises the question of whether the fresh air necessary to maintain a healthy indoor environment is being provided. In the case of multifamily buildings where the air flowing between units carries smoke, dust, biocontaminants, and other pollutants, this question is especially difficult to answer. Air infiltration should only be counted as ventilation when it comes directly into the living space from outdoors without passing through another already polluted environment.

[1] Senior scientist and research engineer, respectively, Ecotope, 2812 E. Madison, Seattle, WA 98112.
[2] Energy Performance of Buildings Group, Energy and Environment Division, Lawrence Berkeley Laboratory, University of California, Berkeley, CA 94720.
This work was sponsored by the Residential Ventilation Consortium operated by the Electric Power Research Institute and also supported by the Assistant Secretary for Conservation and Renewable Energy, Office of Building Technology of the U.S. Department of Energy.

Single-family buildings have been extensively tested, and many methods have been developed to measure airflow in buildings which can be treated as a single zone. Less extensive testing has been done in multifamily buildings where the interzonal nature of airflows complicates the analysis. Most studies of interzonal flows in multifamily buildings have yielded only average results. This paper presents the results of real-time multizone flow measurements made using a multiple tracer gas technique at a multifamily building in Portland, Oregon. The intent of this paper is not to fully analyze the flow data from this test case, but simply to use the case as an example of the type of data which can be collected by this method. The multizone tracer gas measurement system used for this test is described in the next section, followed by an explanation of the experimental instrumentation and procedures used at the test site. The test site and environmental conditions during the test period are discussed in later sections. A summary of the measured data is provided, along with a brief summary and discussion of the findings. The conclusion draws out some of the interesting aspects of the airflow in this building which were discovered during testing. A brief description of the physical characteristics of the test site is given in Table 1.

Multi-Tracer Measurement System

The multi-tracer measurement system (MTMS) used in this study was developed at Lawrence Berkeley Laboratory. A different tracer gas is injected into each zone under study using mass flow controllers. The injection of tracer gases and the measurement of gas concentrations are controlled from a personal computer. This allows for rapid measurement of

TABLE 1—*Physical characteristics of testing site.*

Building Characteristics

Location	Portland, Oregon
Year built	1992
Energy efficiency	Built to Utility Super Good Cents specifications
Number of stories	3
Number of units in building	21
Number of units tested	6
Foundation	Slab on grade
Entry to units	Outside recessed stairwell

Arrangement of Test Units

Two stacks of three units, mirror images. All apartments have south and north facing walls. The test units are in the center of a long row of apartment units, with double fire walls on both sides.

Unit Characteristics

Average floor area	77 m² (829 ft²)
Average volume	183 m³ (6462.6 ft³)
Ceiling height (not including cathedral ceilings)	2.3 m (7.54 ft)

Mechanical Equipment

Automatic exhaust system	Timer-controlled laundry fan
Makeup air	Operable inlet ports in window sashes
Other ventilation fans	Two bath fans, range hood fan
Heating system	Electric resistance wall heaters

Test Period

12:00 noon, February 22, 1992 to 12:00 noon, March 1, 1992

all tracer gases in all zones. Flows between all zones and the outside can then be calculated from the measures of tracer concentrations. At the same time that the concentrations are being measured, temperature and pressure measurements are taken in each zone, and environmental weather data are collected to assist in the flow analysis.

The most demanding aspect of the MTMS is the accurate measurement of tracer gas concentrations. This is accomplished with a residual gas analyzer quadrupole mass spectrometer. Air is selected for analysis from the various zones by a manifold and pumped into a vacuum chamber which houses the spectrometer. The air is ionized, and the resulting positive ions are separated by their charge-to-mass ratio by a variable electric field. The electric field is controlled to select out mass particles which correspond to the various tracer gases being measured. The ions corresponding to the desired tracer gas are then directed at an electron multiplier for detection.

The spectrometer has an absolute accuracy of about 1% of any given reading; however, at the working pressures, electronic noise produces errors of about ± 20 parts per billion (ppb). Most tracer gases can be calibrated to $\pm 1\%$ down to 2 parts per million (ppm) where the noise dominates the uncertainty. Typically, tracer gas concentrations of about 20 to 50 ppm are used in each zone in multifamily buildings so that the flow of air to neighboring zones can be accurately measured. Reference *1* provides a more detailed description of the MTMS.

Methodology and Data Analysis

Different tracer gases were injected into each of six zones using six separate mass flow controllers. Each apartment unit was treated as a separate zone. Mass flow was measured to within 1% accuracy. The gases used in this analysis were helium, sulfur hexafluoride (SF_6), and four types of refrigerants: F12, F13B1, F22, and F116.

In this study, the constant injection method was used whereby the tracers are injected into the zones at a constant rate throughout the test period. Tracer gases were transported to each zone via tubing from the MTMS and injected at three different locations in each zone through a manifold arrangement. Each injection point was attached to an oscillating fan which ran continuously during the test period. These fans served to disperse the tracer into the zone and to improve mixing. The air was sampled from four different locations in each zone and gathered together in a manifold to provide a single air sample to the measurement equipment. Sampling at multiple locations also helps to reduce errors associated with incomplete mixing.

In addition to taking concentration measurements in each of the apartment units, outdoor air was sampled well away from the building at four-minute intervals. This allowed for the measurement of any background levels of the tracer gases in the environment. The MTMS is temperature and pressure sensitive, so a calibration chamber was set up outside and the MTMS was recalibrated throughout the test period using as a baseline the background level of helium in the atmosphere. The calibration chamber consisted of a 6 m (20 ft) length of pipe with a fan located on the downstream end, continuously pulling air through the pipe. Every three hours, helium and one of the other tracer gases were injected into the upstream end of the pipe at a known flow rate, and the concentrations were measured at the other end. The helium reading was then adjusted for the background level and used to calculate the flow rate through the pipe. The reading for the other tracer could then be recalibrated. This resulted in a new calibration for each gas approximately every 18 h. The mean calibration for each tracer gas over the duration of the test period was used to correct the concentration measurements. The relative error in the flow calibrations (standard deviation divided by the mean) was calculated for each gas using all of the calibrations derived during the test period. This error ranged from 2% to 9% for the six gases. The average relative error of all gases together was about 7%. The propagation of error through the multizone matrix equations is very complex, a detailed discussion of which is beyond the scope of this paper. In general, errors in the concentration determination will lead to comparable errors in the calculated flows.

Reference 2 contains a detailed accounting of multizone flow theory and calculations.

Temperatures and Pressures

Temperature readings were taken in each apartment continuously during the testing period. The temperatures were measured with thermocouple wires running from the center of each zone to a personal computer. Temperatures in every zone were recorded each time that any zone was sampled.

Pressure transducers were installed across all of the interior walls between zones, from the top zones to the attic, and at the floor of the ground floor zones to the outside. This allows for a calculation of total stack and wind pressures. These measured pressures can then be used with flow data from the MTMS to help evaluate irregularities in the data.

Data Analysis

The concentration readings were converted to flows using a matrix deconvolution program developed by the primary author. The time series concentrations were filtered using Savitsky-Golay filters [3] to estimate a smoothed concentration and its derivative at each time step. These filters are equivalent to fitting a quadratic by least squares to each successive group of nine points and calculating the predicted central value and its derivative. The flows were then calculated by inverting the matrix equations, resulting in 36 interzone flows at 4-min intervals. The flows to and from outdoors were calculated from flow balances on each zone. The flows were then averaged to 15-min data.

An integral number of days was chosen for this analysis, starting some time after the initial setup to allow for equilibration of the site and instrumentation, and ending before any significant interference from other testing began. We used eight complete days of data: from noon on February 22 to noon on March 1, 1992. For this analysis, the 15-min data were averaged by time-of-day to produce a single average daily profile. This procedure maximizes the signal-to-noise ratio for studying the effects of periodically operated fans.

Test Building Description

Building

The tests were done in a multifamily building in the vicinity of Portland, Oregon, immediately after the building was constructed, but before occupancy. The test period was a full eight days in February and March 1992. The building is three stories high and contains 21 apartment units. All of the apartments have exterior entrances accessed by recessed stairwells. The building is of standard wood frame construction, with a slab-on-grade foundation. The north wall of the ground floor units is buried about 1 m below grade. The floor of every unit has about 4 cm of lightweight concrete and carpeting.

Only six units in the building were tested. These constituted two stacks of three units, accessed by a single exterior stairwell in the center of the building. The units are separated from the rest of the building on both sides by double fire walls. It was assumed for the purpose of this testing that negligible airflow occurs across these fire walls. Therefore, for airflow evaluation and all other references in the remainder of this paper, the six test units are considered as a separate building. For the purposes of this report, the units were arbitrarily numbered from top to bottom, 1 to 6.

All of the test units have identical 77 m² floor plans. Each unit consists of two bedrooms, two bathrooms, a kitchen, living room, and dining room. There is also a small entry area and a hallway with a utility closet which houses the clothes washer and dryer. A rough sketch of the floor plan is included as Fig. 1. The top floor units have cathedral ceilings in the living room, dining room, and entry, so they have slightly larger volumes. The volume of each unit was calculated from interior dimensions, including all interior partition walls and closets. The volume of each unit on the first two floors is 178.5 m³ and the volume of each top floor unit is 192.5 m³. The volume used to characterize the entire test building was their sum, 1099 m³.

Heating System

The building is heated by 750 W fan-forced electric resistance wall heaters located in each bedroom, living room, and dining room. The heaters are controlled by wall-mounted

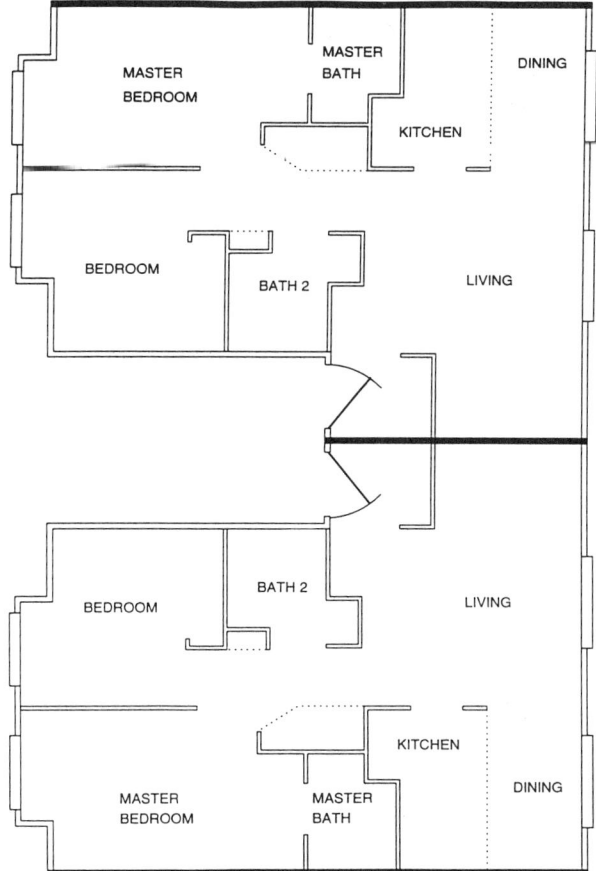

FIG. 1—*Typical floor plan—Site B (not to scale).*

thermostats which were set at a constant temperature for the entire test period. The average measured indoor temperature for all of the units during the testing was 26.1°C.

Ventilation System

This building was built under an energy conservation incentive program called the Super Good Cents program. The 1988 Super Good Cents Specifications [4] have since become the basis for the new Washington State Energy Code. The specifications require high levels of insulation and air-sealing procedures during construction. The air-sealing procedures are intended to reduce natural infiltration levels and in turn reduce heating energy requirements. The sealing specifications include caulking around doors and windows, sealing all plumbing and electrical penetrations, and employing polyvinyl vapor barriers or airtight drywall construction methods. Compliance to tightness specifications is checked by visual inspection rather than performance testing. The field test crew found that the entire building had a leakage rate of about 6.35 air changes per hour at 50 Pa pressure.

Tightly built apartments can have problems associated with poor indoor air quality due to the low level of natural ventilation. To attempt to deal with this the specifications require that automatic ventilation systems be installed in each apartment. In these apartments the automatic ventilation system consists of an exhaust fan in the laundry closet, rated to deliver 85 m³/h at 25 Pa. These fans are controlled from 24 h timers which can be set by the occupants to come on at regular intervals to ventilate the apartments.

The apartments are also equipped with exhaust fans in each bathroom and a range hood exhaust in the kitchen. These are designed to provide intermittent spot ventilation to exhaust high levels of pollutants at their source, such as smoke, moisture, and odors. The fans are operated by manual switches and cannot be relied upon to provide overall ventilation of the apartment, since they are on only intermittently during the day. These fans were not activated during the test period, so the fans referred to in the remainder of this paper are the laundry fans described above which were set to operate automatically during the test period.

To test the ventilation effectiveness of various operation strategies during the test period, the timers were set so that periodically all of the fans would operate at the same time, while at other times they would operate separately. All of the ventilation fans were set to operate from 1 to 3 a.m. and from 1 to 3 p.m. every day. In addition, each fan in units 2, 4, and 6 was set to operate for a single 2 h period each day by itself. The fan in unit 2 was set to operate by itself from 5 to 7 a.m., the fan in unit 4 operated alone between 5 and 7 p.m., and the fan in unit 6 operated alone between 9 and 11 a.m. This allowed for comparison of airflows between periods when all fans were on, periods with only one fan running, and times when all fans were off.

The design specifications require fresh makeup air to be delivered to the primary living spaces. This apartment building meets that part of the specifications by providing slot vents in the tops of the vinyl window frames in each bedroom and the living room. These vents consist of sixteen 1.27 cm-diameter holes which can be opened or closed via a small occupant-controlled slider. The total free area provided by the vents is about 0.002 m² in each bedroom and living room. The vents were closed at the start of the test period and all of them were opened about halfway through the test period. No statistically significant change in airflow could be detected between the periods with opened or closed slot vents. Therefore, the data were averaged over the combined periods of open and closed slot vents.

Environmental Conditions

Environmental conditions of wind and temperature create pressure differences across the building walls which drive air infiltration. These units are very well shielded to the wind due to the fact that they have other apartment units on both sides of them. They are exposed only on the front and rear surfaces, as well as the ceiling of the top floor units. Wind speed and wind direction were monitored at the site and averaged over 15-min intervals throughout the testing period. The maximum site wind speed was 5.8 m/s and the average wind speed was less than 1 m/s. Wind speeds measured at the National Weather Service site at the Portland International Airport during the same period showed a maximum of 11.3 m/s with an average speed of 3.5 m/s.

The outdoor temperature at the site was measured with a radiation-shielded and aspirated thermocouple, and averaged 10.9°C. Average indoor temperature of the six units, measured with thermocouples, ranged from 24.5 to 27.5°C with an overall building average of 26.1°C. The difference between indoor and outdoor temperatures averaged 15.2°C, typical of average heating season values in the Pacific Northwest. The environmental conditions during the test period are summarized in Table 2.

A diurnal variation can be observed in the natural infiltration rate of the entire building in this study, with slightly greater rates at night and in the early mornings when the outside temperature is at its minimum.

Results

Zone and Building Flows from Outdoors

The average ventilation rate from natural driving forces (stack and wind effects) alone was about 240 m³/h for the entire building, or 0.22 air changes per hour (ACH). The average ventilation rate during the two periods with all of the ventilation fans on (1 to 3 a.m. and 1 to 3 p.m.) was 448 m³/h (0.41 ACH). The most widely accepted American standard for residential ventilation rates was put forward by the American Society of Heating, Refrigerating, and Air-Conditioning Engineers (ASHRAE) [5]. This standard states that residential buildings should have a minimum of 0.35 ACH of ventilation or 25.5 m³/h per occupant, whichever is greater. Since these units have two bedrooms, it is reasonable to assume that they could easily house three occupants. To meet the minimum ASHRAE standard, these

TABLE 2—*Environmental conditions during 8-day test period.*[a]

Temperatures	°C	°F
Site indoor average	26.1	78.9
Site outdoor average	10.9	51.6
NWS (Portland) average	10.3	50.5
Site ΔT	15.2	27.3

Wind speeds	m/s	mph
Site maximum	5.8	13.0
Site average	0.9	2.0
NWS maximum	11.3	25.3
NWS average	3.5	7.8

[a] Data summarized from 12:00 p.m. 2/22 to 12:00 p.m. 3/1, 1992.

units would therefore require ventilation rates of about 76 m³/h or an average of 0.42 ACH. By this standard, the apartments are seriously underventilated when only natural infiltration provides ventilation. Even with all of the ventilation fans operating continuously, these apartments on average do not quite meet the minimum standard assuming three occupants per unit.

Figure 2 shows the average flow of air from outside into each unit. This information is summarized in Table 3. It should be noted that the top and bottom floors received more outside air than the middle floor units. This is most likely due to a combination of factors. The lower floor units do not receive an appreciable fraction of their air from other units. Since they are at the bottom of the stack, the pressure forces across their walls bring in mostly outside air. The middle floor units have very little surface area exposed to the outside since they are completely surrounded by other units. While they may have an amount of leakage area similar to that of bottom floor units, they receive some of their infiltration air from the surrounding units, lowering the total amount of outdoor air. The upper floors, on

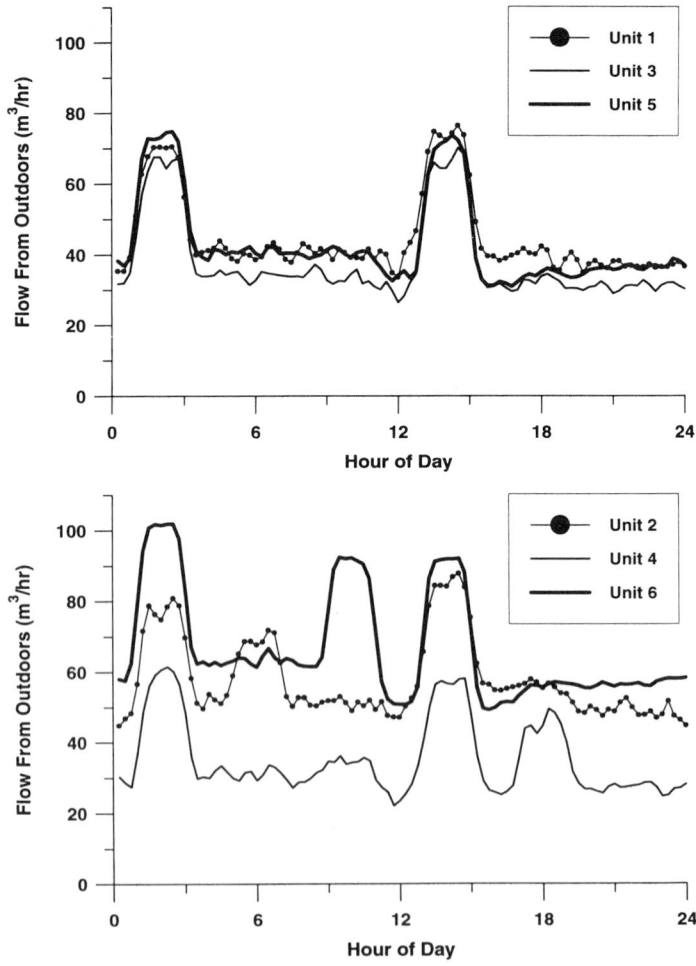

FIG. 2—*Daily unit-by-unit infiltration (8-day average).*

TABLE 3—*Airflow from outdoors.*

Unit	Flow From Outdoors (m^3/h)		Flow From Outdoors (ACH)	
	All Fans Off	All Fans On	All Fans Off	All Fans On
1	39.0	72.0	0.20	0.37
2	50.8	81.7	0.26	0.42
3	31.6	66.2	0.18	0.37
4	27.7	58.3	0.16	0.33
5	35.7	72.7	0.20	0.41
6	55.3	96.7	0.31	0.54
Average	40.0	74.6	0.22	0.41

the other hand, have their entire ceiling exposed to the attic, which is leaky due to penetrations for fans and lights.

The average flow rates with fans on and off were determined by averaging flows over specific periods each day. The fan periods were all set to be two hours in duration; however, due to the effects of data smoothing, averaging, and slight variations in when the fans came on, the data sweep out a smooth curve rather than a sharp-edged peak at each fan period. Therefore, to define the period with all fans on, only 1½ h of each fan period was used. The average of these values was compared with the average flow during two hour periods immediately preceding and following the fan operation.

Interzone Flows

There was a significant amount of air movement between apartments in this building during the test period. These interzone flows were dominated by temperature-driven stack effects. This can be seen by noting that the largest interzone flows were from lower level apartments to the units directly above (see Fig. 3). There was no indication of significant air movement from higher floor units to units below. Flow data also indicated a small amount of flow across some walls between units at the same height. There was an average flow of air from unit 6 to unit 5 of about 6 m^3/h throughout the testing period. Other horizontal interzone flows were typically less than 2 m^3/h. When a single zone exhaust fan was on, flow into that unit from the neighboring unit at the same height was seen to increase; however, horizontal interzone flow was not significantly altered when all exhaust fans were operated at once. Figure 1 shows all of the average interzone flows for periods with all fans off and periods with all fans on. Only the flows greater than 2 m^3/h (1.17 ft^3/min) are shown. This figure also shows the difference in flow between periods with all fans on and off. Note that when the fans were on, almost all of the interzone flows were reduced while the flow from outdoors increased significantly.

Typical average upward flow from lower units to units directly above was in the range of 10 to 20 m^3/h. This flow was significantly altered when single zone fans were operating, and only slightly affected when all fans operated at once. This can be explained by considering the different effects which a single fan would have on the building versus having all of the zone fans on at once. A single fan would be expected to drop the pressure of its zone relative to all other zones and the outside. This would not only cause more outside air to flow through the zone, but would also cause more air flow from neighboring units. If, on the other hand, all of the exhaust fans operate at once and are all moving approximately the same amount of air, then it would be expected that they would have the effect of lowering

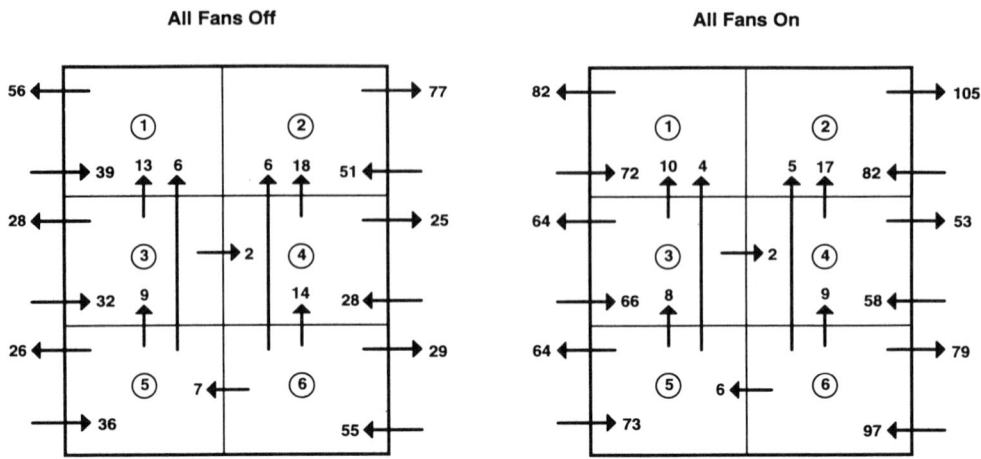

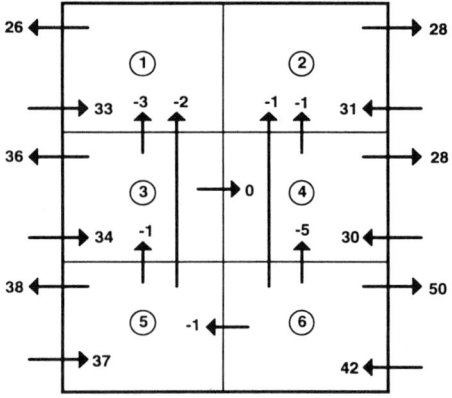

FIG. 3—*Interzone airflows: all fans off versus all fans on (8-day average). Schematic view of south elevation of test building. All flows in m³/h.*

the pressure of the entire building with respect to outside, but would have very little impact on the pressure across the interior walls separating the zones. This was indeed corroborated by the flow data.

These interzone flow data are shown graphically in Figs. 4 and 5. Figure 4 shows the average horizontal interzone flows for the lower four units. For clarity, the flows between units 1 and 2 are not shown in Fig. 4; these flows were about 1 m³/h (0.59 ft³/min). Note that most of the flows were quite small, except for flow from unit 6 to unit 5. Also note that this flow dropped significantly when the fan in unit 6 was on by itself between 9 and 11 a.m. Corresponding to this, flow from unit 5 to unit 6 increased at the same time. Figure 5 shows vertical interzone flows from lower units to the units directly above. All of these flows were much larger than the horizontal interzone flows. This graph also shows how the flows changed due to the different periods of fan operation.

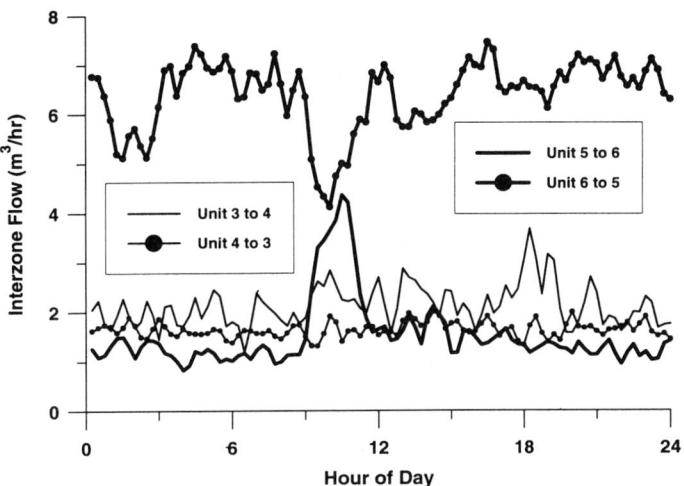

FIG. 4—*Horizontal interzone flow profiles (8-day average)*.

Interestingly, measurements also indicated some significant flows from the ground floor units directly into the top floor units. This can be explained by stack effects within the building. There are some pathways through interior walls and floor cavities for air to travel directly between these units. This is most likely happening in common walls with plumbing and electrical utilities running from the ground floor to the top. The average rate of this flow was about 6 m³/h, larger than most of the horizontal interzone flows.

There are two stack effects working simultaneously on these units. Pressure measurements indicated that there was positive upward pressure across the ceilings between lower units and the units above. Since interzone flow was dominant in the upward direction, we can

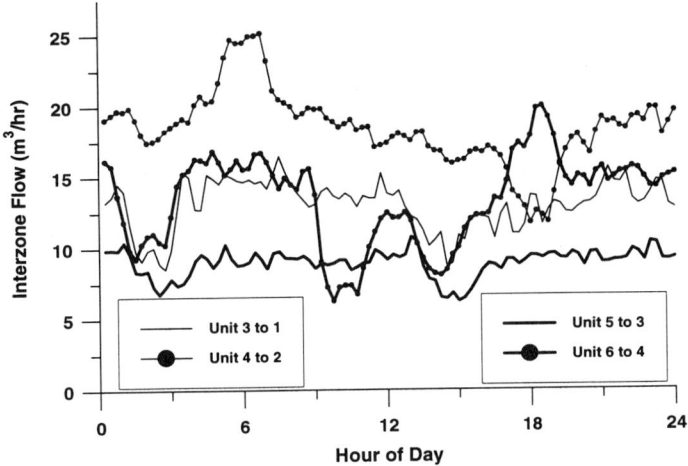

FIG. 5—*Vertical interzone flow profiles (8-day average)*.

surmise that the stack effect is working on the entire building as a single stack. Pressure and flow measurements also indicate that each unit forms its own stack, with outdoor air entering through leaks low in the outside walls and exiting near the ceiling.

Total flow through each of the separate zones is shown in Fig. 6. This clearly shows all of the fan periods, and shows that the top floor units received a higher overall flow rate than the lower units. Table 4 summarizes the total flow through each zone with all fans on and all fans off. Table 4 is identical to Table 3, except that it summarizes total flow through the zones as opposed to just the flow of outdoor air.

Figure 7 and Table 5 show the percentage of the total airflow through each unit that is outdoor air. These data show that under natural driving forces 85% to 95% of the air moving through the ground floor units was fresh outdoor air, while about 65% of the air moving through the upper units was outdoor air. These percentages remained approximately constant when only one zone fan was operating by itself; however, with all fans operating, the percentage of outdoor air in the upper and middle apartments increased significantly.

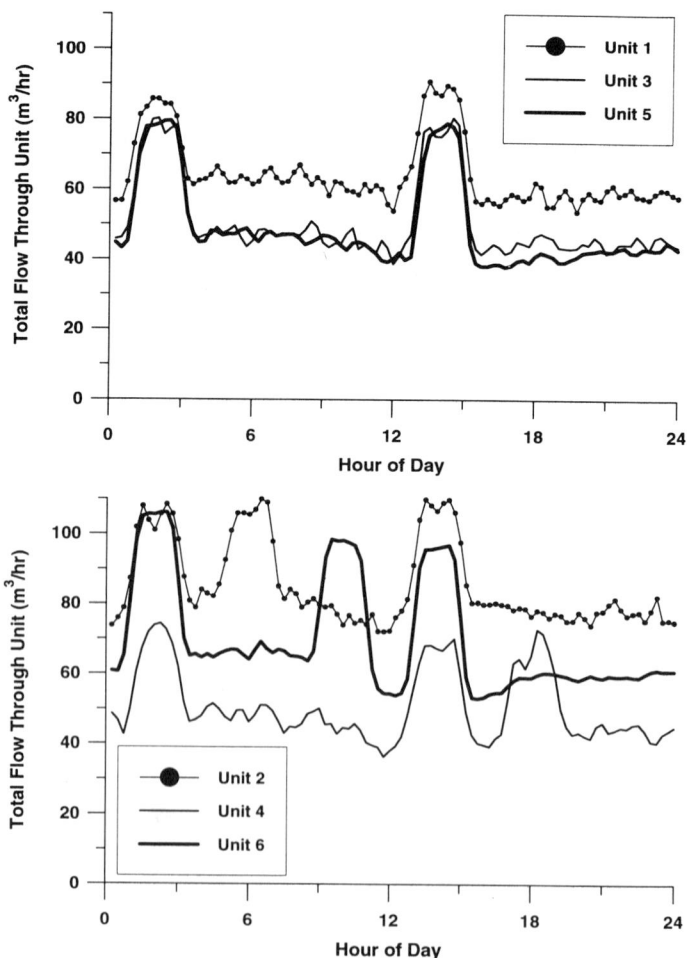

FIG. 6—*Daily profile of total flow through each unit (8-day average).*

TABLE 4—*Total airflow through zones (adjacent zones + outdoors).*

	Total Zone Flow (m³/h)		Total Zone Flow (ACH)	
Unit	All Fans Off	All Fans On	All Fans Off	All Fans On
1	59.0	86.6	0.31	0.45
2	77.9	107.0	0.40	0.56
3	44.3	77.4	0.25	0.43
4	43.4	69.9	0.24	0.39
5	42.1	78.0	0.24	0.44
6	58.4	100.9	0.33	0.57
Average	54.2	86.6	0.30	0.47

Interzone airflow can have a significant impact on indoor air quality. If the lower floor units of this building have strong sources of pollution, the quality of the air in the upper floor units will be negatively impacted.

Fans

The fans increased the pressure across the envelope of each unit by 0.6 to 1.4 Pa. The volumetric flow rate of the fans was measured using a flowhood. The results are given in the second column of Table 6. This should not be confused with the added airflow through the apartment caused by the operation of the fan or the added flow of outdoor air, which are given in the third and fourth columns, respectively. The exhaust fans interact with pressure differences and flows created by stack and wind effects, and the leakage area in the apartments. Some of the flow through the exhaust fan would have been leaking out of the apartment anyway under natural driving forces. Therefore, the added zone flow is always less than the measured fan flow. Furthermore, in a multifamily building, the added zone flow could be coming from outside or it could be coming from neighboring apartment units. It

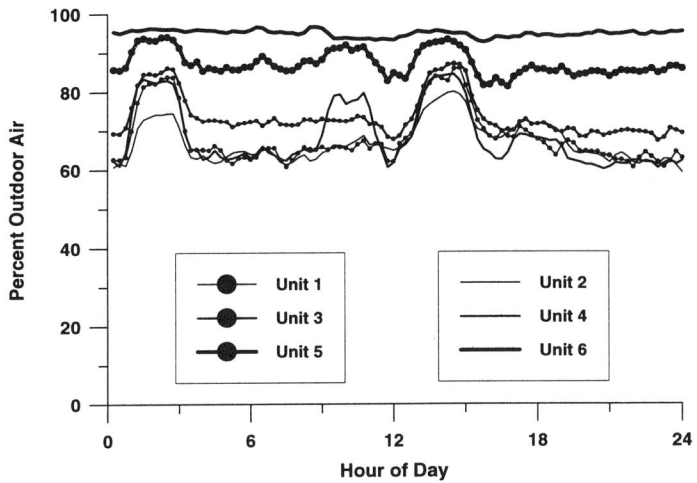

FIG. 7—*Percent of outdoor air in total zone flow.*

TABLE 5—*Percentage of zone flow from outdoors.*

Unit	All Fans Off	All Fans On	Zone Fan Only
1	66%	83%	...
2	65%	76%	65%
3	71%	86%	...
4	64%	83%	68%
5	85%	93%	...
6	95%	96%	94%
Building	100%	100%	...

is important to make this distinction because it could have a large impact on indoor air quality. If most of the air infiltrating into an apartment is coming from a neighboring unit, then the air quality could be severely compromised depending on what is happening in the other unit. For this reason, airflow will be considered as *ventilation* air only when it comes directly into the unit from outside.

The second to last column of Table 6 gives the ratio of the fan flow, measured with a flowhood, to the total added zone flow, measured by the MTMS. Note that this ratio is similar for apartments at the same level, demonstrating the importance of the stack effect and vertical leakage area. This ratio is smallest for the top floor apartments, indicating that exhaust fans in these units had less of an impact on the ventilation rate than on the other floors. The ratio is largest for the middle floor units. About 70% of the fan flow in these units goes toward increasing the flow of outdoor air through the unit. The average ratio of the added flow through the zone to the measured fan flow is 0.55. This finding is consistent with a theoretical model, developed for a single zone, which predicted a typical ratio of added zone flow to fan flow of 0.5 [6].

The final column of Table 6 gives the fan ratio as it relates to the addition of *outdoor* airflow to the zone. Note that the added flow of outdoor air is larger than the added total flow through the zone with the fans on. This is due to the fact that the interzone flows are decreased when all fans are running at the same time. The fans are therefore somewhat more effective at increasing outdoor flow than they are at increasing total zone flow. The ratio of added outdoor flow to measured fan flow for the building is 0.60.

TABLE 6—*Effect of all fans on flow through zones* (m^3/h).[a]

Unit Number	Measured Flow Through Fan	Increase in Total Zone Flow	Increase in Flow from Outdoors	Δ Total Flow/Fan Flow	Δ Outdoor Flow/Fan Flow
1	70.7	27.6	33.0	0.39	0.47
2	65.9	29.1	30.9	0.44	0.47
3	50.1	33.1	34.6	0.66	0.69
4	40.6	26.5	30.6	0.65	0.75
5	62.9	35.9	37.0	0.57	0.59
6	69.1	42.5	41.4	0.61	0.60
Average	59.9	32.5	34.6	0.55	0.60

[a] Fan rating: 85 m^3/h (50 ft^3/min) at 25 Pa (0.1 in. H_2O).

The flow data demonstrate that ventilation is more effective when all fans in the building are operating together. The amount of outdoor air delivered to the apartments is significantly reduced if the fans are run separately.

Conclusion

This testing provided us with a very complete picture of the real-time airflow in a multifamily building. The most significant observations are as follows:

- The test building is very tight, resulting in little natural infiltration. The average ventilation rate for these units was just 0.22 ACH, well below the ASHRAE recommended minimum standard.
- When the ventilation fans in all of the units are operating at the same time, the average ventilation rate is about 75 m³/h or 0.41 ACH. This is very close to the ASHRAE minimum ventilation standard if assuming an occupancy of three people per unit.
- There is a significant amount of airflow from the lower units to the upper units. This indicates that air pollutants in the lower floors will be carried into the neighboring apartments upstairs.
- When individual exhaust fans are on, the interzone air leakage increases, increasing the potential for transfer of indoor air pollution.
- The ventilation fans are much more effective in bringing outdoor air into the units when the fans in all of the apartments are on at the same time. Interzonal airflow is also reduced when all fans operate at once.
- The percentage of outdoor air in the total airflow through the apartments when all fans are off ranges from an average of 90% for the bottom floor units to 66% for the top floor units. These percentages change very little when a single ventilation fan is on. However, when all ventilation fans are on, the percentage of outdoor air increases to an average of 95% for the bottom floor units and 80% for the top floor units.

The authors believe that the test building is typical of many new multifamily buildings in the Pacific Northwest. The combination of building techniques and ventilation and control systems is leading to potentially serious problems of underventilation and cross-unit contamination. The data uncovered in the testing described above could help in the design of better ventilation systems for multifamily buildings.

References

[1] Sherman, M., Feustel, H., and Dickerhoff, D., "Description of a System for Measuring Interzonal Air Flows Using Multiple Tracer Gasses," LBL-26538, Lawrence Berkeley Laboratory, University of California, Berkeley, CA, 1989.
[2] Sherman, M., "On the Estimation of Multizone Ventilation Rates from Tracer Gas Measurements," *Building and Environment,* Vol. 24, No. 4, 1989, pp. 355–362.
[3] Davis, J. C., *Statistics and Data Analysis in Geology,* Wiley, New York, 1986.
[4] Bonneville Power Administration. Super Good Cents Technical Specifications. April 1988.
[5] "Ventilation for Acceptable Indoor Air Quality," ASHRAE Standard 62-1989, American Society of Heating, Refrigerating and Air-Conditioning Engineers, 1989.
[6] Palmiter, L. and Bond, T., "Interaction of Mechanical Systems and Natural Infiltration," Presented at the AIVC Conference on Air Movement and Ventilation Control Within Buildings, Ottawa, Canada, Sept. 1991.

Discussion

David Saum[1] *(written discussion)*—What range of pressure differences were measured? Were these pressures within a range that could be used as a commissioning test?

J. M. Heller (authors' closure)—The paper includes a note about the average pressure differences which were measured during this testing. A more detailed report about this test site will be forthcoming by the primary author. It is the opinion of the authors that it would be difficult to use such pressure measurements as a commissioning test because there are too many variables influencing these measurements.

---

[1] Geomet Technologies, Germantown, MD 20874.

W. Stuart Dols[1] and Andrew K. Persily[1]

A Study of Ventilation Measurement in an Office Building

REFERENCE: Dols, W. S. and Persily, A. K., **"A Study of Ventilation Measurement in an Office Building,"** *Airflow Performance of Building Envelopes, Components, and Systems, ASTM STP 1255,* Mark P. Modera and Andrew K. Persily, Eds., American Society for Testing and Materials, Philadelphia, 1995, pp. 23–46.

ABSTRACT: The National Institute of Standards and Technology has conducted a study of ventilation and ventilation measurement techniques in the Bonneville Power Administration (BPA) Building in Portland, Oregon. The project involved the use of the following outdoor air ventilation measurement techniques: tracer gas decay measurements of whole-building air change rates, the determination of air change rates based on peak carbon dioxide (CO_2) concentrations, the determination of percent outdoor air intake using tracer gas (sulfur hexafluoride and occupant-generated CO_2), and direct airflow rate measurements within the air handling system. In addition, air change rate measurements made approximately three years apart with an automated tracer gas decay system were compared. Airflow rates were measured in the air handling system ductwork using pitot tube, hot-wire anemometer, and vane anemometer traverses, and good agreement was obtained between the different techniques. While accurate determinations of percent outdoor air intake were achieved using tracer gas techniques, the use of CO_2 detector tubes yielded unreliable results. Reliable determinations of ventilation rates per person were made based on SF_6 decay and direct airflow rate measurements, but the use of peak CO_2 concentrations led to overestimations of building air change rates. The measured values of the whole-building air change rates, and their dependence on outdoor air temperature, did not change significantly over a three-year period. The whole-building air change rate under minimum outdoor air intake conditions was determined to be twice the outdoor air intake rate provided by the minimum outdoor air intake fans due to leakage through the main outdoor air intake dampers.

KEYWORDS: airflow, building performance, carbon dioxide, commercial building, indoor air quality, measurements, office building, tracer gas, ventilation

Building ventilation systems are designed to provide sufficient levels of outdoor air to the building, to remove contaminants generated within the space, and to provide an environment that is thermally acceptable to the building occupants. The design of these systems is based on ventilation standards that specify minimum levels of ventilation for occupant health and comfort. It has become increasingly apparent that design values for ventilation rates are not always realized in practice both when the building is constructed and after the building has been in operation for some time [1]. This realization, along with increased concerns about indoor air quality, has led to the need for on-site assessment of building ventilation rates. The requirement for on-site assessment includes the need for practical and reliable procedures for making field measurements of building ventilation rates that are accessible to a range of engineering practitioners.

[1] Mechanical engineer and group leader, respectively, Building and Fire Research Laboratory, National Institute of Standards and Technology, Building 226, Room A313, Gaithersburg, MD 20899.

In order to assess different approaches for evaluating ventilation system performance, the National Institute of Standards and Technology (NIST) has conducted a study of ventilation and ventilation measurement techniques in the Bonneville Power Administration (BPA) Building in Portland, Oregon. The project involved two ventilation measurement issues: comparison of ventilation measurement techniques and changes in building ventilation rates over time. Six ventilation measurement methods were studied in the comparison: (1) whole-building ventilation rates determined by automated tracer gas (sulfur hexafluoride, SF_6) decay, (2) whole building ventilation rates based on peak carbon dioxide (CO_2) concentration, (3) percent outdoor air using SF_6 as a tracer gas, (4) percent outdoor air using an automated infrared CO_2 monitoring system, (5) percent outdoor air using CO_2 detector tubes, and (6) direct airflow rate measurements in heating, ventilating and air-conditioning (HVAC) ducts. Also, two sets of building ventilation rate measurements obtained with the automated tracer gas system were compared. One set of measurements was made between September 1987 and July 1989 [2] and the other was collected as part of the present study from July 1991 to February 1992.

During this project whole-building air change rates were monitored continuously with an automated tracer gas decay system. Percent outdoor air intake rates were also monitored continuously with an automated CO_2 monitoring system and the SF_6 system. During this period, daily peak values of CO_2 were used to calculate ventilation rates per person. Two weeks of intensive measurements of direct airflow rates and CO_2 concentrations were performed simultaneously with the automated monitoring in order to compare various methods of assessing whole-building ventilation rates. These methods were compared in terms of accuracy, ease of use, and the level of effort required to implement them.

Background

Building Description

The BPA building is a seven-story office structure with a one-story basement and a two-story underground parking garage. The conditioned office space within the building has a floor area of approximately 34 600 m² (372 000 ft²) and a volume of 134 000 m³ (4.73 × 10⁶ ft³), assuming an average ceiling height of 3.8 m (12.5 ft), including the return air plenum. A breezeway connects this building to another office building on the first-floor level, and a kitchen and dining room are attached at this level also (floor plans are contained in Ref 2). A penthouse mechanical room houses the main HVAC systems, consisting of three large variable air volume (VAV) systems, one serving the center of the building and the others serving the east and west sides. These three systems serve approximately equal volumes of the building. There are also several smaller air handling systems located on and serving the B1 level.

Sketches of the three main air handling systems are shown in Ref 3. Each system consists of two "cold" supply fans that work in parallel, one "hot" supply fan, a return fan and a minimum outdoor air handling unit. The design supply air capacity of each system is approximately 47 200 L/s (100 000 cfm) and the minimum outdoor air intake fan capacity is 2000 L/s (4200 cfm) per system, which is about 4% of the supply air capacity. Based on the building volume, the minimum design outdoor air intake rate is 0.16 air changes per hour (ACH) or 0.17 L/s · m² (0.034 cfm/ft²), and the maximum supply airflow capacity is 3.8 ACH or 4.1 L/s · m² (0.81 cfm/ft²). An estimate of 2000 building occupants yields minimum and maximum per-person ventilation rates, based on design airflow rates, of 3 L/s (6.3 cfm) per person and 70 L/s (150 cfm) per person. This building was designed to comply with the American Society of Heating, Refrigerating, and Air-Conditioning Engineers (ASHRAE) Standard 62-1981, which contained a minimum outdoor air intake requirement of 2.5 L/s (5 cfm) per person in office space with no smoking present [4] and a default occupancy density of 7 people/100 m². These values correspond to an air change rate of approximately 0.18 ACH for an office building. ASHRAE Standard 62-1989 contains a minimum outdoor air requirement for office space of 10 L/s (20 cfm) per person [5], which corresponds to an air change rate of about 0.72 ACH.

During building occupancy, the minimum outdoor air fans run continuously to provide the design minimum of outdoor air, and the supply fans use variable-pitch fan blades to modulate airflow rate based on supply air demand in the occupied space. Supply air demand is controlled by terminal units located above the ceilings of the occupied space, which modulate supply airflow rates depending on the temperature in the zone being served by the terminal unit. As more units open, requiring additional supply airflow, the associated supply fan blades adjust to increase the airflow and maintain a supply static pressure set point in the main supply ducts. An economizer system modulates the outdoor air intake rate through the "cold" supply fan system during mild weather by modulating the main outdoor air intake (mixed-air) damper position.

Measurement Methods

Whole-Building Air Change Rates

Whole-building air change rates were determined using the tracer gas decay method [ASTM Standard Practice for Measuring Air Leakage Rates by the Tracer Dilution Method (E741-83)]. The automated tracer gas decay system injected SF_6 into the supply airstreams of the building's air handlers every three hours. Tracer gas concentrations were then sampled in ten locations every ten minutes. Tracer gas was injected into the Center, East, and West "cold" supply fans, and four air handlers serving the B1 level. An injection tube carried a metered amount of tracer gas to the supply airstream of the individual air handlers. Tracer gas injection flow rates were based on the volume served by each individual air handler. The locations being sampled were the "cold" supplies and returns of the Center, East, and West systems, the returns of the four air handlers serving the B1 level, the outdoor air, and the diagnostic center which contained the test equipment.

Whole-building air change rates were determined by a volume-weighted averaging of the decay rates of the three main return ducts. An automated tracer gas decay system consisting of a gas chromatograph coupled with an electron capture detector (GC-ECD) was used to determine SF_6 tracer gas concentrations with an uncertainty of about 10%. The accuracy of air change rates measured with this tracer gas system is a function of the uniformity of tracer gas concentration within the building and the calibration of the SF_6 analyzer. Based on the assumption of perfect mixing and the calibration of the SF_6 analyzer, the uncertainty of the air change rates is estimated to be about 10% of the measured value. The tracer gas decay technique determines the total air change rate of the building, including both intentional intake through the ventilation system and unintentional air leakage through the envelope. Previous studies have shown that air change rates due to infiltration can be of the same magnitude as the mechanical ventilation rates [6].

Direct Measurement of Ventilation System Airflows

Direct measurements of system supply and outdoor airflow rates were made during the weeks of August 6, 1991 and January 13, 1992. A hot-wire anemometer, a vane anemometer and a pitot tube with a digital manometer were used during the first week in various locations of the three main systems in order to assess the speed and reliability of these methods for measuring airflow rates in this HVAC system. Both the hot-wire and vane anemometers gave direct readings of velocity in metres per second (feet per minute) [m/s (ft/min)], and the digital manometer used with the pitot tube gave velocity pressure readings in Pascals (inches of water) which were converted to m/s (ft/min). Duct traverses were performed using the hot-wire anemometer and pitot tube in the main supply air ducts, the minimum outdoor air ducts, and the economizer outdoor air intake ducts. Traverses were also performed inside the cold supply fan housings (fan boxes) using the hot-wire anemometer and the vane anemometer. Measurement uncertainties for these airflow rates, based on the uncertainty of the measurement devices alone, were less than 3%. This uncertainty does not include measurement errors due to the use of traverse locations which do not conform with recommended guidelines [7].

The first week of measurements was performed in order to compare velocity measurements with different devices at the same location, to compare measurements of the same airflow rate at different locations, to compare measured airflow rates with design airflow rates, and to use direct airflow rate measurements to determine percent outdoor air intake. The second week of testing focused on the determination of mechanical ventilation rates under minimum outdoor air intake conditions. The results of the first week of measurements revealed that direct traverses of the fan boxes and the minimum outdoor air intake ducts with hot-wire anemometers provided a reasonable means to obtain the desired airflow rates to determine whole building ventilation rates. Therefore, during the second week, supply airflow rates were measured in the cold fan housings immediately downstream of the cooling coils, and duct traverses were performed downstream of the minimum outdoor air intake fans. During two days of the second week, the three main ventilation systems were operated at minimum outdoor air intake.

Percent Outdoor Air Intake Rate

Percent outdoor air intake rates were determined using tracer gas and direct flow measurement techniques. Tracer gas techniques involved a tracer gas balance (SF_6 or CO_2) at the air handler. Based on the measured values of the supply, return, and outdoor air tracer gas concentrations (C_S, C_R, and C_O, respectively), percent outdoor air intake was determined by

$$\%OA = (C_R - C_S)/(C_R - C_O) \qquad (1)$$

Percent outdoor air intake rates were determined using four different methods: tracer gas balances employing the automated SF_6 and automated CO_2 systems, tracer gas balances based on CO_2 detector tube measurements, and direct airflow measurements. SF_6 was measured using the GC-ECD described previously, and CO_2 was measured using an infrared absorption analyzer with an uncertainty of 12.5 part per million (ppm). The CO_2 detector tubes contain a substance which changes color when exposed to CO_2 and have graduated markings on the side that indicate the concentration based on the length of substance that changes color. A handheld volumetric piston pump is used to draw the air sample into the tube. Two sampling strategies were employed using the detector tubes [8]. One sampling strategy was to measure the supply, return, and outdoor air concentrations successively, and then have three people read each tube. The other strategy was to have only one person read the tubes. Uncertainty in the measured CO_2 concentration is assumed to equal 33 ppm for the individual concentration readings using three pump strokes to perform a single measurement. The value of 33 ppm is based only on the resolution of the graduated markings on the detector tube, assuming the user can resolve the tube readings within 100 ppm. Erroneous graduated markings on the detector tubes caused by calibration errors could lead to larger uncertainties. Uncertainty in percent outdoor air intake measurements, $\Delta\%OA$, is based on the propagation of uncertainty in using Eq 1. Each tracer gas concentration measurement has an associated uncertainty due to the measurement uncertainty of the sampling equipment. The uncertainty in percent outdoor air measurements is given by

$$\Delta\%OA = 100 \times \left[\frac{\Delta C_R^2 + \Delta C_S^2}{(C_R - C_O)^2} + \frac{(C_R - C_S)^2(\Delta C_R^2 + \Delta C_O^2)}{(C_R - C_O)^4}\right]^{1/2} \quad (2)$$

where

ΔC_R = uncertainty in return air concentration measurement,
ΔC_S = uncertainty in supply air concentration measurement, and
ΔC_O = uncertainty in outdoor air concentration measurement.

Larger uncertainties will occur when differences between the return and outdoor concentrations are relatively small. Also, the more uncertain the tracer gas concentration measurements the greater the uncertainty in percent outdoor air intake measurements.

Direct airflow measurements to determine percent outdoor air intake were mostly performed under minimum outdoor air intake conditions. The percent outdoor air intake is the ratio of the measured outdoor air intake rate and the measured supply airflow rate. The supply airflow rates were measured upstream of the supply fans inside the fan housings. Under minimum outdoor air intake, the outdoor air intake rates were measured downstream of the minimum outdoor air intake fans. The percent outdoor air intake under economizer operation was determined by the direct airflow method on only one occasion because it required a great deal of time to perform a traverse of the main outdoor air intake ducts. During the traverses of the main outdoor air intake ducts, the supply airflow rate rarely remained constant, interfering with several attempts to make these measurements.

Ventilation Rate per Person

Three methods were used to determine the outdoor air intake rate per person: tracer gas decay, peak CO_2 levels, and direct measurement of the total supply airflow rate multiplied by the percent outdoor air intake rate (multiplicative method). The measurement of peak CO_2 levels was performed using the automated CO_2 system, CO_2 detector tubes, and air sample bags filled using a portable pump and analyzed with the automated CO_2 analyzer. The determination of the per-person ventilation rate by the multiplicative method employed the various methods to determine percent outdoor air intake described previously. The tracer gas and CO_2 methods determine the ventilation rate due to both mechanical ventilation and envelope leakage. The multiplicative method accounts for only outdoor air intake through the mechanical system.

Whole-building ventilation rates determined using the tracer gas decay method were based on the average of the decay rates measured in the return air ducts of the three main air handlers. In order to convert whole-building ventilation rates in air changes per hour to ventilation rate per person, the whole-building air change rate is multiplied by the building volume and divided by the number of building occupants, which is approximately 2000. The building volume was determined from a scale set of drawings, and the number of occupants was obtained from a computerized personnel directory.

Building ventilation rates were also estimated by measuring equilibrium or peak values of CO_2 inside the building and the outdoor concentration. This method is based on a mass balance of CO_2 within the building with a constant ventilation rate, a uniform and constant CO_2 generation rate inside the building (that is, constant occupancy), and a constant outdoor CO_2 concentration [9]. Under these assumptions the building ventilation rate is related to the equilibrium CO_2 concentration inside the building by

$$Q_p = G_p/(C_{eq} - C_o) \tag{3}$$

where

Q_p = per-person building ventilation rate (m^3/s per person),
G_p = per-person CO_2 generation rate (assumed equal to 5.3×10^{-6} m^3/s per person),
C_{eq} = indoor CO_2 concentration at equilibrium, and
C_o = outdoor CO_2 concentration.

This method is also based on the assumption that the CO_2 concentration is the same throughout the building and that it has attained equilibrium. The uncertainty of the ventilation rate per person determined using Eq 3 is given by Eq 4 and depends on the uncertainty in the measured indoor and outdoor CO_2 concentrations, ΔC_{eq} and ΔC_o. However, Eq 4 does not account for the uncertainty in the CO_2 generation rate:

$$\Delta Q_p = Q_p \frac{(\Delta C_{eq}^2 + \Delta C_o^2)^{1/2}}{C_{eq} - C_o} \tag{4}$$

Determination of the ventilation rate per person using CO_2 detector tubes is also based on Eq 3 and the associated assumptions. Detector tubes were used to measure ventilation rates per person on January 15 and 16, 1992. On the 15th, concentrations were measured in the return air ducts of the three main air handlers, and on the 16th they were measured in several locations within the occupied space.

Peak CO_2 measurements were performed within the occupied space using portable hand-held sample pumps and air sample bags and the infrared CO_2 detector of the automated CO_2 system. This was done only on January 16, 1992 at the same time that the CO_2 detector tube measurements were being performed. Samples were collected over a period of about one hour, which spanned the time at which the peak indoor concentration was expected to occur. The expected peak time was based on the automated CO_2 data collected the previous day and verified by the automated data collected during the test. Typically, the CO_2 concentration reached a peak at around 11:00 a.m. Air samples were collected at three locations on floors 1 through 7, and an average of these sample concentrations was used as the equilibrium concentration. Outdoor air samples were also collected before and after the interior samples were collected, and the average of these values was used in the calculation.

When determining ventilation rates per person using the multiplicative method, supply airflow rates were measured using a hot-wire anemometer in the supply fan housings. Percent outdoor air intake rates were measured using the SF_6 and CO_2 automated systems, CO_2 detector tubes, and direct airflow measurements of the minimum outdoor air intake fans. The latter case corresponds to the direct measurement of the outdoor air intake rate at the minimum outdoor air handler units. The uncertainty in ventilation rates per person, ΔQ_p, determined using the multiplicative method is dependent on the uncertainty of the supply airflow rate measurement, ΔQ_s, the uncertainty of the percent outdoor air intake measurement, $\Delta\%OA$, and is given by

$$\Delta Q_p = \frac{\left[\left(\Delta Q_s \cdot \frac{\%OA}{100}\right)^2 + \left(\frac{\Delta\%OA}{100} \cdot Q_s\right)^2\right]^{1/2}}{\text{Number of occupants}} \tag{5}$$

CO_2 Buildup Analysis

While the peak CO_2 measurement technique requires steady-state conditions to exist, one can also analyze the buildup in CO_2 concentrations to determine ventilation rates. The buildup method is based on the transient analysis of CO_2 as the concentration increases or builds up in the morning. The technique is based on a single-zone mass balance expressed in volumetric terms as

$$V\, dC/dt = Q(C_o - C) + G \tag{6}$$

where

  V = building volume,
  C = interior CO_2 concentration,
  t = time,
  Q = airflow rate into and out of building,
  C_o = outdoor CO_2 concentration, and
  G = generation rate of CO_2 within the building.

Assuming that Q, C_o, and G are constant, and setting C' equal to the difference between the indoor and outdoor CO_2 concentration, the solution to Eq 6 can be expressed as

$$C' = C'_{eq} + (C'_{init} - C'_{eq})e^{-It} \tag{7}$$

where

  C'_{eq} = equilibrium value of C', G/Q,
  C'_{init} = $C - C_o$ at $t = 0$, and
  $I = Q/V$, building air change rate.

Based on the time history of the CO_2 concentration inside the return air ducts, nonlinear regression techniques were used to solve for the parameters C'_{eq}, C'_{init}, and I. The value of I can be converted to the ventilation rate per person based on the building volume and the number of occupants. In addition, an average CO_2 generation rate per person can be determined by

$$G_p = C'_{eq}IV/(\text{Number of occupants}) \tag{8}$$

Results

Whole-Building Air Change Rates

The automated tracer gas system was used to measure whole-building air change rates in a previous study from September 1987 to July 1989 [2] and in the present study from July 1991 to January 1992. This enabled an assessment of changes in the operation and performance of the ventilation system since the previous study was performed. Figure 1 shows the

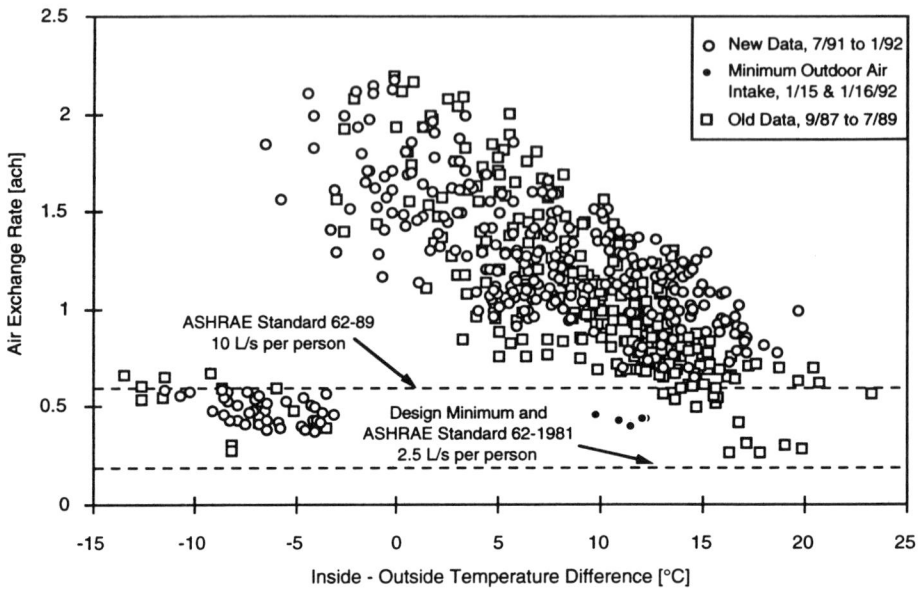

FIG. 1—*Building air change rates for old and new data.*

whole-building air change rates as determined by the automated SF_6 system for the old and new sets of data. The new data are very similar to the data collected in the previous study, indicating that for these weather conditions, that is, outdoor temperatures between 12 and 32°C (54 and 90°F), the building ventilation system is operating as it was during the period of September 1987 to July 1989. These two data sets provide the first opportunity to assess changes in the ventilation characteristics of a building over such a long period of time.

Figure 1 also shows the design minimum air change rate and the air change rates corresponding to the minimums recommended in ASHRAE Standard 62-1981 (2.5 L/s per person, 5 cfm per person) and ASHRAE Standard 62-1989 (10 L/s per person, 20 cfm per person) based on the building occupancy of 2000 people. The average air change rate measured using tracer gas decay under minimum intake for the new data set is 0.45 ACH, about twice the design minimum of 0.19 ACH. The maximum measured air change rate shown in Fig. 3 is 2.2 ACH, about one half of the design capacity.

Direct Measurement of Ventilation System Airflows

Direct measurements of system supply and outdoor airflow rates were made during the weeks of August 6, 1991 and January 13, 1992. The results are presented in Tables 1 and 2 respectively. These tables list the date and time the measurements were performed, the measurement location, the measurement instrument used, the design airflow rate capacity at that location in the system, and the percent difference from design (% DIFF DESIGN = (measured flow − design flow)/design flow) of each measured airflow rate. Measurement locations in individual supply air submain ducts are denoted by their diameters in Table 1. Because the supply fans modulate supply airflow based on cooling demand, the measured supply airflow rates are not expected to equal their design capacities unless the fan is running at full capacity. The minimum outdoor air handler units (AHU-10,11&12) are always run at full capacity; therefore, the differences between the measured and design values are of in-

TABLE 1—*Direct airflow measurements (week 1).*

Date	Time	Fan System	Measurement Location	Measurement Method	Design Flow Rate [L/s]	Measured Flow Rate [L/s]	% DIFF DESIGN	% DIFF METHOD
8/6/91	9:30	SFC12	OA Duct	hot wire	53081	46520	-12%	4%
8/6/91	9:30	SFC12	OA Duct	pitot	53081	48285	-9%	
8/6/91	11:00	SFC12	fan box	hot wire	53081	56397	6%	3%
8/6/91	11:00	SFC12	fan box	vane	53081	58050	9%	
8/6/91	11:00	AHU-10		hot wire	1982	2156	9%	
8/6/91	14:30	SFC12	1.12 m OD	hot wire	16416	11124	-32%	16%
8/6/91	16:00	SFC12	1.12 m OD	pitot	16416	13090	-20%	
8/6/91	15:00	SFC12	1.42 m OD	hot wire	29840	21887	-27%	7%
8/6/91	16:00	SFC12	1.42 m OD	pitot	29840	23406	-22%	
8/6/91	16:00	SFC12	fan box	hot wire	53081	57605	9%	7%
8/6/91	16:00	SFC12	fan box	vane	53081	61839	16%	
8/6/91	15:00	AHU-10		pitot	1982	2425	22%	
8/6/91	16:40	SFC56	0.76 m OD	hot wire	6603	6759	2%	2%
8/6/91	16:40	SFC56	0.76 m OD	pitot	6603	6628	0%	
8/6/91	16:30	SFC56	1.22 m OD	hot wire	20069	17130	-15%	10%
8/6/91	16:30	SFC56	1.22 m OD	pitot	20069	15465	-23%	
8/6/91	16:50	SFC56	1.42 m OD	hot wire	27433	25880	-6%	3%
8/6/91	16:50	SFC56	1.42 m OD	pitot	27433	25037	-9%	
8/6/91	16:40	SFC56	Duct Total	hot wire	54105	49754	-8%	5%
8/6/91	16:40	SFC56	Duct Total	pitot	54105	47121	-13%	
8/6/91	17:00	SFC56	fan box	hot wire	54105	57387	6%	1%
8/6/91	17:00	SFC56	fan box	vane	54105	58050	7%	
8/7/91	9:20	SFC12	1.12 m OD	hot wire	16416	12923	-21%	26%
8/7/91	8:30	SFC12	1.12 m OD	pitot	16416	9924	-40%	
8/7/91	9:30	SFC12	1.42 m OD	hot wire	29840	22004	-26%	15%
8/7/91	8:40	SFC12	1.42 m OD	pitot	29840	18877	-37%	
8/7/91	8:30	SFC12	OA Duct	hot wire	53081	45243	-15%	18%
8/7/91	10:00	SFC12	OA Duct	pitot	53081	53946	2%	
8/7/91	10:15	AHU-10		hot wire	1982	2375	20%	4%
8/7/91	10:15	AHU-10		pitot	1982	2473	25%	
8/7/91	14:40	SFC56	0.76 m OD	hot wire	6603	6575	0%	0%
8/7/91	13:20	SFC56	0.76 m OD	pitot	6603	6583	0%	
8/7/91	14:50	SFC56	1.22 m OD	hot wire	20065	15761	-21%	6%
8/7/91	13:50	SFC56	1.22 m OD	pitot	20065	16703	-17%	
8/7/91	15:15	SFC56	1.42 m OD	hot wire	27433	22921	-16%	15%
8/7/91	13:40	SFC56	1.42 m OD	pitot	27433	26544	-3%	
8/7/91	14:50	SFC56	Duct Total	hot wire	54105	45257	-16%	10%
8/7/91	13:40	SFC56	Duct Total	pitot	54105	49831	-8%	
8/8/91	9:20	SFC12	1.12 m OD	hot wire	16416	9724	-41%	
8/8/91	9:30	SFC12	1.42 m OD	hot wire	29840	19925	-33%	
8/8/91	9:25	SFC56	Duct Total	hot wire	54105	48209	-11%	
8/8/91	10:35	SFC34	Duct Total	hot wire	47856	35020	-27%	
8/8/91	11:30	AHU-12		hot wire	1982	1838	-7%	
8/8/91	12:00	SFC56	fan box	vane	54105	59813	11%	
8/8/91	12:05	SFC56	Duct Total	hot wire	54105	43939	-19%	
8/8/91	14:25	SFC56	fan box	vane	54105	56088	4%	
8/8/91	16:30	SFC56	fan box	vane	54105	58750	9%	
8/8/91	16:40	SFC56	Duct Total	hot wire	54105	44471	-18%	
8/8/91	14:30	AHU-12		pitot	1982	2101	6%	

terest. The airflow rate of AHU-10 was measured six times and always yielded a higher airflow rate than the design value of 2000 L/s (4200 cfm). Measured values were on the average 20% higher than the design capacity. AHU-11 was measured three times, and the airflow rates were approximately 20% lower than design. AHU-12 was measured five times and yielded airflow rates within 7% of the design capacity.

The difference between measurements taken at the same location with different instruments was divided by their mean value and is presented in the column labeled "%DIFF

TABLE 2—*Direct airflow measurements (week 2).*

Date	Time	Fan System	Measurement Location	Measurement Method	Design Flow Rate [L/s]	Measured Flow Rate [L/s]	% DIFF DESIGN
1/14/92	10:50	SFC12	fan box	hot wire	53081	30534	-42%
1/14/92	11:00	SFC34	fan box	hot wire	47856	35041	-27%
1/14/92	10:40	SFC56	fan box	hot wire	54105	28847	-47%
1/14/92	13:40	SFC56	fan box	hot wire	54105	29605	-45%
1/14/92	16:10	SFC12	fan box	hot wire	53081	28499	-46%
1/14/92	16:00	SFC34	fan box	hot wire	47856	35544	-26%
1/14/92	16:20	SFC56	fan box	hot wire	54105	26386	-51%
1/15/92	9:25	SFC12	fan box	hot wire	53081	57157	8%
1/15/92	9:40	SFC34	fan box	hot wire	47856	38591	-19%
1/15/92	9:15	SFC56	fan box	hot wire	54105	51522	-5%
1/15/92	15:00	SFC12	fan box	hot wire	53081	28356	-47%
1/15/92	15:50	SFC34	fan box	hot wire	47856	22849	-52%
1/15/92	16:10	SFC56	fan box	hot wire	54105	24948	-54%
1/16/92	8:15	SFC12	fan box	hot wire	53081	54329	2%
1/16/92	8:30	SFC34	fan box	hot wire	47856	37538	-22%
1/16/92	8:00	SFC56	fan box	hot wire	54105	40171	-26%
1/16/92	11:50	SFC12	fan box	hot wire	53081	56048	6%
1/16/92	12:00	SFC34	fan box	hot wire	47856	42268	-12%
1/16/92	11:40	SFC56	fan box	hot wire	54105	56058	4%
1/16/92	14:15	SFC12	fan box	hot wire	53081	55201	4%
1/16/92	14:30	SFC34	fan box	hot wire	47856	38847	-19%
1/16/92	14:00	SFC56	fan box	hot wire	54105	55965	3%
1/15/92	13:50	AHU-10		hot wire	1982	2441	23%
1/16/92	9:30	AHU-10		hot wire	1982	2560	29%
1/15/92	11:30	AHU-11		hot wire	1982	1556	-22%
1/16/92	8:40	AHU-11		hot wire	1982	1589	-20%
1/16/92	11:40	AHU-11		hot wire	1982	1644	-17%
1/15/92	14:15	AHU-12		hot wire	1982	1865	-6%
1/16/92	9:10	AHU-12		hot wire	1982	1864	-6%
1/16/92	11:30	AHU-12		hot wire	1982	1838	-7%

METHOD" in Table 1. The comparisons are made between the hot-wire measurements which are presented in the same row as the %DIFF METHOD and the measurement which appears in the row below the hot-wire value. Airflow rates measured using the hot-wire anemometer and the pitot tube, and the hot-wire anemometer and vane anemometer, were generally within 10% of each other.

In order to evaluate the measurement of the same airflow rate at different locations, measurements of the West air handler system (SFC-5&6) taken inside the cold supply fan box were compared with the sum of the individual airflow rates measured in the three submain ducts of the West system. The difference between these measurements divided by the average measured airflow rate ranged from 15% to 30%.

Percent Outdoor Air Intake Rate

Percent Outdoor Air: SF_6 and CO_2 Automated—Percent outdoor air intake rates, based on data collected with the SF_6 tracer gas system during the previous study, were determined for the Center air handling system for the period of September 1987 to July 1989 and are presented in Fig. 2 along with the new measurements. An average maximum outdoor air intake rate of 83% occurred when outdoor temperatures were between 15% and 25°C (59°F and 77°F), and a minimum of about 8% occurred when the outdoor temperature was above about 25°C (77°F). Due to the operation of the economizer cycle, the percent outdoor air intake was modulated when the outdoor temperature was below 12°C (54°F).

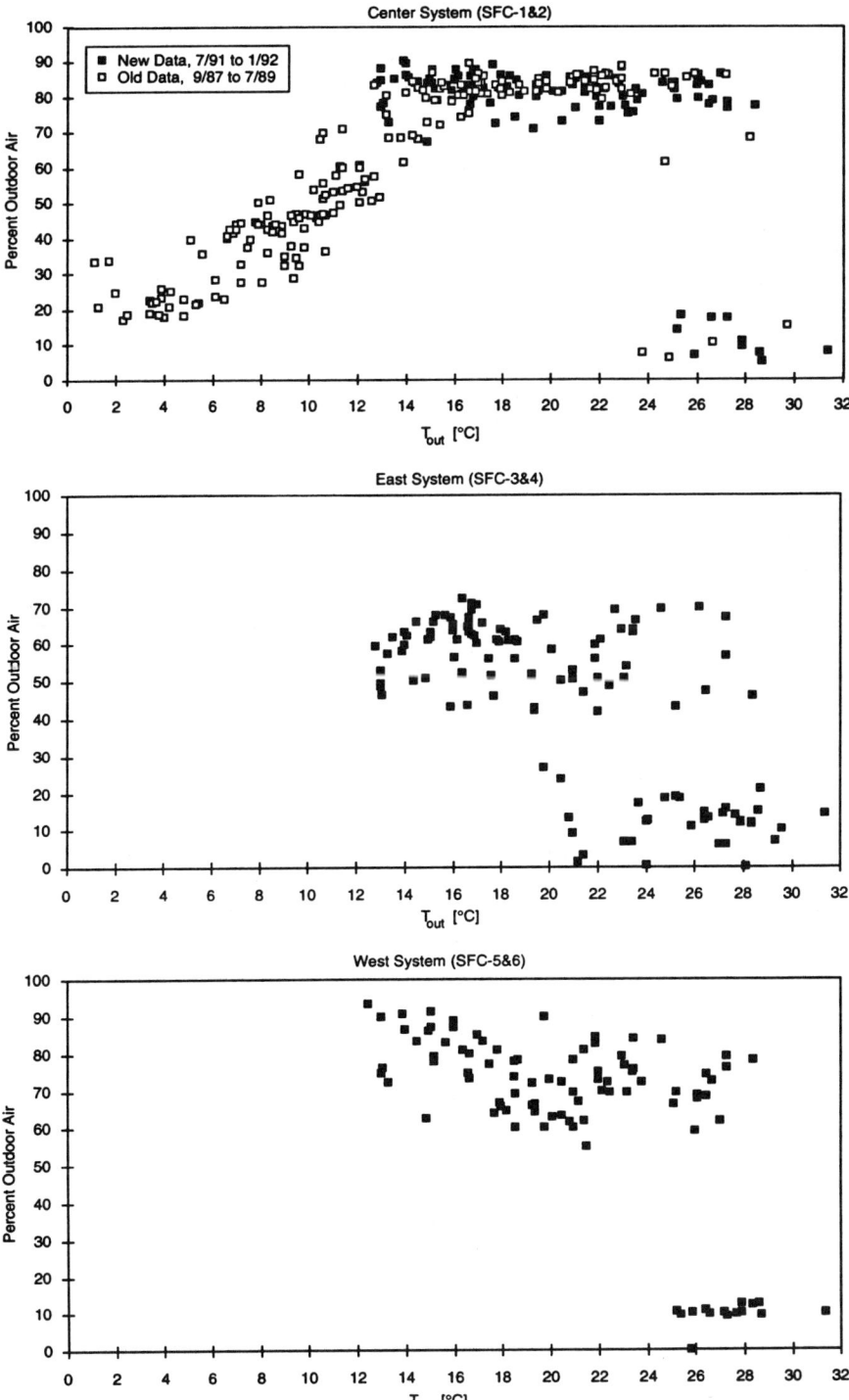

FIG. 2—*Percent outdoor air intake (SF$_6$ automated).*

For this study, percent outdoor air intake was continuously monitored using both the automated SF_6 tracer gas system and the automated CO_2 system simultaneously. Figure 2 shows percent outdoor air intake rates for all three main fan systems as determined by the automated SF_6 system. Figure 3 shows percent outdoor air intake measured with the automated CO_2 system. All three fan systems have approximately the same minimum outdoor air intake rate of about 10% compared to a design value of 4%. The average maximum rates of the Center, East, and West systems (SFC-1&2, SFC-3&4, and SFC-5&6) are 82%, 60%, and 74% respectively based on the SF_6 data.

Percent Outdoor Air: CO_2 Detector Tubes—The results of the percent outdoor air intake determinations based on CO_2 detector tube measurements are presented in Table 3, which gives the individual detector tube readings divided by the number of pump strokes used to take the sample. The percent outdoor air based on the readings obtained by each individual are shown, along with the mean and standard deviation of these three readings. The averages of the concentration readings taken by the three people were also used to calculate the percent outdoor air intake rates given in bold in Table 3. The values to the right of the bold values are the measurement uncertainties given in percent outdoor air. The determinations on August 8, 1991 and January 15, 1992 were all done by a single person, and uncertainty estimates are shown for each value of percent outdoor air.

The percent outdoor air calculations based on one individual's readings are quite variable and subject to significant uncertainty. This uncertainty is due to the low resolution of the detector tubes used in this study, the difficulty in reading the tubes, and calibration errors. In some cases the single-person results were quite unreasonable, that is, less than 0% or greater than 100%.

Percent Outdoor Air: Direct Airflow Measurement—Percent outdoor air intake rates were determined under both minimum and maximum percent outdoor air intake conditions during the first week of direct airflow measurements and under minimum outdoor air intake conditions during the second week. These values of percent outdoor air intake were determined by dividing the airflow rate measured in the outdoor air intake ducts by the supply airflow rate, with both values being obtained by duct traverses. Table 4 lists the results of these determinations along with the mixed-air damper status for the fan system being measured as obtained from the HVAC control system. Very few measurements were made with the mixed-air dampers open because of difficulties in accessing some of the ductwork and due to the modulation of the mixed-air damper positions during traverses.

Ventilation Rate Per Person

Ventilation Rate Per Person: SF_6 Automated—Table 5 shows the whole building air change rate determined by the tracer gas decay method and the ventilation rate per person based on the measured air change rate, 2000 building occupants, and a building volume of 114 000 m^3 (4 030 000 ft^3). The measurement uncertainty associated with the ventilation rate as determined by the tracer gas decay method is approximately 10% of the indicated values. The values obtained during minimum outdoor air intake were approximately 8 L/s (16 cfm) per person which is higher than the recommended minimum outdoor air intake given in ASHRAE Standard 62-1981 (2.5 L/s per person, 5 cfm per person) and slightly lower than the value in ASHRAE Standard 62-1989 (10 L/s per person, 20 cfm per person). These measurements include both intentional outdoor air intake through the ventilation system and unintentional air leakage through the building envelope.

Ventilation Rate Per Person: Peak CO_2 Automated—Building ventilation rate estimates based on equilibrium analysis were determined using the peak values of the average building CO_2 concentration. These concentrations were measured in the return ducts of the main air

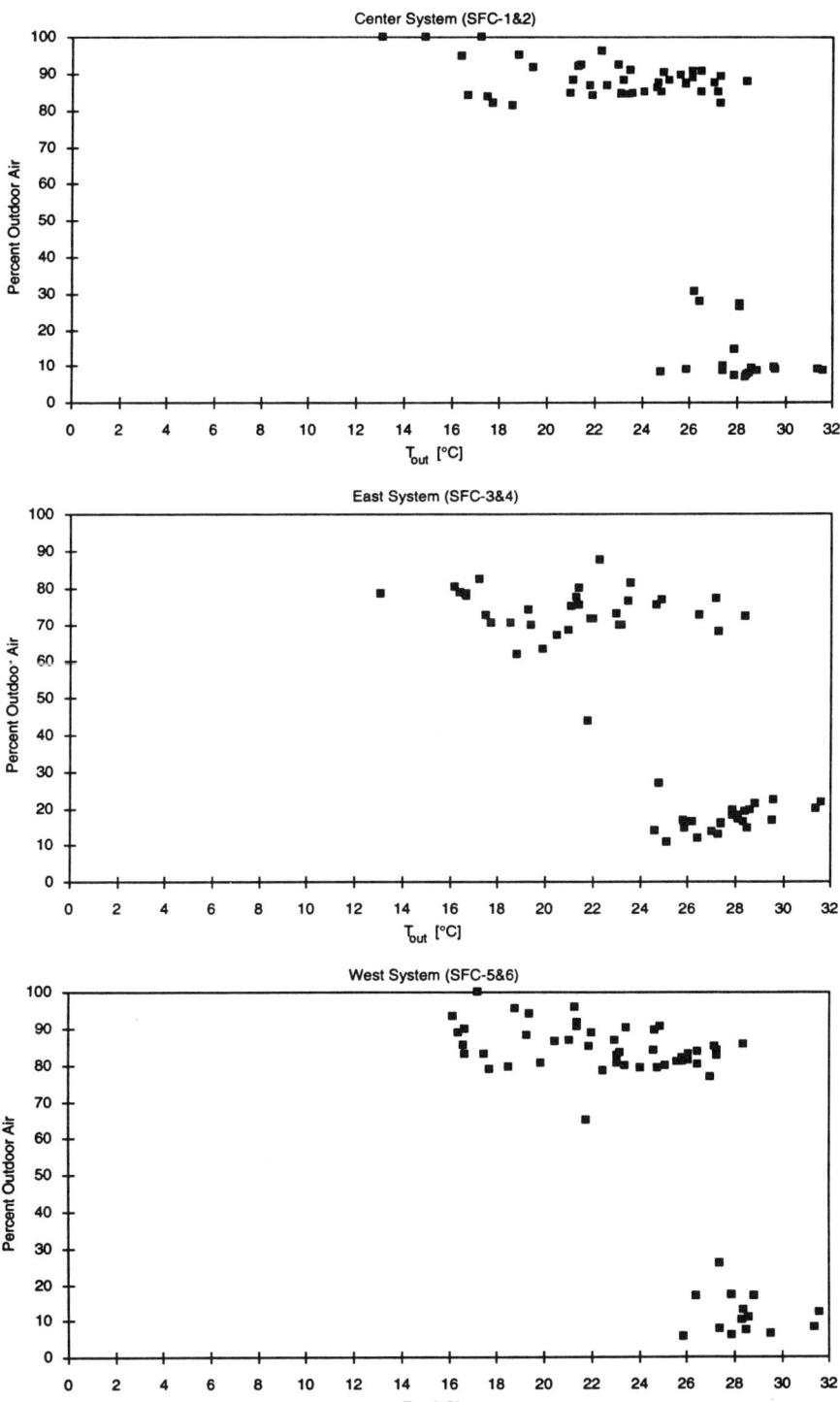

FIG. 3—*Percent outdoor air intake (CO_2 automated).*

TABLE 3—*Percent outdoor air by CO_2*
detector tubes.

Date	Time	System	CO₂ Concentrations			%OA	Person
			Supply	Return	Outdoor		
8/6/91	10:10	SFC12	467	533	417	57%	1
			467	567	417	67%	2
			317	433	300	87%	3
						70%	<-- Avg
						13%	<--Std
		Avg-->	417	511	378	**71%**	± 43%
8/6/91	10:40	SFC12	467	533	417	57%	1
			467	567	433	75%	2
			433	500	400	67%	3
						66%	<-- Avg
						7%	<--Std
		Avg-->	456	533	417	**67%**	± 48%
8/7/91	8:30	SFC12	417	500	417	100%	1
			417	500	433	125%	2
			383	467	400	125%	3
						117%	<-- Avg
						12%	<--Std
		Avg-->	406	489	417	**115%**	± 100%
8/7/91	10:35	SFC12	467	500	467	100%	1
			467	500	500	ERR	2
			433	500	433	100%	3
						ERR	<-- Avg
						ERR	<--Std
		Avg-->	456	500	467	**133%**	± 237%
8/7/91	13:30	SFC56	400	500	433	150%	1
			400	500	467	300%	2
			400	500	400	100%	3
						183%	<-- Avg
						85%	<--Std
		Avg-->	400	500	433	**150%**	± 126%
8/7/91	15:30	SFC56	667	767	467	33%	1
			733	700	433	-12%	2
			667	700	433	12%	3
						11%	<-- Avg
						19%	<--Std
		Avg-->	689	722	444	**12%**	± 17%
8/8/91	11:00	SFC56	667	600	417	-36%	± 27%
8/8/91	11:30	SFC56	567	700	467	57%	± 23%
8/8/91	17:25	SFC56	550	750	525	89%	± 28%
8/8/91	18:05	SFC56	550	600	500	50%	± 53%
1/15/92	11:00	SFC12	900	1067	475	28%	± 8%
1/15/92		SFC34	967	1067	475	17%	± 8%
1/15/92		SFC56	1000	1267	475	34%	± 6%
1/15/92	14:00	SFC12	933	1125	454	29%	± 7%
1/15/92		SFC34	983	1050	454	11%	± 8%
1/15/92		SFC56	1050	1483	454	42%	± 5%

handlers using the automated CO_2 system. Table 6 shows the calculated ventilation rate per person based on Eq 3, the measurement uncertainty in the calculated ventilation rate, and the building air change rate corresponding to this ventilation rate. Under minimum percent outdoor air intake, the peak CO_2 method yielded a ventilation rate of about 15 L/s (30 cfm) per person and about 50 L/s (100 cfm) per person under maximum percent outdoor air intake conditions.

The measurement uncertainties shown in Table 6 are based on the accuracy of the infrared CO_2 monitor utilized by the automated system and the propagation of uncertainty in Q_p given by Eq 4. The uncertainties given in the table do not include other sources of error such as in the number of building occupants, variations in the number of building occupants during the measurements, variations in the ventilation rate, non-constant outdoor CO_2 concentrations, nonuniformities in the CO_2 concentrations within the building, and the indoor CO_2 concentration not being at equilibrium.

Ventilation Rate Per Person: Peak CO_2 Detector Tubes—Peak CO_2 was measured with detector tubes on January 15 and 16, 1992. Measurements were performed in the three main return air ducts on the 15th and in several locations in the office space on the 16th. The

TABLE 4—*Percent outdoor air by direct flow method.*

Date	Time	System	% Outdoor Air by Direct Flow Method	Mixed-air Damper Status [%open]
8/6/91	10:10	SFC-1&2	87%	100%
	10:40	SFC-1&2	87%	100%
	16:40	SFC-5&6	3%	0%
8/7/91	15:00	SFC-5&6	4%	0%
8/8/91	11:30	SFC-5&6	3%	0%
	16:30	SFC-5&6	4%	0%
1/15/92	9:25	SFC-1&2	4%	5%
	9:40	SFC-3&4	4%	0%
	9:15	SFC-5&6	4%	0%
1/16/92	8:00	SFC-1&2	5%	4%
	8:15	SFC-3&4	4%	0%
	8:30	SFC-5&6	5%	0%
1/16/92	11:50	SFC-1&2	5%	4%
	12:00	SFC-3&4	4%	0%
	11:40	SFC-5&6	3%	0%
1/16/92	14:15	SFC-1&2	4%	4%
	14:25	SFC-3&4	4%	0%
	14:05	SFC-5&6	4%	0%

calculated ventilation rates per person under minimum outdoor air intake conditions were approximately 9 L/s (18 cfm) per person on both days with an associated uncertainty of approximately 8% of the measured value.

Ventilation Rate Per Person: Peak CO_2 Air Sample Bags—Peak CO_2 was determined in the office space on January 16, 1992 using air sample bags and the infrared CO_2 analyzer of the automated system. The interior sample concentrations ranged from 450 to 1200 ppm, and the outdoor concentration was approximately 550 ppm. The ventilation rate per person based on the average of these interior measurements was approximately 16 L/s (32 cfm) per person. For comparison, a value of 13 L/s (26 cfm) per person was obtained using the automated CO_2 system to measure the return air concentrations at the main air handlers. The uncertainties associated with these ventilation rates are approximately 5% of the measured values.

Ventilation Rate Per Person: Multiplicative Method—Table 7 displays the results of the determinations of per-person ventilation rate based on the multiplicative method during the week of January 13, 1992. Of particular interest are the four sets of measurements performed under minimum outdoor air intake conditions. Excellent agreement was obtained between the per-person ventilation rates at minimum intake as determined using the percent outdoor air intake rates based on the automated SF_6 and automated CO_2 systems and the results obtained based on SF_6 decay. The ventilation rates per person obtained by the direct measurement of the minimum outdoor air intake fan airflow rates were approximately one-half the rates obtained using the multiplicative methods based on the automated SF_6 and CO_2 systems. Results based on the CO_2 detector tubes were inconsistent with the results of the

TABLE 5—*Ventilation rate per person*
(SF$_6$ automated).

Date	Hr	SF$_6$ Decay		
		ach	per person	
			[L/s]	[cfm]
7/26/91	9	1.7	27	57
7/29/91	9	1.8	29	62
7/30/91	9	1.8	29	61
7/31/91	9	2.0	32	67
7/31/91	15	0.5	8	16
8/01/91	9	1.9	30	63
8/12/91	9	1.7	27	57
8/12/91	17	0.5	8	17
8/13/91	9	1.9	30	64
8/14/91	9	1.8	28	59
8/14/91	16	0.6	9	19
8/15/91	9	2.1	33	71
8/15/91	15	0.6	9	19
8/16/91	10	1.6	25	52
8/19/91	9	1.9	31	65
8/19/91	15	0.6	9	19
8/20/91	9	1.8	29	61
8/20/91	15	0.6	9	18
8/21/91	8	1.6	26	54
8/21/91	15	0.6	9	20
1/15/92	11	0.4	7	15
1/16/92	11	0.4	7	14

other methods. The relative uncertainties associated with these ventilation rates are dependent upon the uncertainty in the percent outdoor air intake measurement technique used.

CO$_2$ Buildup Analysis

The buildup analysis was performed with the data collected on January 15 and 16 under minimum outdoor air intake conditions between 9:00 and 11:00 a.m. The results are presented in Table 8. A plot of the data collected on January 15 along with the nonlinear curve fit to the data is shown in Fig. 4. Table 8 lists the air change rate and the per person CO$_2$ generation rate based on the curve fit and the whole-building air change rate determined by SF$_6$ decay.

Discussion

Measurement Results

Direct Measurement of Ventilation System Airflows—In making direct measurements of ventilation system airflow rates using duct traverses, the impact of practical considerations on the use of these techniques was noted. As expected, the physical configuration of the HVAC system and the manner in which it is operated can limit which airflows can be measured, when they can be measured, and the accuracy of the measurements. None of the airflow rate measurement locations were consistent with handbook recommendations for pitot

TABLE 6—*Ventilation rate per person*
(*peak CO_2 automated*).

Date	Hr	Peak CO_2 Automated		
		per person	%	ach
		[L/s]	[cfm]	Uncertainty	
7/26/91	9	50	106	17%	3.2
7/26/91	14	41	87	14%	2.6
7/29/91	9	45	96	15%	2.9
7/29/91	15	15	33	5%	1.0
7/30/91	9	49	103	16%	3.1
7/30/91	16	21	44	7%	1.3
7/31/91	9	52	111	18%	3.3
7/31/91	15	18	38	6%	1.1
8/01/91	9	54	115	18%	3.4
8/12/91	9	43	90	14%	2.7
8/12/91	17	37	78	12%	2.3
8/13/91	9	45	95	15%	2.8
8/13/91	14	40	84	13%	2.5
8/14/91	9	45	95	15%	2.8
8/14/91	16	20	41	7%	1.2
8/15/91	9	47	100	16%	3.0
8/15/91	15	15	31	5%	0.9
8/16/91	10	45	95	15%	2.8
8/19/91	9	43	92	14%	2.7
8/19/91	15	20	43	7%	1.3
8/20/91	9	52	109	17%	3.3
8/20/91	15	17	37	6%	1.1
8/21/91	8	49	103	16%	3.1
8/21/91	15	18	37	6%	1.1
8/22/91	9	45	95	15%	2.8
0/22/91	15	1C	04	5%	1.0
1/15/92	11	14	30	5%	0.9
1/16/92	11	13	27	4%	0.8

tube or hot-wire traverses [*7,10*]. All of the traverse locations represented a compromise with recommended practice, but the results obtained appeared to be consistent and reasonable. In some cases, ducts of interest were entirely inaccessible. Along with these physical constraints, there are also time constraints associated with these airflow measurements. Throughout the course of a day, changes in system demands affect system airflow rates. This effect

TABLE 7—*Ventilation rate per person*
(*multiplicative method*).

Date	Time	%OA	Ventilation Rates [L/s per person]				
			Total Supply x %OA				SF_6 Decay
			CO_2 Auto	CO_2 Tube	SF_6 Auto	Direct Flow	
1/14/92	16	max	34	--	30	--	--
1/15/92	9	min	6	20	7	3	7
1/15/92	16	max	28	10	25	--	16
1/16/92	8	min	7	--	6	3	7
1/16/92	12	min	7	--	5	3	7
1/16/92	14	min	6	--	5	3	7

TABLE 8—*Ventilation rate per person (CO_2 buildup).*

| Date | Time | Regression Results | | SF$_6$ Decay [ach] |
		I [ach]	G$_p$ [m^3/s per person]	
1/15/92	9:00-11:00	0.59	4.06 x 10^{-6}	0.44
1/16/92	9:00-11:00	0.40	3.48 x 10^{-6}	0.43

was particularly evident during the first week of measurements when the weather was quite warm, that is, near to the temperature at which the main outdoor air intake (mixed-air) dampers close due to the operation of the economizer cycle. Under these conditions, the building operated with the mixed-air dampers wide open until the outdoor air temperature increased to the level at which these dampers shut and the intake was reduced to its minimum value. Several duct traverses were interrupted by this sudden change in airflow rate. Such system effects can sometimes be anticipated, but this requires an understanding of how the system is intended to work and, more importantly, how the system is actually working.

The measurements of ventilation system airflow rates using pitot tube, hot-wire anemometer, and vane anemometer duct traverses were generally consistent with each other even though the duct configurations were not consistent with standard recommendations. Measurements of the same airflow rate using these different devices were generally within 10% of each other.

Percent Outdoor Air Intake Rate—Table 9 compares the results of the measurements based on SF$_6$ and CO_2 balances. Percent outdoor air intake rates based on the automated SF$_6$ and CO_2 test results are generally in good agreement with each other as seen in Figs. 2 and 3. All three fan systems have approximately the same minimum outdoor air intake rate of about 10% compared to the design value of 4%. As seen in Table 9, the determination of percent outdoor air intake based on detector tube readings was often quite inaccurate, particularly at low values of percent outdoor air intake.

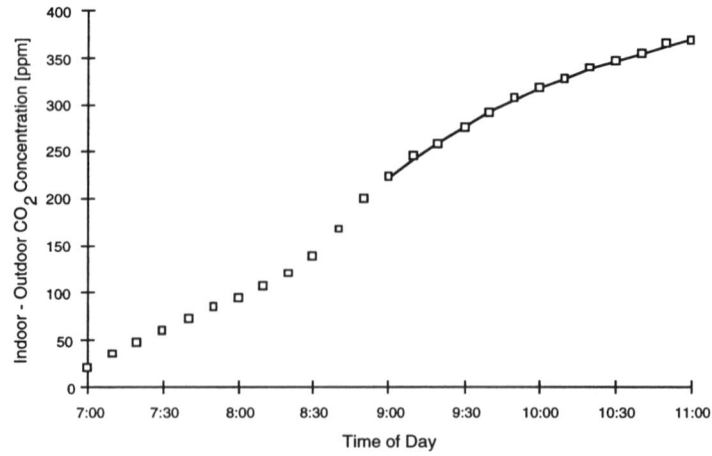

FIG. 4—*Nonlinear curve fit of CO_2 buildup data (1/15/92).*

TABLE 9—*Comparison of percent outdoor air measurements by method.*

Date	Time	System	Percent Outdoor Air			
			CO_2 Tubes (1 person)	CO_2 Tubes (3 person)	CO_2 Auto	SF_6 Auto
8/6/91	10:10	SFC-1&2	57%	71%	87%	--
	10:40	SFC-1&2	57%	67%	82%	--
8/7/91	8:30	SFC-1&2	100%	115%	90%	--
	10:35	SFC-1&2	100%	133%	90%	--
	13:30	SFC-5&6	150%	150%	88%	--
	15:40	SFC-5&6	33%	12%	11%	--
8/8/91	11:00	SFC-5&6	-36%		8%	--
	11:30	SFC-5&6	57%		12%	--
	17:25	SFC-5&6	89%		13%	--
	18:05	SFC-5&6	50%		11%	--
1/15/92	9:25	SFC-1&2	28%		7%	11%
	9:15	SFC-5&6	34%		8%	16%
1/15/92	13:50	SFC-1&2	29%		10%	11%
	14:10	SFC-3&4	11%		10%	10%
	14:00	SFC-5&6	42%		9%	16%
1/16/92	8:00	SFC-1&2	--		8%	5%
	8:15	SFC-3&4	--		14%	7%
	8:30	SFC-5&6	--		13%	14%
1/16/92	11:50	SFC-1&2	--		8%	8%
	12:00	SFC-3&4	--		9%	9%
	11:40	SFC-5&6	--		9%	12%
1/16/92	14:15	SFC-1&2	--		6%	4%
	14:25	SFC-3&4	--		12%	6%
	14.05	3FC-5&0	--		7%	12%

The CO_2 detector tubes used in this study did not yield reliable measurements of percent outdoor air intake due to the difficulty in obtaining consistent readings and the low resolution of the tubes. The single-person CO_2 tube method yielded results which were as much as several hundred percent different from the values obtained by the other tracer gas methods. For example, on August 8, 1991 and January 15, 1992 the automated SF_6 and CO_2 systems yielded percent outdoor air intake rates under minimum outdoor air intake conditions of approximately 10%, while the CO_2 tube method yielded results from −36% to 89%.

As compared with the CO_2 concentration measurements using the automated system, the measurements of CO_2 concentrations with detector tubes in this study were associated with significant measurement uncertainties. Large variations between readings taken by individuals occurred because the line separating the reacted and nonreacted chemical is diffuse, making it difficult to read the tubes consistently. Graduations on the tubes are in increments of 500 ppm, which is very coarse for resolving concentrations typical of indoor and outdoor air.

The three-person method seemed to be useful only to determine a rough estimate of percent outdoor air intake, but it still yielded some unreasonable results. Percent outdoor air values obtained using the three-person CO_2 detector tube sampling method correlated better with the outdoor air damper status than the values obtained by the single-person method, which did not correlate at all.

The percent outdoor air intake rates measured with the direct flow technique under minimum outdoor air intake conditions were approximately 5%, which is about one-half the intake rates determined with the automated SF_6 and CO_2 systems. The reason for this difference is that the direct airflow measurements accounted only for intake through the minimum outdoor air intake fans and did not account for leakage of outdoor air through the mixed-air dampers when they were in the fully closed position. This damper leakage was verified by performing velocity measurements in the economizer outdoor air intake ducts under minimum outdoor air intake conditions when the mixed-air dampers were closed. Based on these results it is seen that under minimum outdoor air intake conditions, outdoor air leakage through the mixed-air dampers is approximately equal to the intentional minimum outdoor air intake rate.

Ventilation Rate per Person—Table 10 presents per-person ventilation rates from SF_6 decay and peak CO_2 concentrations determined with the automated CO_2 measurement system. Under minimum outdoor air intake conditions, the CO_2-based values are roughly twice those determined by SF_6 decay. At higher air change rates, as expected, the difference is not quite as large, though still significant. Under minimum percent outdoor air intake the tracer gas decay method yielded a ventilation rate of approximately 8 L/s (16 cfm) per person and the peak CO_2 method yielded about 15 L/s (30 cfm) per person. Under maximum outdoor air intake conditions the tracer gas decay method yielded about 30 L/s (60 cfm) per person and the peak CO_2 method about 50 L/s (100 cfm) per person.

As seen in other studies [9], overprediction by the peak CO_2 approach appears to occur because the CO_2 concentrations are not at equilibrium at the time of the measurements. This and most other office buildings are only occupied by an approximately constant number of people at best from about 9:00 a.m. until 12:00 noon and from 1:00 p.m. until 5:00 p.m. In the test building, this period of constant occupancy is further shortened by the implementation of flex-time schedules. Under constant occupancy, three hours are required to reach 95% of the steady-state concentration at an air change rate of 1.0 ACH and approximately

TABLE 10—*Comparison of ventilation rate per person measurements by method.*

Date	Hr	SF_6 Decay L/s per person	Peak CO_2 Auto L/s per person
7/26/91	9	27	50
7/29/91	9	29	45
7/30/91	9	29	49
7/31/91	9	32	52
7/31/91	15	8	18
8/01/91	9	30	54
8/12/91	9	27	43
8/12/91	17	8	37
8/13/91	9	30	45
8/14/91	9	28	45
8/14/91	16	9	20
8/15/91	9	33	47
8/15/91	15	9	15
8/16/91	10	25	45
8/19/91	9	31	43
8/19/91	15	9	20
8/20/91	9	29	52
8/20/91	15	9	17
8/21/91	8	26	49
8/21/91	15	9	18
1/15/92	11	7	14
1/16/92	11	7	13

six hours at a rate of 0.5 ACH. Therefore, in this and other office buildings, it is unlikely that the peak CO_2 concentration is an actual equilibrium value, leading to overpredictions of per-person ventilation rates based on peak CO_2 analysis.

The limited number of ventilation rate per-person determinations based on peak CO_2 measured with detector tubes agreed with the SF_6 measurements, but this agreement was fortuitous. The peak CO_2 approach with detector tubes has the same tendency to overpredict based on the use of pre-equilibrium concentrations. However, inaccuracies in the CO_2 concentration measurements using the detector tubes just happened to balance out the pre-equilibrium overprediction.

Under minimum outdoor air intake, the multiplicative method of determining ventilation rate per person using percent outdoor air based on the automated SF_6 and CO_2 measurements agreed with the ventilation rate per person based on SF_6 decay. This is interesting because the multiplicative method does not account for envelope infiltration, while the SF_6 decay method does. The ventilation rate per person based on direct airflow rate measurement accounts for only intake through the minimum outdoor air intake fans, and the results of these determinations were about one-half the results based on SF_6 decay. Based on these results, the amount of envelope infiltration appears to be minimal under conditions of minimum outdoor air intake. However, there appears to be outdoor air leakage into the building through the mixed-air dampers at a rate approximately equal to the intake through the minimum outdoor air intake fan, and this is the reason for the agreement between the multiplicative and tracer gas decay methods.

Measurement Issues

The approaches to ventilation assessment studied in this project have both advantages and disadvantages. The amount of effort associated with each technique, and the completeness and quality of the information obtained, is in general a function of the building being studied and the resources available to those performing the assessment. Building layout and HVAC system configuration are two important factors that impact the required level of effort. Additional resource requirements including the initial cost of the measurement equipment; the cost associated with installation, calibration, and maintenance; the number of measurements to be conducted; and the time for data analysis. The number of measurements is an important consideration when deciding between an automated monitoring system and a manual approach. This decision must involve a balance between the amount of time required to make the manual measurements and the installation time of an automated system. A discussion of some of these measurement issues follows, including estimates of the level of effort associated with each technique.

SF$_6$ Automated and CO$_2$ Automated—The initial setup of the automated SF_6 and CO_2 systems for measuring the concentrations at the main air handlers required about 10 person-weeks. However, these systems were installed for a different project, and if they were installed to conduct this study alone the installation would have required only about four person-weeks. Once the system was installed, the automated measurements required little effort to keep them running. Maintaining these systems involves calibrating the detectors; maintaining the sample pumps, valves, and controllers within the systems; changing diskettes used to store the data; and replacing compressed gas cylinders. Calibration of the SF_6 system takes about one hour, and the CO_2 system about 15 min. The CO_2 system does not require a tracer gas injection system, is easier and quicker to calibrate, and requires less maintenance than the SF_6 system.

Direct Flow—Direct airflow measurements require a detailed inspection of the HVAC system in order to identify the most suitable measurement locations. Once these locations are determined, the layout of traverse points must be determined and holes drilled in the ducts if necessary. As is often the case, traverse locations for performing these measurements in accordance with recommended practice [7] were unobtainable in this HVAC system. Traverses were performed in the only accessible locations with several different instruments, and the various approaches and measurement locations yielded similar results. Total supply airflow rates measured by performing duct traverses of the submain ducts of all three systems, some of which were inaccessible, required approximately 120 min to perform, whereas measurements taken inside all three fan boxes required only about 40 min to perform. Duct traverse measurements required about 60 min for all three minimum outdoor air intake fans.

CO_2 *Tubes*—Each measurement performed with a detector tube required approximately 10 min using three pump strokes per sample. A total of 60 min was required to measure per-person ventilation rates, based on peak CO_2 concentrations, at all three main air handlers. A similar amount of time was required to determine the percent outdoor air intake for the three air handlers. CO_2 detector tubes have the advantages of being portable and not requiring any installation time. The manufacturers of the CO_2 detector tubes used in this study state that their tubes require no calibration; however, the tubes which were used in this study often disagreed with values determined with the infrared monitor used in the automated system. Results based on the CO_2 tube measurements are subject to user interpretation and can yield very unreliable results when compared with a calibrated CO_2 monitor. This study was based on the use of only one brand of detector tube. There are other manufacturers of CO_2 detector tubes, and a more detailed study of some of these tubes is presented in Ref *11*.

CO_2 *Sample Bags*—Air sample bags and portable pumps have the advantage over an automated system of requiring no installation time, except perhaps the drilling of access holes in the ductwork. In these tests, it took about one minute to collect the air samples and another minute to measure the CO_2 concentration. Another option is to use a portable CO_2 monitor at the measurement site. The monitor must be calibrated periodically, requiring about 15 min. This calibration time is relatively insignificant with performing a large number of measurements.

CO_2 *Buildup*—This method was based on data collected by the automated CO_2 sampling system. Because the data were collected automatically, the time required for this method was primarily associated with the nonlinear regression analysis. The buildup method also requires that the CO_2 concentration data fit the model given by Eq 7. The limited amount of data analyzed in this study fit the model well, but as seen in other studies, this is not always the case [9].

Conclusions

This study of ventilation assessment in an office building concentrated on two issues, changes in building ventilation characteristics over time and a comparison of different approaches to ventilation evaluation. In the study, whole-building air change rates were measured over several months using the tracer gas decay technique. The results of these measurements were compared with a similar data set collected about three years earlier, and the comparison showed no significant changes in the ventilation rates of this building over time. The measured ventilation rates were always above the design minimum of 3 L/s (6 cfm) per person, in compliance with the recommendations of ASHRAE Standard 62-1981, the standard on which the design was based. The minimum ventilation rates were below the minimum levels recommended in ASHRAE Standard 62-1989 of 10 L/s (20 cfm) per person.

Ventilation rates per person ranged from about 7 L/s (14 cfm) per person to about 36 L/s (72 cfm) per person. Also, the measured air change rates indicate that the outdoor air intake controls are operating in accordance with design.

The measurements of ventilation system airflow rates using duct traverses showed that the minimum outdoor air intake was within 10% or 20% of design for the three minimum outdoor intake fans. Although the traverse locations were not in accordance with standard recommendations, the results of traverses using pitot tubes, hot-wire anemometers, and vane anemometers were generally within 10% of each other. In conducting these measurements, several logistical difficulties were identified such as inaccessible ductwork and modulations in system airflow rates during the traverses.

The percent outdoor air intake at the building air handlers was determined reliably using SF_6 and CO_2 balances in the supply, return, and outdoor airstreams. However, when CO_2 detector tubes were used to perform these balances, the percent outdoor air determinations were inaccurate and unreliable. These problems were due to the inaccurate determinations of CO_2 concentration using the detector tubes. The determination of percent outdoor air by dividing the outdoor air intake rate measured at the minimum outdoor air intake fans by the measured supply airflow rate yielded values approximately one-half those obtained from the tracer gas mass balance under minimum intake conditions. This difference is due to outdoor air leakage at the mixed-air dampers under minimum outdoor air intake.

Ventilation rates per person were determined from whole building SF_6 decay tests and equilibrium analysis of peak CO_2 concentrations. The values obtained from the peak CO_2 analysis were about 50% to 100% above the values obtained from SF_6 decay, depending on the ventilation rate. This overestimation presumably occurred because CO_2 concentrations in this building did not attain equilibrium due to insufficiently long periods of constant CO_2 generation, that is, constant occupancy. When the percent outdoor air was determined by an SF_6 or CO_2 mass balance and multiplied by the supply airflow rate, the resultant ventilation rate was in good agreement with the rate obtained by SF_6 decay.

All of these approaches to ventilation evaluation have certain advantages and disadvantages, but none of them yields a complete characterization of the ventilation system performance. In order to obtain a complete understanding, a combination of methods must be considered and a certain investment of resources is required. In many situations, resource limitations result in the ability to perform only a partial evaluation of ventilation. With proper planning and careful consideration, however, a partial evaluation can still yield useful and reliable information.

References

[1] Persily, A. K., "Ventilation Rates in Office Buildings" in *Proceedings,* ASHRAE/SOEH Conference IAQ 89, The Human Equation: Health and Comfort, American Society of Heating, Refrigerating, and Air-Conditioning Engineers, Atlanta, GA, 1989.

[2] Grot, R. A., Persily, A. K., Hodgson, A. T., and Daisey, J. M., "Environmental Evaluation of the Portland East Federal Office Building Preoccupancy and Early Occupancy Results," NISTIR 89-4066, National Institute of Standards and Technology, 1989.

[3] Dols, W. S. and Persily, A. K., "A Study of Ventilation Measurement in an Office Building," NISTIR 92-4905, National Institute of Standards and Technology, 1992.

[4] ASHRAE Standard 62-1981, "Ventilation for Acceptable Air Quality," American Society of Heating, Refrigerating, and Air-Conditioning Engineers, 1981.

[5] ASHRAE Standard 62-1989, "Ventilation for Acceptable Air Quality," American Society of Heating, Refrigerating, and Air-Conditioning Engineers, 1989.

[6] Persily, A. K. and Norford, L. K., "Simultaneous Measurements of Infiltration and Intake in an Office Building," *ASHRAE Transactions,* Vol. 93, Part 2, 1987, pp. 42–56.

[7] *Industrial Ventilation. A Manual of Recommended Practice,* 20th ed. American Conference of Governmental Industrial Hygienists, 1988.
[8] Ancker, K., Göthe, C., and Bjurström, R., "Evaluation of CO_2 Detector Tubes for Measuring Air Recirculation," *Environmental International,* Vol. 15, 1989, pp. 605–608.
[9] Persily, A. K. and Dols, W. S., "The Relation of CO_2 Concentration to Office Building Ventilation," *Air Change Rate and Airtightness in Buildings, ASTM STP 1067,* M. H. Sherman, Ed., American Society for Testing and Materials, Philadelphia, 1990, pp. 77–92.
[10] *ASHRAE Handbook of Fundamentals,* American Society of Heating, Refrigerating, and Air-Conditioning Engineers, 1989.
[11] Norbäck, D., Ancker, K., and Johanson, G., "Field Evaluation of CO_2 Detector Tubes for Measuring Outdoor Air Supply Rate in the Indoor Environment," *Indoor Air,* Vol. 2, No. 1, 1992, pp. 58–64.

Discussion

Stephen N. Flanders[1] (written discussion)—How was the SF_6 tracer gas distributed in the building and how was the uniformity of concentration tested?

W. S. Dols (authors' closure)—Tracer gas was injected into the supply airstreams of the three main air handlers which serve the entire building. Each air handler serves a zone which is approximately one third of the total building volume. Each of these zones is made up of one third of floors 1 through 7. Because the three zones are about the same size, tracer gas was injected at approximately equal rates and allowed to mix for about twenty minutes prior to calculating the decay rates in the return airstreams of the three main air handlers. Uniformity of tracer gas was verified by sampling the return airstreams and selected locations within the occupied space every ten minutes during the decay.

David Saum[2] (written discussion)—How typical is this building? Have there been any air quality complaints?

W. S. Dols (authors' closure)—There are many different types of office buildings and mechanical ventilation systems, and this variety makes it difficult to describe the test building or any other building as typical. The test building was certainly not unusual.

Aside from a few isolated complaints primarily associated with thermal comfort, there haven't been any major indoor air quality complaints in this building.

---

[1] USA CRREL, Hanover, NH 03755.
[2] Geomet Technologies, Germantown, MD 20874.

C. Y. Shaw[1] *and J. T. Reardon*[1]

Changes in Airtightness Levels of Six Office Buildings

REFERENCE: Shaw, C. Y. and Reardon, J. T., **"Changes in Airtightness Levels of Six Office Buildings,"** *Airflow Performance of Building Envelopes, Components, and Systems, ASTM STP 1255,* Mark P. Modera and Andrew K. Persily, Eds., American Society for Testing and Materials, Philadelphia, 1995, pp. 47–57.

ABSTRACT: This paper reports the changes in the overall airtightness values of six tall Canadian office buildings. These buildings, which are 10 to 25 stories high, were built between 1964 and 1974 and were previously tested shortly after their construction. The results indicate that as a result of various retrofit measures applied, except for one building, the building envelopes are more airtight now than 20 years ago. The improvement in the overall airtightness value at 50 Pa ranges from 0 to 43% of its original value. Details are discussed in the paper.

KEYWORDS: pressurization, air leakage, measurement, office buildings, retrofit

Air leakage affects the energy consumption and thermal comfort of a building. Moist air leaking outward in winter can also cause significant damage to a building's exterior walls. Controlling air leakage is, therefore, essential to prolong the life of building envelopes, reduce building energy use, and maintain thermal comfort.

With a long severe winter, most Canadian buildings at some time will likely be retrofitted to improve their airtightness. As a result, the airtightness of old buildings can be very much different from when they were new. About 20 years ago the Institute for Research in Construction tested eight new office buildings in the Ottawa area to measure their air leakage characteristics. As these and many other buildings of similar age are still in use, a follow-up fan pressurization test was conducted on six of these buildings to determine the changes in their airtightness characteristics. This information will be useful to assess the potential for retrofitting old office buildings in Canada. It also provides a realistic basis for establishing an achievable airtightness criterion for office buildings. This paper presents the test results.

Test Buildings

Six office buildings in the Ottawa area identified as Buildings A, B, D, E, F, and G were tested [1]. The buildings, 10 to 25 stories high and built between 1964 and 1974, were previously tested between 1970 and 1974 [2,3]. A detailed description of the buildings is given in Table 1.

---

[1] Research Officers, Institute for Research in Construction, National Research Council, Ottawa, Canada.

TABLE 1—Description of test buildings.

				Building			
	A	B	D	E	F	G
Year constructed	1970	1964	1971	1968	1973	1974
Year-first test	1970	1971	1971	1974	1974	1974
Year-second test	1991	1991	1991	1991	1991	1991
No. of typical floors	9	17	20	21	16	25
Floor plan m × m	51 × 64	27 × 43	23 × 28	25 × 48	25 × 56	37 × 44
ft × ft	166 × 210	88 × 140	75 × 93	83 × 158	83 × 183	123 × 143
Floor height, m	4	3.4	3.2	3.2	3.2	3.2
ft	13	11	10.5	10.4	10.6	10.6
Wall area per floor m²	908	466	328	466	525	524
ft²	9776	5016	3528	5013	5656	5639
Window area (% wall area)	38	33	26	35	52	26
Ratio of roof to total wall	31%	12%	8%	11%	15%	11%
Window type	Fixed sealed double glazing	Openable sealed double glazing	Fixed sealed double glazing	Fixed sealed double glazing	Fixed sealed double glazing	Fixed sealed double glazing
Wall construction	Precast concrete 203 mm (8 in.) tile 51 mm (2 in.) insulation air space 152 mm (6 in.) tile plaster	Precast concrete panel 51 mm (2 in.) tile insulation	Metal panel air space 51 mm (2 in.) insulation 508 mm (20 in.) concrete	Metal panel 51 mm (2 in.) insulation	Precast concrete panel 25 mm (1 in.) insulation	Precast concrete panel 25 mm (1 in.) insulation

Measurement Method

The test method used in the present study was identical to the previous one. The locations of airflow measuring stations were identical to the previous tests (this was possible because most of the total pressure averaging tubes used in the previous tests were not removed from the supply air ducts). As shown in Fig. 1, the test method [2,3] involves pressurizing the test building using the building's supply air systems with 100% outside air and measuring the corresponding pressure differences across the building envelope. All the return and exhaust fans are turned off during the test. Supply airflow rates are varied to create four or five different pressure differentials between 0 and 80 Pa. These airflow rates are measured at the supply air ducts of each supply air system using a pair of total pressure averaging tubes and a static pressure probe. Airflow rate measurements are accurate to within 6.5% of the measured values [2].

The corresponding pressure differences across the building envelope are measured at the ground and roof levels using an electronic manometer with a strip chart recorder (accurate to within 5% of the measured values). The average of the measured pressure differentials at the ground and roof levels for each airflow rate is used to represent the mean pressure difference across the building envelope. Prior to and immediately after the test, the supply air fans are turned off and the pressure differences across the envelope are measured. The before and after two readings obtained at each location (ground and roof) are averaged to give the base readings of the pressure difference for the two locations. These base readings are then subtracted from subsequent pressure difference measurements to minimize wind and stack effects which exist when tests are conducted under different inside-outside air temperatures and wind conditions [4]. The air leakage characteristic of the test building is then determined by plotting the measured airflow rates against the measured average pressure differences across the building envelope.

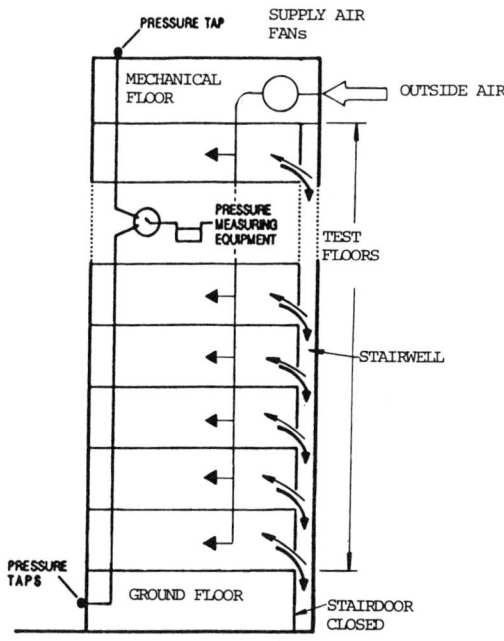

FIG. 1—*Test setup.*

Results and Discussion

Figures 2 and 3 show the previous and current measured overall airtightness values of the six buildings. The results indicate that the overall airtightness values of the six buildings at 75 Pa varied from 2.6 to 5.2 L/s · m² (0.5 to 1 cfm/ft²). The corresponding values measured 20 years ago varied from 2.2 to 6.1 L/s · m² (0.44 to 1.2 cfm/ft²) [3]. For comparison, the overall airtightness values of eight U.S. office buildings at 75 Pa measured by Persily and Grot [5] ranged from 1.1 to 5.2 L/s · m² (0.22 to 1 cfm/ft²).

Except for Building A, all data lie within a narrow band. Building A is only half as high as the other buildings and its roof area is about twice that of other buildings. The leakage through the roof and basement, therefore, may play a larger role in Building A's overall air leakage than in the other buildings. For this reason, Building A may not be representative of the same category of tall buildings as the others, which typically have a smaller roof-to-wall area ratio (see Table 1). Ignoring Building A, the data were fitted to the standard air leakage equation with a flow exponent of 0.65 [3]

$$Q = CA(\Delta P)^{0.65}$$

where

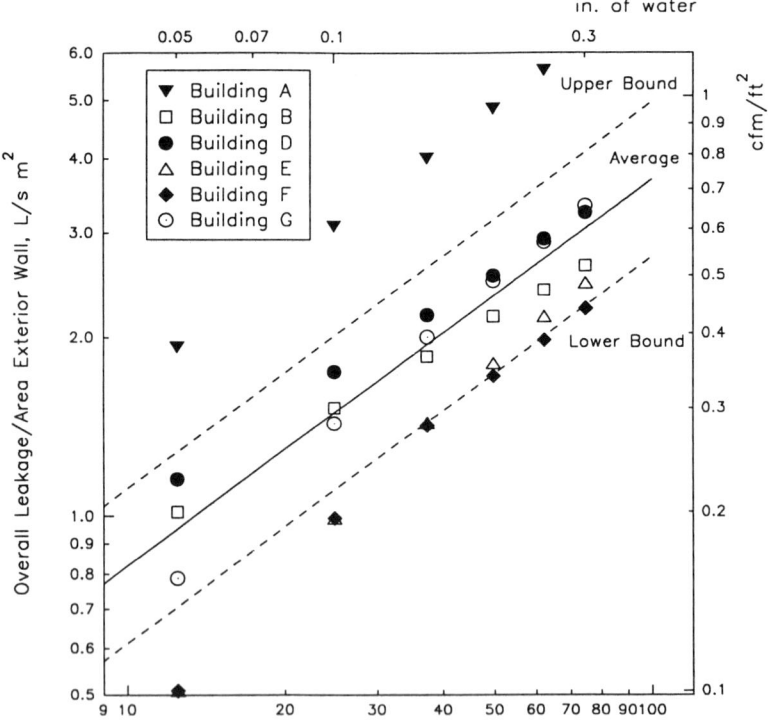

FIG. 2—*Overall airtightness values, all buildings, previous.*

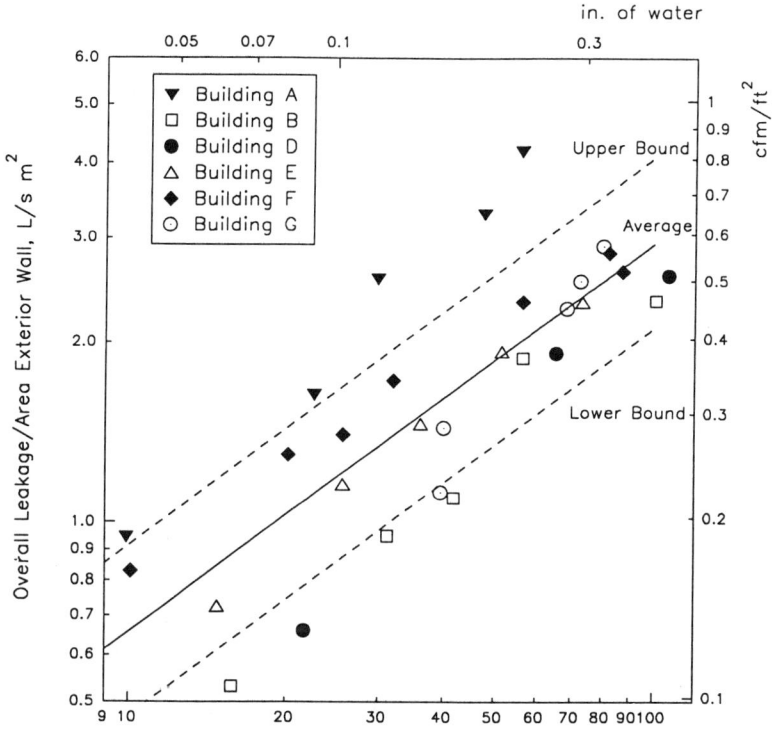

FIG. 3—*Overall airtightness values, all buildings, present.*

Q = overall airtightness value, L/s [cfm],
C = flow coefficient, L/s · m² · (Pa)^{0.65} [cfm/ft² · (in. H$_2$O)^{0.65}],
A = area of exterior wall, m² [ft²], and
ΔP = pressure difference across envelope, Pa [in. H$_2$O].

Three curves were obtained representing the lower and upper limits and the average value. The values of the flow coefficient in L/s · m² · (Pa)^{0.65} [cfm/ft² · (in H$_2$O)^{0.65}] were

	Lower Limit	Average	Upper Limit
Present study	0.106 [0.753]	0.147 [1.045]	0.205 [1.457]
Previous study	0.137 [0.974]	0.185 [1.315]	0.249 [1.770]

For comparison, the corresponding flow coefficients for the previous study are also given [2,3]. The above coefficients can be used by designers to estimate air infiltration rates for heating and cooling load or energy consumption calculations.

The results indicate that, except for Building F, the building envelopes are now more airtight than 20 years ago. Discussions with the property managers and building engineers indicated that except for Building F, Buildings B and D have been extensively retrofitted to improve airtightness and the other three have been partially retrofitted. The applied retrofit measures and their effects on the overall airtightness level are discussed below.

Building A: A new vapor barrier with 100-mm-thick (4 in.) rigid insulation was installed for the 10th floor and part of the 9th floor of this 10-storey building. As this was the leakiest building, Fig. 4 shows that even a partial retrofit measure made quite an improvement in its envelope airtightness. As shown, at 50 Pa, the overall air tightness value with its estimated uncertainty was reduced from 4.85 ± 0.01 (0.95 ± 0.002) to 3.65 ± 0.23 (0.72 ± 0.05) L/s · m² (cfm/ft²), an improvement of 25% [6].

Building B: All windows were recaulked and resealed. All vertical columns were sealed from the inside. Figure 5 shows the previous and current measured overall airtightness values of this building. It indicates that these two retrofit measures caused a change in the overall airtightness value from 2.17 ± 0.02 (0.43 ± 0.004) to 1.36 ± 0.19 (0.27 ± 0.04) L/s · m² (cfm/ft²) at 50 Pa, an improvement of 37%.

Building D: This building underwent the most extensive retrofit. A new curtain wall cladding system was installed. The overall airtightness values before and after the retrofit are shown in Fig. 6. At 50 Pa, the overall airtightness value reduced from 2.54 ± 0.02 (0.50 ± 0.004) to 1.44 ± 0.16 (0.28 ± 0.03) L/s · m² (cfm/ft²), an improvement of 43%.

Building E: Although all joints in the curtain wall were recaulked in 1990, Fig. 7 shows that the overall airtightness level was practically unchanged. As the airtightness level would likely deteriorate with time, this retrofit measure appeared to be sufficient to offset the effect of aging.

Building F: No retrofit measures were applied to this building. Figure 8 indicates that the overall airtightness level of this building deteriorated with time. At 50 Pa, the airtightness value increased from 1.73 ± 0.06 (0.34 ± 0.01) to 2.13 ± 0.12 (0.42 ± 0.02) L/s · m² (cfm/ft²). For this building, the overall airtightness value at 50 Pa was 23% more (i.e., leakier) now than 20 years age.

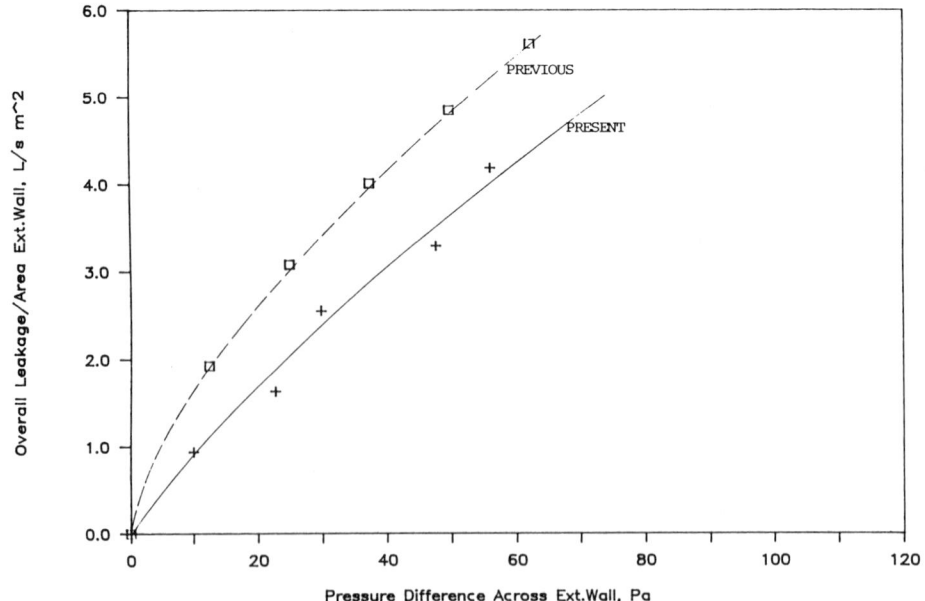

FIG. 4—*Overall airtightness values: Building A.*

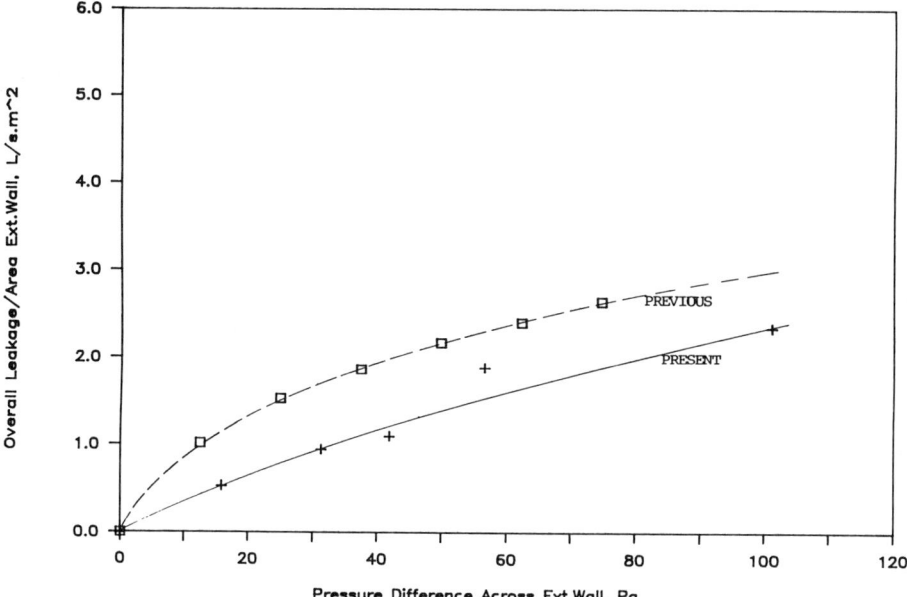

FIG. 5—*Overall airtightness values: Building B.*

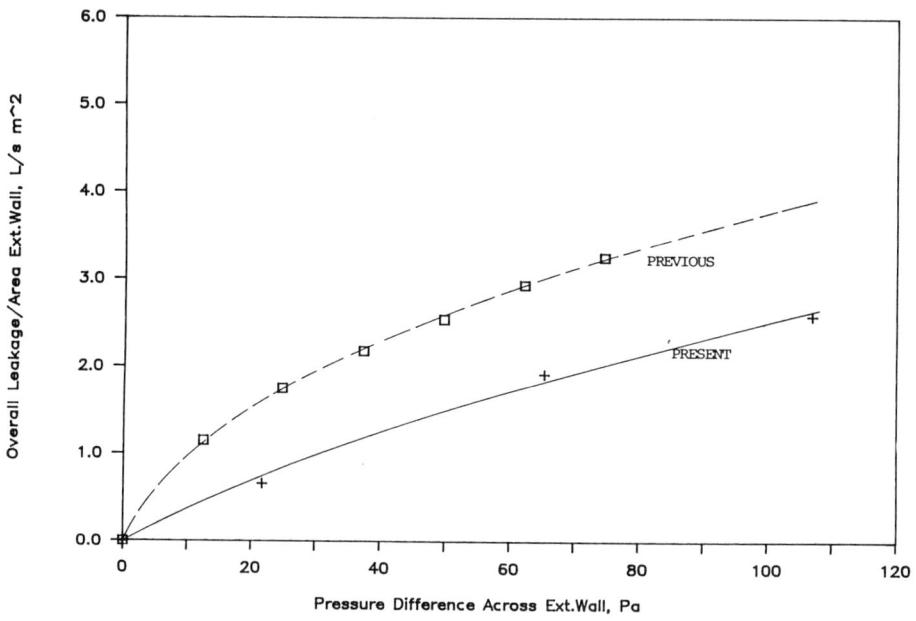

FIG. 6—*Overall airtightness values: Building D.*

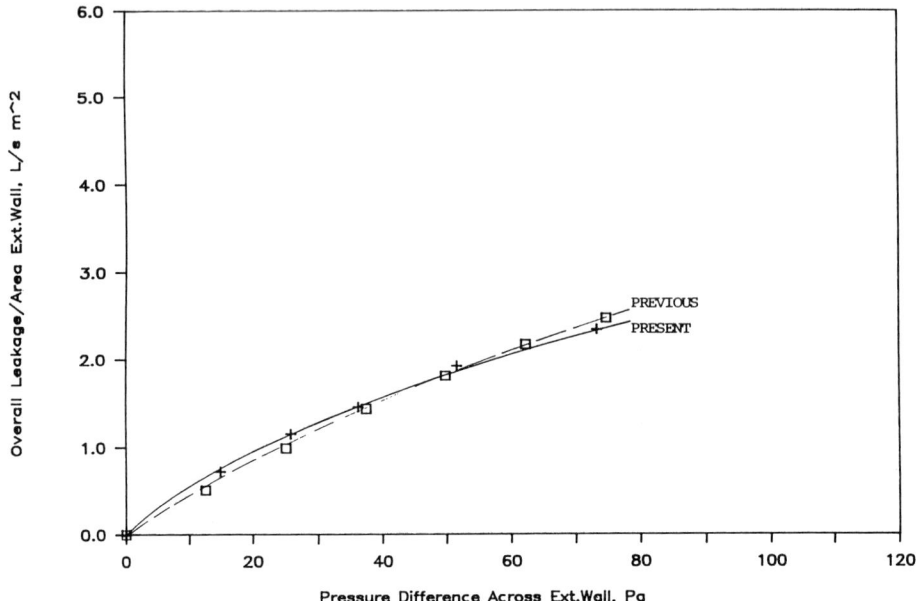

FIG. 7—*Overall airtightness values: Building E.*

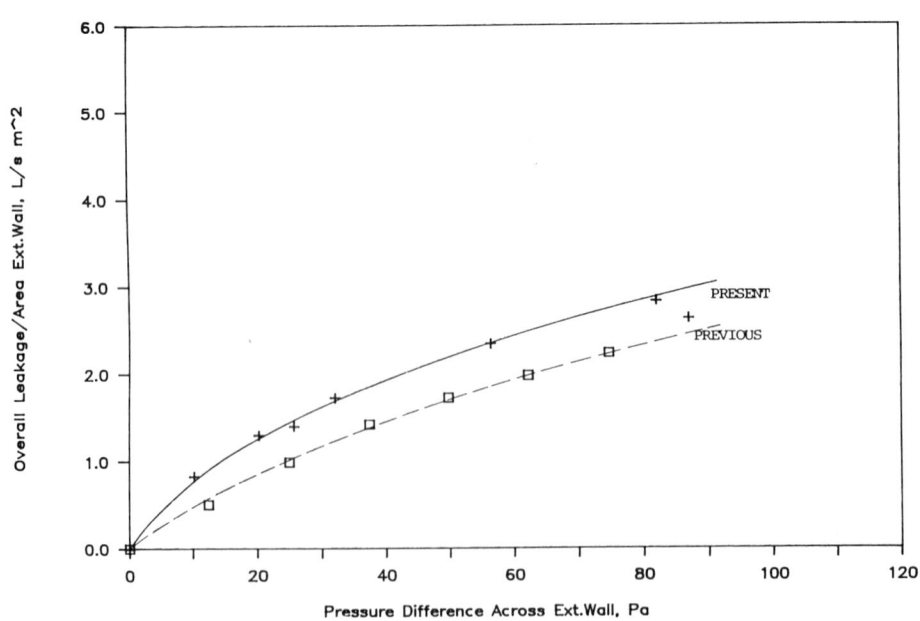

FIG. 8—*Overall airtightness values: Building F.*

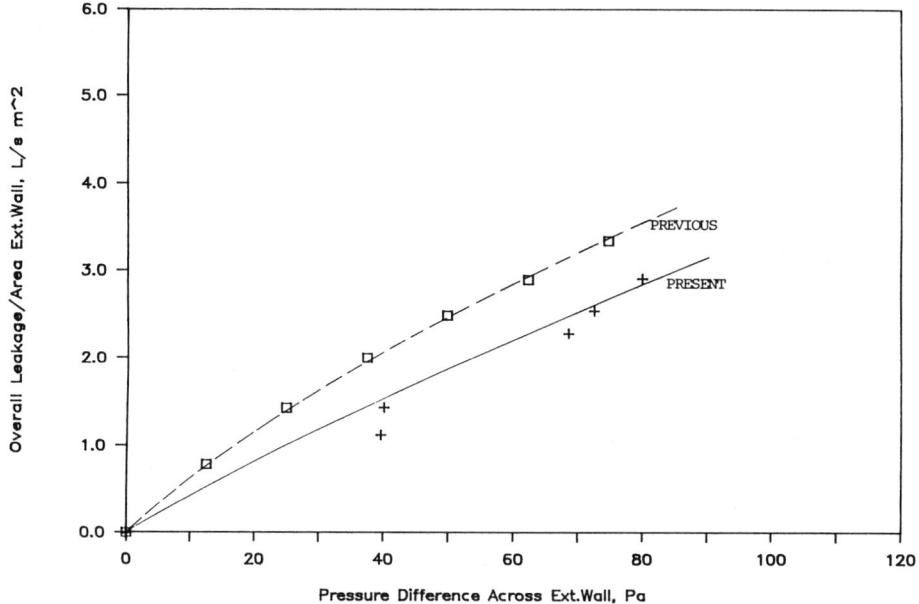

FIG. 9—*Overall airtightness values: Building G.*

Building G: A new roof was installed. Figure 9 indicates that with a new roof, the overall airtightness value at 50 Pa reduced from 2.49 ± 0.04 (0.49 ± 0.01) to 1.8 ± 0.14 (0.35 ± 0.3) L/s · m² (cfm/ft²), a 28% improvement.

To give some indication of the effect of air sealing on the building's energy consumption, the energy consumption data of Building D before and after the retrofit are given in Table 2. The heating energy consumption data after retrofit were adjusted by multiplying the re-

TABLE 2—*Energy consumption data of Building D before and after retrofit.*

		Heating Energy Consumption	
	Degree Days (DD)	(kWh)* Uncorrected	(kWh) Corrected for DD
BEFORE RETROFIT			
Dec. 1983	837.3	540 089	...
Jan. 1984	950.4	511 314	...
Feb. 1984	637.4	406 309	...
	Total	1 457 712 kWh	
AFTER RETROFIT			
Dec. 1988	815.3	402 914	413 786 [= (837.3/815.3) *kWh]
Jan. 1989	807.1	435 955	513 358
Feb. 1989	772.6	447 908	369 527
	Total	1 296 671 kWh	
	Energy reduction = (1 296 671 − 1 457 712)/1 457 712		
	= −11%		

* kWh = kilowatt hours = 3.6 MJ.

corded consumption in kilowatt hours (kWh) by the ratio of before retrofit to after retrofit degree days. For Building D, the recorded energy consumption for three winter months (i.e., December, January and February) before and after the retrofit was 1 457 712 kWh and 1 296 671 kWh, respectively. This translated to a corrected energy reduction of 11% with respect to the energy consumption before retrofit.

In Canada, a large number of old buildings are in need of repair to improve airtightness, extend building life, and reduce energy consumption. There is a need to establish an achievable airtightness criterion for building retrofit. To do so, it is necessary to know what are the commonly applied retrofit measures and their results. This study will provide such information.

Summary

Fan pressurization tests were conducted on six Canadian office buildings to determine the change in their airtightness levels 20 years after they were first tested. Of the six buildings, five have been retrofitted to improve airtightness. The results are summarized as follows:

1. Building F is now 23% leakier than 20 years ago. As no major work has been done on the envelope, aging is probably the main cause for the decrease in its envelope airtightness.
2. Buildings A, B, D, E and G are more airtight now than 20 years ago. The maximum improvement in the overall airtightness value at 50 Pa was 43% of its original value (Building D). The energy reduction caused by the retrofit for Building D was estimated to be 11%. Therefore, old office buildings can and should be retrofitted to improve their airtightness. Significant improvement can be realized in the overall airtightness with a corresponding reduction in energy consumption due to air infiltration.
3. The overall airtightness values of these buildings at 50 Pa vary from 1.36 to 3.65 L/s · m^2 (0.27 to 0.72 cfm/ft^2). These results can be used by designers to estimate air infiltration rates for heating and cooling load or energy consumption calculations. They also provide a realistic basis for establishing an achievable airtightness criterion for office buildings in cold climates, particularly for buildings which are to be retrofitted.

Acknowledgments

The authors wish to acknowledge the financial support from Public Works Canada and, particularly, the assistance of Dr. M. S. Cheung. The authors also wish to thank the property managers and building engineers of the test buildings for their cooperation and assistance during the tests. The energy consumption data were supplied by Mr. B. P. Solomon of Carleton University and the field tests were conducted by R. A. G. Evens, R. J. Magee, H. J. Weichert, B. Belwa and J. Egan.

References

[1] Shaw, C. Y., Reardon, J. T., and Cheung, M. S., "Changes in Air Leakage Levels of Six Canadian Office Buildings," *ASHRAE Journal,* Feb. 1993.
[2] Shaw, C. Y., Sander, D. M., and Tamura, G. T., "Air Leakage Measurements of the Exterior Walls of Tall Buildings," *ASHRAE Transactions,* Vol. 79, 1973.
[3] Tamura, G. T. and Shaw, C. Y., "Studies on Exterior Wall Air Tightness and Air Infiltration of Tall Buildings," *ASHRAE Transactions,* Vol. 82, 1976.

[4] Shaw, C. Y., Gasparetto, S., and Reardon, J. T., "Methods for Measuring Air Leakage in High-Rise Apartments," *Air Change Rate and Airtightness in Buildings, ASTM STP 1067,* M. H. Sherman, Ed., American Society for Testing and Materials, Philadelphia, 1990.

[5] *ASHRAE Handbook, Fundamentals,* Chapter 23, American Society of Heating, Refrigerating, and Air Conditioning Engineers, Atlanta, GA, 1989.

[6] "Determination of the Airtightness of Building Envelopes by the Fan Depressurization Method," CAN/CGSB-149.10-M86, Canadian General Standards Board, 1986.

Arthur C. Kozik,[1] Paul Oppenheim,[1] and David Hintenlang[2]

The Effect of Varying Levels of Induced Duct Leakage on Differential Pressures in a Florida House

REFERENCE: Kozik, A. C., Oppenheim, P., and Hintenlang, D., **"The Effect of Varying Levels of Induced Duct Leakage on Differential Pressures in a Florida House,"** *Airflow Performance of Building Envelopes, Components, and Systems, ASTM STP 1255,* Mark P. Modera and Andrew K. Persily, Eds., American Society for Testing and Materials, Philadelphia, 1995, pp. 58–67.

ABSTRACT: The objective of this research is to quantify the effect of varying levels of induced duct leakage on differential pressures and subsequent radon levels in a typical Florida residence. The heating and air conditioning (HAC) system of a residence was modified with the addition of a fresh air ventilation system. This modification provided the ability to simulate either a return leak or a supply leak in the HAC system. The amount of leakage induced can be varied using flow control dampers and an auxiliary fan located in the ventilation system ductwork.

The house was pressurized and depressurized by adding 0.06 m³/s of ventilation air into the return side and removing 0.05 m³/s of ventilation air from the supply side of the air handling system, respectively. The subsequent changes in house pressure and ventilation rates reduced the indoor radon concentrations from 1865 Bq/m³ to 170 Bq/m³ under pressurized conditions and to 305 Bq/m³ under depressurized conditions.

KEYWORDS: duct leakage, radon, ventilation

There are many factors that affect indoor radon concentrations that are not yet fully understood. The migration of radon into a structure is a function of a number of variables [1] which include: the radon source strength in the soil, the permeability of the soil, the cumulative size of penetrations in the house to soil interface, pressure differentials across the house to soil interface, ventilation rates, and atmospheric pressure fluctuations [2–4]. The objective of this study is to examine how the operation of the heating, ventilating, and air conditioning (HVAC) system affects pressure differentials across the shell and soil/structure interface, and primarily how the effects of creating controlled duct leaks influence house pressures and radon entry.

Differential pressures across a house shell are a function of the natural forces of winds, thermal buoyancy, and the mechanical forces and duct leakage associated with air moving equipment [5,6]. Air moving equipment in a structure can include the blower of the HVAC system, exhaust fans, attic fans, clothes dryers, and kitchen exhaust fans. A blower door can be used to simulate the net pressurization or depressurization created by these mechanical

---

[1] Graduate research assistant and associate professor, respectively, University of Florida M. E. Rinker School of Building Construction, Gainesville, FL 32611.

[2] Associate professor, University of Florida Department of Nuclear Engineering Sciences, Gainesville, FL 32611.

devices. We observed, however, that the flow rates produced by the blower door were unstable at the low flow rates required to simulate the driving forces produced by the HVAC system alone. Therefore, a system permitting controlled infiltration/exfiltration was added to the HVAC system to study varying levels of air duct leakage effects on the house and was subsequently compared to the results of blower door tests. Differential pressures were measured across the slab and the house shell while varying levels of air duct infiltration or exfiltration were maintained.

Laboratory Facility

The experimentation for this project was conducted in an unoccupied house located in Gainesville, FL. The house is part of the Florida Radon Research Project (FRRP) sponsored by the U.S. Environmental Protection Agency (EPA) and the Florida Department of Community Affairs. The house has been used as a dedicated testing facility since the summer of 1991. The house is a single-story concrete block and wood frame structure representative of houses built in the region between 1965 and 1973. There are three bedrooms, two bathrooms, a family room, a living room, a kitchen, and a two-car garage. The washer and dryer are located in the garage (floor plan, Fig. 1). The house foundation consists of a floating slab-on-grade contained by a concrete block stem wall. The total living area is approximately 163 m². The ventilation rate of the house with the air handler operating under the original configuration is 0.3 air changes per hour (ACH), which is typical for the Florida Housing stock [2]. Detailed house characteristics are given in Table 1.

The air handling system in the house is composed of a central forced air gas furnace and split air conditioner. The furnace is located in the garage and the condensing unit is located outdoors. The details of the forced air system are presented in Fig. 1 and Table 2. The supply air ductwork is a single rectangular trunk line constructed of 2.54 cm (1 in.) fiberglass duct board with round insulated sheet metal ducts feeding the individual spaces. A centralized return is used in the original duct system configuration. Duct integrity was evaluated using a protocol developed as part of the FRRP, Test Method for Determining HAC Duct System Leakage [7] for evaluating Florida-style houses. The method consists of mechanical depressurization of the structure and its HAC duct system and measurements of the resulting airflow

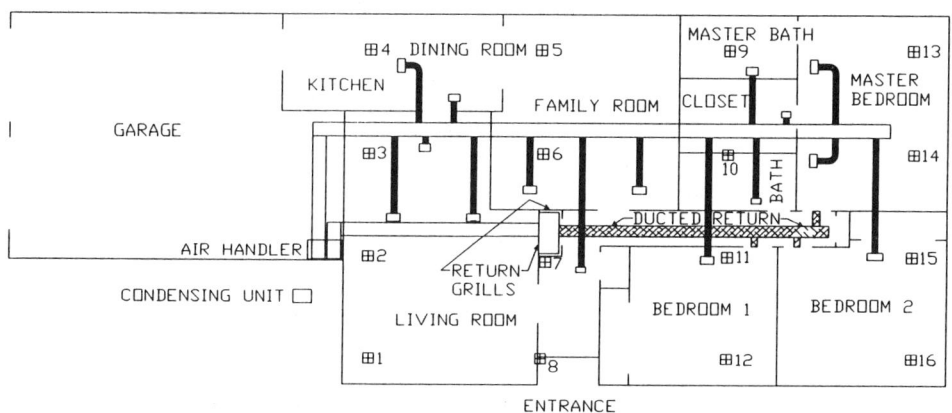

FIG. 1—*Plan view of research house.*

TABLE 1—*Laboratory house characteristics.*

Location	2418 NW 64th Terrace, Gainesville FL
Year Constructed	1970 (approximately)
Style	single floor, slab on grade
Floor area	163 m²
Volume	420 m³
Construction	
Exterior Walls	20 cm (8 in.) concrete block with R-5 insulation 88 m²
Adjacent Wall	Wood frame with R-11 insulation 7 m²
Roof	Ceiling with R-19 insulation under a ventilated attic space
Windows	Single pane with metal frame 4 m² North, 14 m² E/W
Doors	Solid core wood doors and metal frame sliding door in the family room
Floor	Concrete slab on grade with no edge insulation
House Leakage	8.6 ACH50

rates at several indoor-outdoor static pressure differences. The first portion of the procedure is with the duct system as operated and the second is with the duct system sealed from the interior air. The air leakage characteristics of the HAC duct system are then evaluated from the relationship of the various airflow rates and pressure differentials.

The air handling system in the house was modified in July 1992 to allow the introduction or removal of air from the distribution system. These modifications allowed for effective

TABLE 2—*Forced air system.*

Gas Furnace	General Electric, Model BLU100C942A1, 29 kW input, 23 kW bonnet capacity
Air Conditioner	General Electric, Model BGTA742B1B, 5 kW condensing unit
Blower	370 W direct drive
Distribution System	Supply trunk constructed of 2.5 cm (1 in.) fiberboard 48 cm by 41 cm to 30.5 cm by 30.5 cm
	Runouts: 10 cm to 20 cm round
	Conventional central return: 48 cm × 41 cm
	Ducted return
	Main trunk
	Master Bedroom 20 cm round
	Bedroom 1 15 cm round
	Bedroom 2 15 cm round

pressurization or depressurization of the house by forced infiltration/exfiltration. The system modifications were made in the garage of the house, outside the living space.

Round snaplock aluminum ductwork (25.4 cm [10 in.]) connects the existing return and supply ductwork of the furnace to a fabricated aluminum box housing two 40 W auxiliary in-line duct fans, one to provide additional airflow in through the supply duct, and one to provide additional airflow out through the exhaust duct. The new aluminum ductwork connects to the 2.54 cm (1 in.) ductboard of the existing return and supply ducts by way of ductboard taps, followed by 25.4 cm guillotine dampers that allow the modified portion of the system to be isolated from the original duct system.

The new ductwork is attached to the fan box through 25.4 cm to 15.2 cm (10 in. to 6 in.) reducers. The fans are operated independently of one another via variable speed controllers. Following the fan box are two fabricated boxes, each containing one adjustable flow control damper. These are used in conjunction with the variable speed fans to control the airflow being introduced or removed from the original distribution system. The 25.4 cm ducts then continue 6 m across the garage and exit through an existing door opening. The return tap exits the garage approximately 1.2 m above the supply tap to minimize re-entrainment in the event the ducts are operated simultaneously.

Experimental Methodology

Data Acquisition

The system being used for data collection in the project house consists of a stand-alone personal computer interfaced to a data acquisition and control system. The data currently being collected include ambient and indoor conditions. The ambient conditions being measured include temperature, relative humidity, barometric pressure, wind speed and direction, and average rainfall. The indoor conditions being measured include indoor radon concentrations, temperature, humidity, and house shell pressure differentials.

The pressure transducers used to measure house to sub-slab pressure differentials are Setra model C-264 low range transducers with a full scale of ± 25 Pa and a minimum sensitivity of ± 0.25 Pa. The transducers are laid out on a grid pattern in the house as displayed in Fig. 1. These pressure transducers are set up to measure differential pressures between the region immediately below the slab and the region immediately above the slab. The pressure transducers used to measure indoor to outdoor pressure differentials are located in the north and east wall. A Pylon model AB-5 was used as a continuous indoor radon monitor, and counts were conducted at 10-min intervals with an uncertainty of $\pm 10\%$. A hand-held anemometer, an electronic digital manometer (EDM), and a flow hood (volumetric flow) were used to take duct system measurements on the existing and modified system.

Testing Procedures

A set of five test conditions was used to quantify the effect that operation of the ventilation system has on supply and return airflows, and house pressurization and depressurization. The five tests conducted were:

> Test 1, House pressurization: auxiliary fan operating; return side duct leak open; supply side duct leak closed.
> Test 2, House depressurization: auxiliary fan operating; return side duct leak closed; supply side duct leak open.
> Test 3, House normal: return side duct leak closed; supply side duct leak closed.

Test 4, House pressurization: auxiliary fan assembly removed; return side duct leak open; supply side duct leak closed.

Test 5, House depressurization: auxiliary fan assembly removed; return side duct leak closed; supply side duct leak open.

During each of these tests, differential pressure data, radon data, and barometric pressure data were examined. For all tests the HVAC system blower was switched to the on position and the compressor was actuated by thermostatic control.

Differential pressures and radon data are reported for two zones of the house, the kitchen and master bedroom. The two pressure test points chosen for comparison, pressure transducer 13 (PT-13) in the master bedroom zone and pressure transducer 4 (PT-4) in the kitchen zone, were selected because of similar locations in reference to the slab edge, displayed in Fig. 1. The results from both zones are reported because there is a noticeable difference in the house to subslab differential pressure (ΔP) measurements for the test periods. The difference between the average ΔP measured at PT-4 and PT-13 was 0.5 Pa for both pressurization tests. The difference between the average ΔP measured at both points for the maximum depressurization test was 0.1 Pa. The pressures in the fill below the test points are usually not identical because of the dynamic environmental forces acting on the structure and the compacted sand fill material that limits the rate of pressure equalization under the foundation.

Results and Discussion

The continuous data collected during the tests were analyzed to provide differential pressures across the house shell and barometric pressure as a function of time. By creating a controlled leak in the return or supply side of the air distribution system, the house is effectively pressurized or depressurized, respectively. The modified air handling system provided good control over the small magnitudes of house pressurization and depressurization required for this study. Differential pressures across the house shell of 0.5 to 2.2 Pa were able to be maintained over a period of several days with less than 25% variation during periods of stable weather. The maximum net pressurization that could be achieved was 1.7 Pa, which corresponds to a measured infiltration rate of 0.06 m³/s. The flow rates required to develop this pressure are smaller than predicted from an extrapolation of blower door data. This is not surprising since the blower door data is collected at higher differential pressures (≥ 10 Pa) [6] and permits additional leaks to contribute to the overall leakage rate. Table 3 displays house supply/return quantities and the magnitude of house pressurization for the five test conditions. The reduction in supply quantities from Test 1 to Test 2 occurs as a result of the simulated duct leak being moved from the return side to the supply side of the system.

Tests 4 and 5 relied only on the system circulation fan to produce house pressurization and depressurization, respectively. This configuration produced the largest magnitude of pressurization and depressurization. The auxiliary in-line fans used in Tests 1 and 2 actually resulted in smaller airflows into and out of the system than achieved in Tests 4 and 5. Although the fans were expected to increase the airflows, the combination of the restriction in the duct work to the 15.2 cm fan diameter and limited fan capacity limit the available flow. The volumetric airflow rate through the unrestricted ductwork (0.06 m³/s) is greater than the airflow capable of being driven by these fans (estimated to be about 0.03 m³/s).

The maximum house pressurization resulted from a simulated leak of 0.06 m³/s in the return side of the existing distribution system. This was accomplished by opening the return side duct leak with the auxiliary fan assembly removed from the duct. The maximum house depressurization resulted from a simulated leak of 0.05 m³/s in the supply side duct of the

TABLE 3—*Supply and return quantities.*

Room (Register #)	Test #1 (m³/s)	Test #2 (m³/s)	Test #3 (m³/s)	Test #4 (m³/s)	Test #5 (m³/s)
Living Room (1)	0.028	0.024	0.028	0.031	0.022
(2)	0.039	0.033	0.039	0.039	0.032
(3)	0.037	0.033	0.037	0.039	0.033
Kitchen (1)	0.040	0.034	0.042	0.042	0.035
(2)	0.033	0.029	0.033	0.033	0.028
Family Room (1)	0.033	0.031	0.035	0.037	0.031
(2)	0.031	0.028	0.034	0.037	0.032
Foyer	0.009	0.009	0.009	0.007	0.006
Bedroom 1	0.029	0.027	0.031	0.032	0.028
Bedroom 2	0.046	0.042	0.047	0.050	0.042
Master BR (1)	0.032	0.029	0.033	0.034	0.031
(2)	0.031	0.028	0.031	0.032	0.028
Bath 1	0.009	0.007	0.012	0.009	0.009
Master Closet	0.012	0.09	0.012	0.014	0.012
Master Bath	0.028	0.024	0.026	0.027	0.026
Supply Total	0.437	0.387	0.449	0.463	0.395
Return Total	0.297	0.340	0.335	0.281	0.340
Ventilation Air into Return	0.03	...	...	0.06	0.06
Ventilation Air into Supply	...	0.03	...	...	0.05
House Pressure (Pa) (Ref. to ambient)	+1.0	−0.2	+0.5	+2.2	−0.5

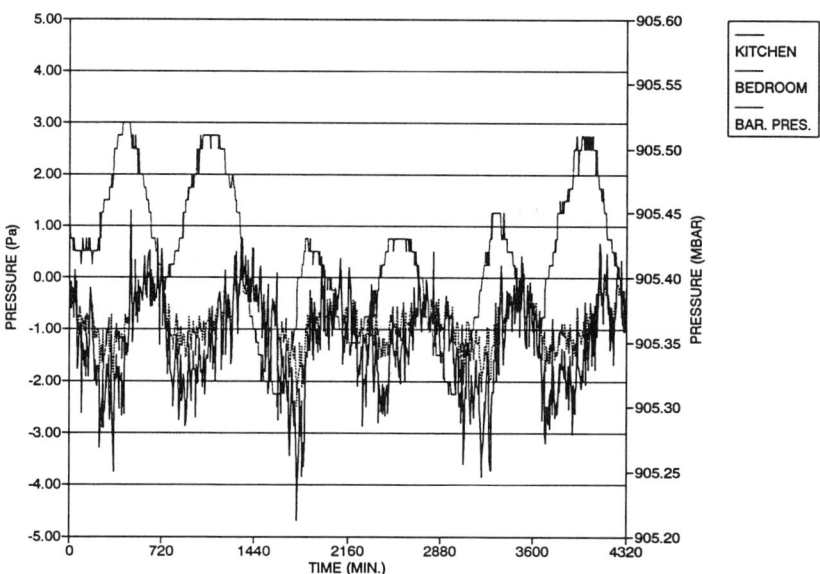

FIG. 2—*9/18 maximum pressurization test.*

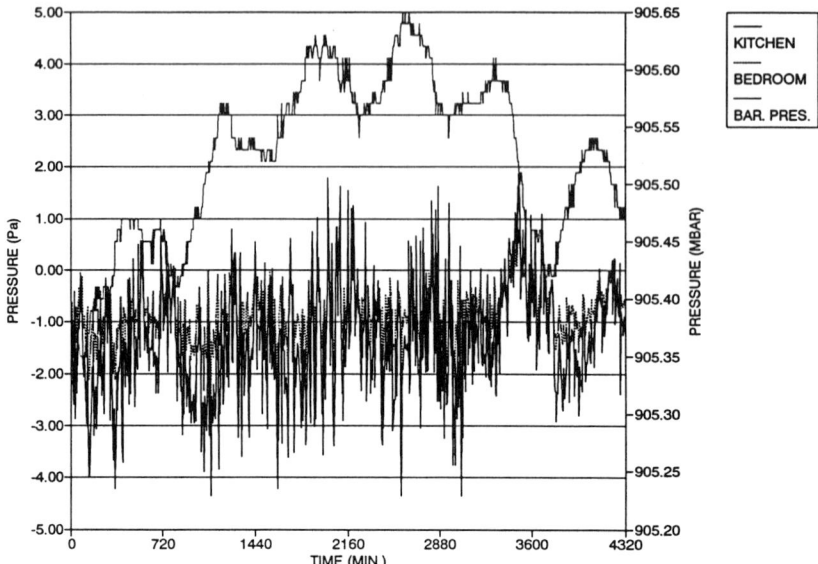

FIG. 3—*9/29 maximum pressurization test.*

existing distribution system. This is also occurred with the auxiliary fan assembly removed from the duct.

Figures 2, 3, and 4 show differential pressure and barometric pressure data over two 3-day periods of maximum pressurization conditions, and a 3-day period of depressurization

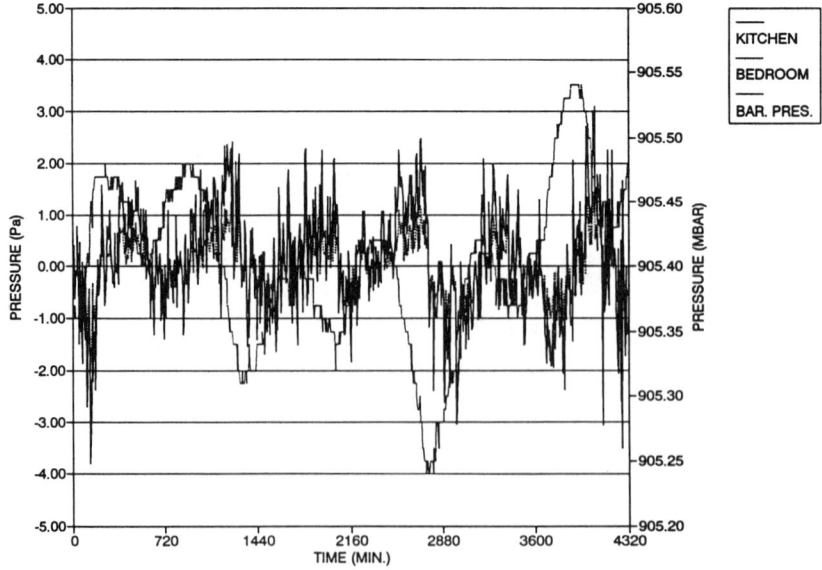

FIG. 4—*Maximum depressurization test.*

conditions, respectively. The average ΔP between indoor and subslab at PT-4 and PT-13 for the pressurization periods were 1.3 Pa and 0.8 Pa, respectively. During both pressurization tests the house was at a positive pressure relative to the subslab region, reducing the rate of radon migration across the slab. The average ΔP at PT-4 and PT-13 during the depressurization test was -0.1 Pa and -0.1 Pa, respectively. These values represent a negative pressure in the house relative to the subslab soil gas which can increase radon entry into the structure. Table 4 displays the average indoor to subslap ΔP, average radon levels, and average barometric pressure readings for each of the three test periods.

Fluctuations in barometric pressure have previously been shown to produce a distinct effect on indoor to subslab differential pressure [4,8–10]. The differential pressures created by the changes in barometric pressure are the driving forces with which whole house pressurization must compete to prevent radon entry through advective flow. This variation in indoor to subslab differential pressure is observed in this structure and demonstrates that even though the indoor pressure is positive relative to ambient (as great as 2.2 Pa) it does not always maintain the indoor pressures above the subslab gas pressures. Consequently, even though radon entry is reduced it is not eliminated.

While the house is under static conditions, HVAC off and interior doors open, the indoor radon concentrations averaged 1865 Bq/m^3. The average radon concentration for the two pressurization tests was 170 Bq/m^3. This reduction in radon levels can be attributed to the increased pressure in the house, which reduces the radon entry rate, along with the dilution of radon concentrations by the 0.6 ACH increase in the ventilation rate. When the house was under a negative pressure the radon concentrations were reduced to 305 Bq/m^3. This reduction of the indoor radon concentrations from the static conditions must be a result of only the increased house ventilation rate, since the radon entry rate would be expected to increase under these conditions.

Conclusion

Return and supply duct leakage of a magnitude of 0.06 m^3/s effectively pressurizes or depressurizes the interior of this residential structure by approximately 1 Pa. The flow rates required to produce this pressure are significantly smaller than predicted by extrapolating the results of blower door testing to the low-pressure regime. Under a return leak situation the interior of the structure becomes pressurized relative to the subslab region and ambient. Under a supply leak situation the house interior becomes depressurized relative to the subslab region and ambient.

The pressurization or depressurization resulting from air duct leakage effectively reduces indoor radon levels. The radon levels in the research house were reduced from an average of 1865 Bq/m^3 under static conditions, down to 170 Bq/m^3 under pressurized conditions and 305 Bq/m^3 under depressurized conditions. Of the two methods studied, the one that delivers the highest degree of positive control and more effectively reduces indoor radon levels is house pressurization. Even small levels of house pressurization offer a solution that can effectively reduce indoor radon concentrations. The practical implementation of this technique should use a controlled source of infiltration air that can be properly filtered and conditioned rather than allowing the uncontrolled entry of air from unknown sources.

TABLE 4—*Average readings for each of the three tests.*

Room	Test Period	Pressurization 9/18 (Test 1)			Pressurization 9/29 (Test 1)			Depressurization 9/21 (Test 2)		
		ΔP House to Subslab, Pa	Barometric Pressure, kPa	Radon, Bq/m³	ΔP House to Subslab, Pa	Barometric Pressure, kPa	Radon, Bq/m³	ΔP House to Subslab, Pa	Barometric Pressure, kPa	Radon, Bq/m³
Kitchen	0 to 1440 min	1.1	100.9	230	1.7	100.9	130	-0.2	100.8	280
	1440 min to 2880 min	1.3	100.8	230	1.3	101.2	115	-0.3	100.7	365
	2880 min to 4320 min	1.4	100.8	190	1.0	101.0	135	0.1	100.9	315
	Avg for 4320 min period	1.3	100.8	215	1.3	101.0	125	-0.1	100.8	320
Bedroom	0 to 1440 min	0.7	100.9	250	1.0	100.9	125	0.0	100.8	295
	1440 min to 2880 min	0.8	100.8	255	0.8	101.2	105	-0.1	100.7	400
	2880 min to 4320 min	0.9	100.8	200	0.6	101.0	125	0.1	100.9	345
	Avg for 4320 min period	0.8	100.8	230	0.8	101.0	120	0.1	100.8	290

References

[1] Nazaroff, W. M. and Nero, A. V., *Radon and Its Decay Products in Indoor Air,* John Wiley & Sons, New York, 1988.

[2] Cummings, J. B., Tooley, J. J. Jr., and Moyer, N., "Radon Pressure Differential Project, Phase I. Florida Radon Research Program," EPA-600/R-92-008 (NTIS PB92-148519), January 1992.

[3] Clements, W. E. and Wilkening, M. H., "Atmospheric Pressure Effects on ^{222}Rn Transport Across the Earth-Air Interface," *Journal of Geophysical Research,* Vol. 79, No. 33, 1974, pp. 5025–5029.

[4] Hintenlang, D. E. and Al-Ahmady, K. K., "Pressure Differentials for Radon Entry Coupled to Periodic Atmospheric Pressure Variations," *Indoor Air,* Vol. 2, 1992, pp. 208–215.

[5] Modera, M. P., "Residential Duct System Leakage: Magnitude, Impacts, and Potential for Reduction," *ASHRAE Transactions 95,* VA-89-5-5, 1989, pp. 561–569.

[6] *ASHRAE Handbook, Fundamentals,* American Society of Heating, Refrigerating and Air-Conditioning Engineers, Atlanta, GA, 1985.

[7] Standard Measurement Protocols, Florida Radon Research Program, Test Method for Determining HAC Duct System Leakage: U.S. EPA, EPA-600/8-91-212, 1991.

[8] Hintenlang, D. E. and Al-Ahmady, K. K., "Influence of Ventilation Strategies on Indoor Radon Concentrations Based on a Semi-Empirical Model for Florida-Style Houses," *Health Physics,* Vol. 66, No. 6, 1994, pp. 427–432.

[9] Narasimhan, T. N., Tsang, Y. W., and Holman, Y., "On the Potential Importance of Transient Air Flow in Advective Radon Entry into Buildings," *Geophysical Research Letters,* Vol. 17, No. 6, 1990, pp. 821–824.

[10] Tsang, Y. W. and Narasimhan, T. N., "Effects of Periodic Atmospheric Pressure Variation on Radon Entry into Buildings," Berkeley, CA, Lawrence Berkeley Laboratory Report LBL-31164, 1991, (submitted to *Journal of Geophysical Research*).

DISCUSSION

David Saum[1] *(written discussion)*—The blower door air tightness measurement of 8.6 ACH 50 Pa appears to be within the typical range, but the tracer gas measurement of 0.3 ACH with the forced airflow on seems to be very tight compared to other measurements reported for FL (Toohy) and CA (Modera) for houses with ducts in the attic. Are these ducts unusually tight?

A. C. Kozik, P. Oppenheim, and D. Hintenlang (authors' closure)—The ducts in this structure are tight, but do not appear to be unusually tight for single family detached housing built in north central Florida. The Florida Solar Energy Center has conducted measurements in central Florida and have observed some comparatively leaky duct installations. We have examined approximately 30 houses in north central Florida and have obtained very similar results for blower door and trace gas measurements of house ventilation. This appears to be a typical house for north central Florida.

---

[1] Geomet Technologies, Germantown, MD 20874.

Ashley D. Williamson,[1] Charles S. Fowler,[1] and Susan E. McDonough[1]

Assessing the Effectiveness of Slab Flooring as a Barrier to Soil Gas and Radon Infiltration

REFERENCE: Williamson, A. D., Fowler, C. S., and McDonough, S. E., "**Assessing the Effectiveness of Slab Flooring as a Barrier to Soil Gas and Radon Infiltration,**" *Airflow Performance of Building Envelopes, Components, and Systems, ASTM STP 1255,* Mark P. Modera and Andrew K. Persily, Eds., American Society for Testing and Materials, Philadelphia, 1995, pp. 68–78.

ABSTRACT: Experimental studies on the entry of soil gas and radon into slab-on-grade buildings have been carried out in instrumented, single-zone test structures. This work, as part of the Florida Radon Research Program, focused on the effectiveness of slab flooring variants as barriers to soil gas/radon entry. A second objective was the study of the role of subslab fill soil as both a potential source of and barrier to radon entry. Studies were made in well-sealed (~600 mm² ELA) unoccupied test buildings placed on well-characterized, radium-bearing sandy soil. The buildings were instrumented with data acquisition systems to continuously monitor indoor radon concentrations, differential pressures at several subsurface locations, weather conditions, and soil moisture. The response of the structures to mechanical depressurization as well as natural driving forces was measured. Limited measurements were made regarding direct diffusive transport of radon through apparently intact concrete slabs, as well as transport through cracks in the floor structure.

KEYWORDS: radon, soil gas entry, infiltration, concrete slabs, slab cracks, slab in stem wall, floating slab

Soil gas infiltration into occupied structures represents a component of overall infiltration whose environmental impact is proportionally much greater than its volumetric contribution. Soil gas entry represents the primary source of several indoor pollutants, including radon. It is generally accepted that the primary driving force for entry of radon and other soil-bound pollutants is pressure driven soil gas infiltration, and that the primary barrier to entry in slab-on-grade buildings is the integrity of the floor slab. Unfortunately, there are no quantitative methods to assess the relative resistance of different portions of the slab barrier. Such measurements would be useful in pinpointing vulnerable features in current slab construction in order to refine strategies for radon-resistant slab design.

The project described in this paper was conducted as part of the Florida Radon Research Program, administered by the Florida Department of Community Affairs with collaboration from the U.S. Environmental Protection Agency. The purpose of the overall program is the development of construction standards for radon-resistant buildings. The objectives of this study include evaluation of elements of the foundation and subslab fill as pathways or barriers

[1] Division head, research physicist, and research environmental engineer, respectively, Southern Research Institute, P.O. Box 55305, Birmingham, AL 35255-5305.

to indoor radon, validation of models for radon transport and entry, and development of procedures to evaluate radon resistance of new and existing buildings.

Experimental

In order to assess the factors affecting radon entry and the effectiveness of slab features as barriers to radon entry, studies were performed in small, single-zone structures rather than full-scale houses. The advantages of these test cells, as they will be called in this paper, are numerous. With unoccupied, instrumented single-zone structures, studies can be designed which eliminate the confounding effects of multizone air transport; ducted heating, ventilation and air-conditioning (HVAC) systems; distributed uncontrolled air leakage and pressure distributions, as well as occupancy patterns. The entire slab is available for measurements, and outdoor air infiltration and building pressure differentials can be independently controlled in order to focus controlled studies on the slab and subslab fill soil.

The test cells used for the first phase of measurements reported here were built in December 1990 by Geomet Technologies, Inc., under contract with the Florida Department of Community Affairs (DCA). They were built over the edge of an old spoil bank of sand tailings from earlier phosphate mining operations currently on the facilities of the Florida Institute of Phosphate Research (FIPR). Their foundations made a 20-by-20-ft (6 × 6 m) footprint. The first of the test cells (referred as TC1) was built with a "floating" slab (FS) with a significant edge crack. TC2 had a slab-in-stem wall (SSW) construction. The foundations of both test cells comprised poured concrete 8-by-16-in. (0.2 × 0.4 m) footings reinforced with two continuous No. 5 reinforcing bars. The trenches for these footings were dug about 16-to-20-in. (0.4 to 0.5 m) below the existing site grade. The soil excavated from the footing location was placed under where the slabs were to be poured. A small amount of additional fill soil was brought in for both slabs. While the composition of this sandy fill is not known, we feel the amount [approximately 2 in. (50 mm) thickness] to be too small to have an applicable impact on this study. Above the footings three courses of concrete block were laid. At both cells, the core holes of both of the lower two courses of blocks were filled with soil and plugged at the top. No polyethylene vapor barrier was placed under either of the 4 in. (0.1 m) slabs poured in these two cells. In TC1, the top course of blocks was poured full of concrete when the slab was placed. The gap between the side of the top course of blocks and the slab was filled with a removable form about ¾ in. (20 mm) thick. The slab was placed against the inside edge of this top course, leaving a perimeter crack all around the inside of the block. In TC2 the top course of blocks was a "chair," "L," or "shoe" block which had a 4 in. (0.1 m) cut in its inside face. In this design, when the slab is placed the concrete spills over the cut side and into the core holes; thus the term "slab-in-stem wall."

The cell superstructures were wood frames with stucco over plywood cladding. A 1 in. (25 mm) layer of spray urethane foam was applied to the inside of the plywood surface, in the walls, and under the roof. While this foam provided some thermal insulation, the primary purpose was to enhance airtightness of the superstructure. Each cell had a steel door opening into the unit with positive magnetic seals on all four edges. Neither cell was built with any other fixed or planned opening.

All additions to the cell were likewise designed to minimize unplanned air infiltration. Each test cell was equipped with a wall-mounted ductless split heat pump for space conditioning. All of the lines for this system were fed through a single penetration which was sealed with spray urethane foam. Later still, when the data logger and weather station were installed, another small penetration was required in each cell through which necessary signal lines could be run. All of these penetrations were sealed with spray urethane or caulk or both and were leak-tested to ensure that the sealing was adequate and complete. Leak tests were performed by depressurizing the whole structure and using chemical smoke on the inside of the structure to identify infiltration routes for sealing.

After sealing as well as possible, a system for independent control of induced airflow and pressure was installed. An exhaust (or supply) port was made in one wall of each cell. On the inside, 3 in. (76 mm) polyvinylchloride (PVC) pipe was run up the wall toward the ceiling. On the outside a 3 in. (76 mm) PVC pipe was run parallel to the grade and either capped or connected through a calibrated orifice to an in-line centrifugal depressurizing (or pressurizing fan). Another 3 in. (76 mm) penetration was made near a corner of the opposite wall in each test cell. A 3 in. (76 mm) ball valve was placed on the outside of each building at the end of a 3 in. (76 mm) PVC pipe run along the inside of this wall near the slab level; small holes were drilled in this pipe all along its length. Figure 1 illustrates the heat pump locations and piping arrangements for TC1 in a depressurizing mode. Lengths of ¼ in. (6 mm) stainless steel tubing were placed through each wall so that pressure lines could be connected to and from either the inside, outside, or between the cells as any given experiment might require. Using the fan system described above, several air leakage measurements were performed in each cell. The measured equivalent leakage area varied from 600 to 1300 mm² (0.9 to 2.0 in.²) in test TC1 and 400 to 800 mm² (0.6 to 1.3 in.²) in TC2. This level of sealing was found adequate for our studies.

A Campbell Scientific, Inc., (CSI) 21X data logger was installed in one of the FIPR test cells for most of the study period. Temperatures were monitored by the internal panel instrumentation of the datalogger and by Type K thermocouples placed at the other points to be monitored. Differential pressures were monitored with pressure transducers of various ranges. Zero checks were performed on the transducers at least once each working day. The data logger also stored information on wind speed and direction, barometric pressure, solar flux density, outside air temperature and relative humidity, and rainfall.

Differential pressures that were monitored included indoor/outdoor and indoor/subslab as well as pressure drop across the flowmetering orifice used to monitor exhaust flow from the building. Indoor radon was monitored with commercial continuous radon monitors (Pylon AB5 and Femto-Tech R210 F). Output from these monitors was collected using either the datalogger or the monitors' internal data storage.

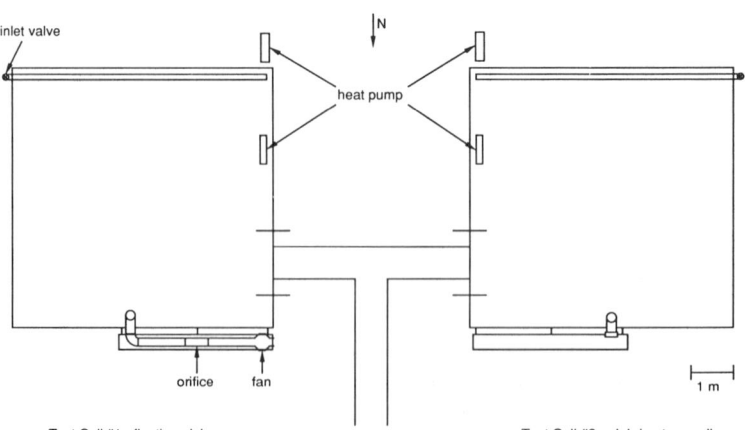

FIG. 1—*Test cell floor plans showing TC1 set for a depressurization run.*

Several variants of a common methodology described below were used to measure soil gas radon entry across the slab barrier. The primary measurement was the whole-building radon entry measurement to assess the resistance of the entire slab. In a whole-building test, indoor radon concentration and indoor-outdoor differential pressures were monitored continuously over a test period. If the study condition included depressurization using the external fan, the exhaust flow rate was continuously monitored as well and gave a direct measure of net air infiltration.

Data from the whole-building entry test were analyzed using a well-mixed first-order mass balance model as follows. The total amount of radon in a given volume is VC, where V is the volume of the space and C the radon concentration in picocuries per litre (pCi/L) or some equivalent units. V is constant for the structure, and C generally varies with time. It is assumed that the radon enters at some fixed entry rate, I, for any given pressure applied in the building. Radon exits by way of the exfiltration of the indoor air, represented by QC, where Q is the exfiltration rate. The radon concentration is also reduced by radioactive decay, but for most of the cases examined in this report the decay term is small compared to the exfiltration term. Therefore, a decay term is not included in this derivation. For the present, Q is assumed also to be constant at a given building condition of closure, applied pressure, etc. The net rate of change in the radon quantity is then

$$\frac{d(VC)}{dt} = V\frac{dC}{dt} = I - QC \tag{1}$$

This therefore is the governing differential equation describing the change in radon. A typical initial condition for solving Eq 1 occurs when the concentration at time zero is zero: $C(0) = 0$. (Other initial conditions are easily accommodated as well.) The solution that fulfills the governing equation and initial condition is

$$C(t) = C_s(1 - e^{-\lambda t}) \tag{2}$$

where C_s is the steady-state (long-time equilibrium) radon concentration, and λ is the ratio of exfiltration rate to volume

$$\lambda = \frac{Q}{V} \tag{3}$$

The radon entry rate, I (assumed to be constant for a given structure condition), must equal the equilibrium radon exfiltration after a sufficiently long time for the inside concentration to stabilize, or $I = QC_s$. Alternatively, the entry rate can be fit to the slope of the rise of $C(t)$ from a near-zero initial value. For depressurized conditions, the exfiltration rate fit from the radon data can be compared with the measured exhaust flow rate to confirm the well-mixed single-zone assumptions used. For passive conditions, the infiltration rate must be inferred from the time dependence of the radon concentration.

An example of the applicability of this first-order entry and mixing model is given in Fig. 2. Here, radon in both test cells is recorded over a period of depressurization (-10 Pa) followed by several days of passive monitoring. The fit to the first-order entry model shown in the figure clearly reproduces the major features of the data. The fitted curves for the initial depressurized portion of the data set used constants $I = 570$ pCi/s and 360 pCi/s, respectively, for TC1 and TC2, with $\lambda = 0.15$ and 0.17 h^{-1}, respectively. The exfiltration flow rate

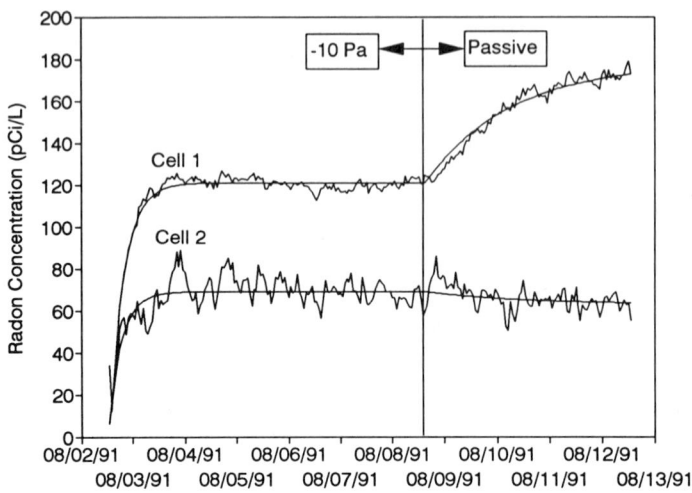

FIG. 2—*Radon concentrations with both cells at −10 Pa and then passive.*

corresponding to these decay terms is 4.7×10^{-3} and 5.2×10^{-3} m^3/s, respectively, in excellent agreement with the 4.7×10^{-3} m^3/s (10 ft^3/min) target exhaust flow rate. The passive period was fit using $I = 143$ and 45 pCi/s, respectively, and $\lambda = 0.026$ and 0.023 h^{-1}, respectively, for cells TC1 and TC2.

An interesting feature of the data in Fig. 2 is the different behavior as the exhaust fans are deactivated, resulting in a drop in steady-state radon in TC2 and a rise in TC1. This indicates that the ratio of radon entry rate to ventilation rate (I/λ) is greater at 10 Pa depressurization than under passive conditions for TC2, while the reverse is true for TC1. This effect is not atypical of our studies in test cells and houses. It illustrates the benefit of the model as a framework to separate the effects of radon entry and infiltration (that is, dilution). Both of these quantities increase with increased depressurization, but have different pressure dependences resulting in radon concentration which may not vary monotonically with pressure.

Many replications of the whole-building radon entry measurements were performed in both structures at various pressure differentials and at different times of the year. The tabulated radon entry rates serve as a figure of merit for the overall radon resistance of the entire slab, including all potential pathways for radon entry. A modified measurement was derived in order to assess the relative importance of these individual "weak points"—cracks and penetrations as well as apparently intact slab. Two sealed plenums were constructed and used to monitor radon entry into smaller zones over selected segments of the slabs using techniques analogous to the whole-building tests described above. The first, and 18-in.-high (460 mm) box made with acrylic plastic sheets joined at the corners by aluminum 90-deg angle bars, was built to cover a 3-by-3-ft (0.9 × 0.9 m) square of intact slab area. The sheets, bars, and holes drilled for joining were all caulked leaktight. The open bottom of the box was sealed to the concrete slab with continuous rope caulking all the way around the perimeter. The box was depressurized and leak-tested with chemical smoke to identify and seal all detectable leaks. With the test cell at neutral to slightly positive pressure and as flushed as much as feasible with low radon outside air, the box was left at neutral pressure, and the radon concentration was monitored with a continuous radon monitor inside the box in a manner directly parallel to the whole-building passive test. Figure 3 gives an example of

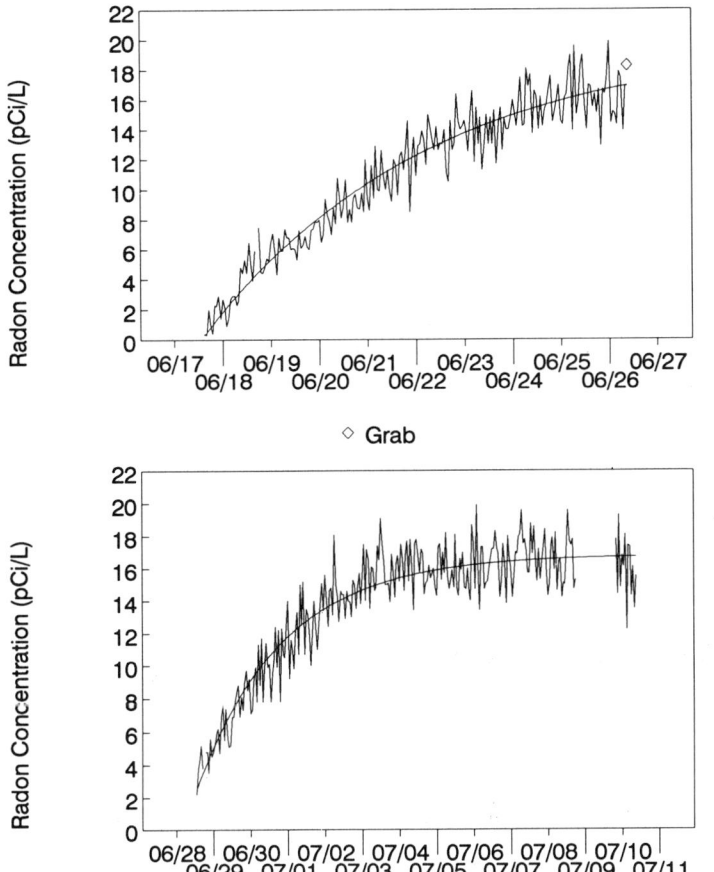

FIG. 3—*Plots of radon concentrations measured under larger plenum over intact slabs in TC1* (top) *and TC2* (bottom).

data from one of these measurements in each test cell. For depressurized plenum operation, a small pump and flowmetering orifice arrangement was used to withdraw air from one end of the box, while a bleed valve at the other end of the plenum was used to simultaneously establish desired values of the exhaust flow rate and the applied pressure differential between the interior and exterior of the box.

A smaller box of similar construction was used to study radon infiltration through cracks. This box was 18 in. (460 mm) long, 12 in. (300 mm) wide, and 12 in. (300 mm) high. It was sealed, leak-tested, and affixed to the slab in the same way as the larger plenum, except that the long dimension of the box was placed over a length of floor crack. The same valve and pumping arrangements were used, though these tests were typically conducted over shorter periods and grab samples of radon were used rather than continuous monitors. In TC1, this was first done over a length of the perimeter crack of the "floating" slab and later over a length of unintentional crack that developed. In TC2, only the unintentional cracks existed; one of these was monitored with the box. Both unintentional slab cracks were measured at points near the centers of their respective cells. The "stress" on the slab crack varied from no depressurization to −20 to −40 Pa. At a given depressurization an air inlet valve was usually opened as necessary to give 12.2 in.³/min (200 cm³/min) flow. Radon grab samples were taken before any depressurization and then at regular 1 to 2 h intervals thereafter until the end of the test (usually several hours). Figure 4 illustrates typical behavior of radon in the small boxes under depressurized and passive conditions. These results are discussed in the following section.

Results and Conclusions

The results of several periods of whole-building radon monitoring are summarized in Fig. 5. The lowest pressure point on each plot corresponds to the means of all the passive condition measurements, namely, 141 ± 61 pCi/s for TC1 and 33 ± 12 pCi/s for TC2. The regression lines for radon entry into cells 1 and 2, respectively, are given by

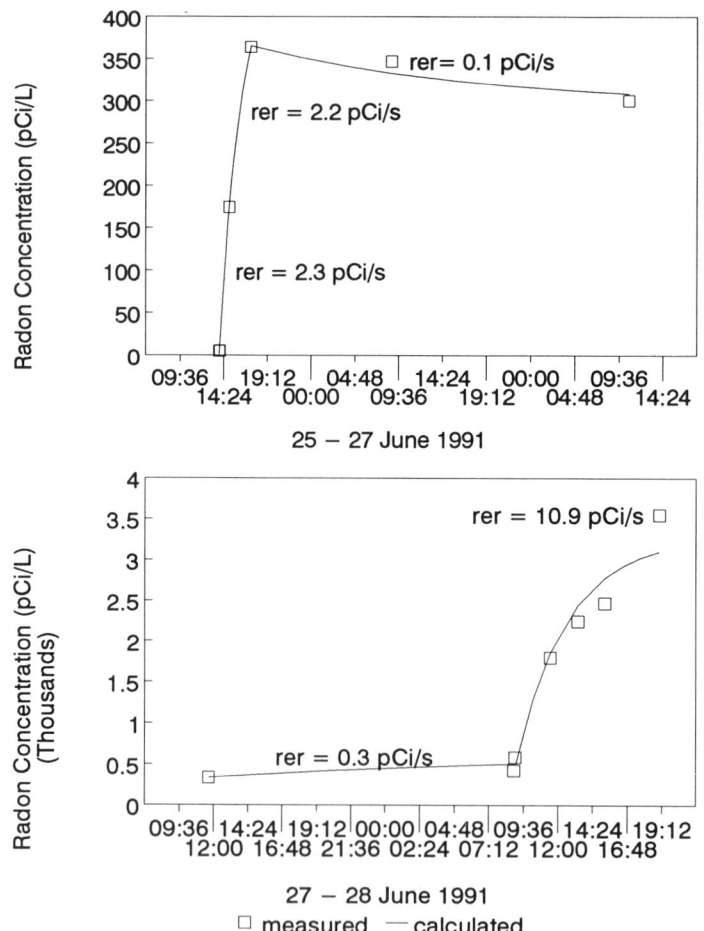

FIG. 4—*Radon concentrations extracted from smaller plenum over settling cracks in TC1 (top) and TC2 (bottom).*

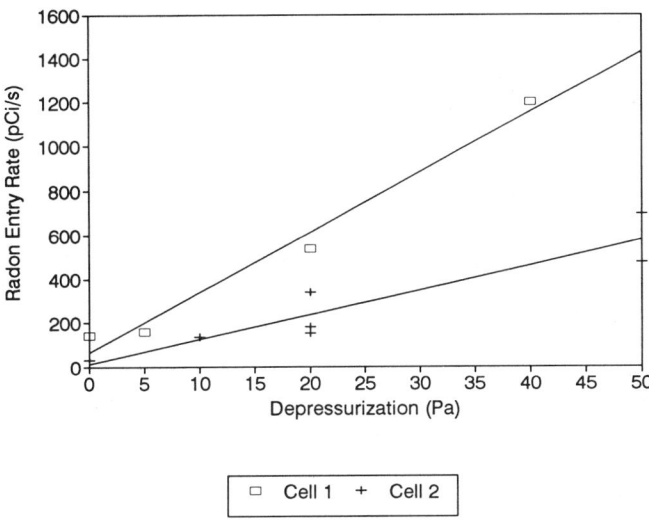

FIG. 5—*Radon entry rates in test cells as a function of applied depressurization.*

$$\text{TC1 entry} = (66 \pm 85) + \Delta P \times (27 \pm 3); \qquad r^2 = 0.98$$

$$\text{TC2 entry} = (12 \pm 96) + \Delta P \times (11 \pm 2); \qquad r^2 = 0.86$$

where the units of the intercept and slope coefficients, respectively, are pCi/s and pCi/s · Pa^{-1}.

The form of the regression equation follows our anticipation of the behavior of radon entry:

$$I = C_1 + C_2 \times \Delta P$$

The intercept term represents radon entry via diffusional or other pressure-independent mechanisms, while the slope term represents advective or pressure-driven entry of soil gas containing radon, which for Darcy flow should be linearly proportional to applied pressure. That is, $Q_s = K_s \times \Delta P$, where Q_s is soil gas flow rate and K_s is constant. Further, if the subslab radon concentration C_s is known, the second term can be decomposed into the product $C_s \times Q_s = C_s \times K_s \times \Delta P$.

Thus, the radon entry fits can provide figures of merit for the effectiveness of a building's substructure to radon entry. The pressure-dependent term C_2 provides a measure of the "conductance" of the building to pressure-driven radon entry. The flow proportionality constant, K_s, provides a normalized measure for soil gas entry, allowing comparison of foundations placed on soils of differing radon source strength. In the present case, there is a factor of 2.5 difference in the slopes (C_2) of the regressions, slightly less than the factor of 4.3 ratio of equilibrium radon concentrations under passive conditions. The means of several subslab radon measurements under TC1 and TC2 are 8550 ($\pm$1500) and 6000 ($\pm$800) pCi/L, respectively. If these concentrations are typical of the entire period of this study, the ratio of the flow rate proportionality terms K_s is 1.7, indicating that in the limit of higher applied pressures, TC1 admits approximately 1.7 times as much soil gas as TC2. This suggests that the slab in stem wall design under TC2 represents a superior barrier compared with the floating slab design of TC1.

The results of the tests of slab sections under plenums provide further insight into specific entry pathways. Table 1 contains a list of the entry rates measured during this study, with extrapolations of the zone measurements to the entire slab surface. The first three columns of Table 1 refer, respectively, to the specific portion of the slab monitored, the applied depressurization of the tests, and the measured radon entry rate. For the whole-building studies, the entry rate shown represents the mean of several passive measurements, or the regression prediction for each depressurization condition. All other entries in this column are the actual results of the measurements shown. The final column, "Whole-Building Equivalent" measurements, represents extrapolation of the 9 ft² (0.84 m²) intact slab segment to the 398 ft² (37 m²) slab area, or the 1.5 ft (0.46 m) crack segment to the roughly 39 ft (12 m) of interior crack or 78 ft (24 m) of perimeter crack.

Comparison of the whole-building equivalent entry rates for different zones and sample pressures allows several conclusions. First, radon entry through the intact slabs, while measurable, does not play a major role in total radon entry for these slabs. In only one measurement (TC2, passive) does this route appear to account for as much as 1% of total radon entry. Second, as might be expected, the perimeter crack in TC1 represents a major route of radon entry. In fact, only this crack proved to be a significant pathway for measurable air infiltration as well as radon entry. The reason 1 Pa was used for the depressurization measurement was that only pressures less than 3 Pa could be sustained at the maximum pump flow rate with the plenum over the crack segment used. While we believe that sustained entry rates for higher applied depressurization would be somewhat higher than the

TABLE 1—*Radon entry rates through elements of slab.*

		Radon Entry Rate, pCi/s	
Element	Depressurization, Pa	Test Zone	Whole-Building Equivalent
TEST CELL 1 (FLOATING SLAB)			
Whole building	Passive	141	141
	10	336	336
	20	606	606
	40	1146	1146
Perimeter crack	Passive	0.6	32
(0.46 m)	1	11	590
Interior crack	Passive	0.14	3.8
(0.46 m)	40	2.3	61
Intact slab	Passive	0.02	0.9
(0.84 m²)	20	0.04	1.8
TEST CELL 2 (SLAB-IN-STEM-WALL)			
Whole building	Passive	33	33
	10	122	122
	20	232	232
	40	452	452
Interior crack	Passive	0.3	8
(0.46 m)	40	11	290
Intact slab	Passive	0.03	1.3
(0.84 m²)	20	0.04	1.8

NOTE: 1 m = 3.28 ft; 1 m² = 10.76 ft².

590 pCi/s observed at 1 Pa, we suspect that the increase would be limited (perhaps to 1000–1500 pCi/s) due to dilution by outdoor air and depletion of soil radon in the vicinity of the crack. The interior cracks appear to be less effective pathways than the larger perimeter crack, although in TC2 this route appears to account for at least 25% to 30% of the predicted total entry. Finally, the whole-building entry rates seem to be of the same magnitude as, but somewhat greater than, the sum of the predicted component entry rates. The reason for the remaining deficit is not clear.

In conclusion, these results demonstrate that for the small, single-zone structures used in this study, radon ingrowth monitoring coupled with a well-mixed, first-order mass balance model is adequate to derive radon entry rates for whole structures. Further, application of this technique to smaller slab elements allows partitioning of the entry among anticipated pathways through the slab feature. Finally, the edge crack formed in the floating slab design represents a major entry pathway. The intact slab of the structures represents only a minor contributor even without a plastic membrane vapor barrier. The transport across intact slabs of structures placed over plastic vapor barriers would be expected to be at least as low as that across slabs placed in the manner of the present study.

Acknowledgment

This work was funded by the Florida Department of Community Affairs under Interagency Agreement 92RD-41-15-00-22-001 and the U.S. Environmental Protection Agency under Cooperative Agreement CR818413.

Discussion

David Saum[1] (written discussion)—One of your figures shows that the indoor radon concentration decreases as the building is depressurized. This is contrary to the conventional wisdom and EPA guidance that depressurization/exhaust fans generally cause increases in radon concentrations. Comments?

A. D. Williamson (authors' closure)—While the results of our study indicate that radon *entry rates* into the test structures uniformly increase with applied depressurization, the radon *concentrations* can decrease at some pressure levels. This can be visualized as a competition between entry and dilution by ventilation, as follows. Using a well-mixed, mass-balance model such as assumed by us and others, the steady-state radon level in a building is approximated by $C_s = I/Q$, using the variables in our paper. Our results are consistent with an entry rate of the expected form $I = I_0 + I_1 \times \Delta P$, consisting of a term linear in applied pressure due to pressure-driven soil gas flow in the Darcy regime, and a constant term attributable to pressure-independent mechanisms such as diffusive transport. (Note that we are using the convention that $\Delta P > 0$ for depressurized or exhaust ventilation.) We can assume an infiltration rate of the form $Q = q_0 + q_1 \times (\Delta P)^n$ where n can be assumed to lie in the range of 0.5 to 1.0 with typical values approximately 0.6. (Here the constant term serves to remove the singularity at $\Delta P = 0$, and is in accordance with residual loss due to

---

[1] Geomet Technologies, Germantown, MD 20874.

radioactive decay as well as residual infiltration even at zero applied pressure). Thus, the radon concentration would be of the form

$$C = \frac{I_0 + I_1 \times \Delta P}{q_0 + q_1 \times \Delta P^n}$$

with all terms positive and n between 0.5 and 1.0. It can be shown that for parameters in this range, as ΔP is increased from 0, the concentration decreases from its passive level of I_0/q_0, passes through a minimum, and increases asymptotically in proportion to $\Delta P^{(1-n)}$. This is completely consistent with behavior seen in the test cells and in several slab-on-grade houses in central Florida, in which radon is noticeably lowered by modest levels of exhaust ventilation, while at higher levels of depressurization (over 10 to 20 Pa) the radon concentration again increases.

Saum (discussion)—Some models (Saum and Modera) suggest that the neutral pressure level (NPL) is an important parameter in determining how depressurization will affect indoor radon concentrations. Is the NPL of these buildings known?

Williamson (closure)—The NPL of the buildings is not known under the conditions studied. A separate study of the NPL of test cell 1 was attempted after the period reported here, with inconclusive results. Part of the difficulty with measuring the NPL is the low average pressure difference to outside—typically in the range of ± 0.5 Pa. With negligible stack-induced pressures, the net pressure differential is apparently dominated by fluctuating wind pressures, and there is no stable neutral pressure plane.

Window Air Leakage

David W. Kehrli[1]

A Description of the New ASTM Test Method E 1424, Used for Measuring Fenestration Air Leakage at Differential Temperatures and Pressures

REFERENCE: Kehrli, D. W., " **A Description of the New ASTM Test Method E 1424, Used for Measuring Fenestration Air Leakage at Differential Temperatures and Pressures,**" *Airflow Performance of Building Envelopes, Components, and Systems, ASTM STP 1255,* Mark P. Modera and Andrew K. Persily, Eds., American Society for Testing and Materials, Philadelphia, 1995, pp. 81–89.

ABSTRACT: A new committee has been developed by ASTM Committee E6 for measuring air leakage rates of fenestration products under imposed conditions of differential pressures and temperatures. This new method is different from the long-standing, internationally referenced ASTM Method E 283[2] in that it is performed under temperature differentials across the test specimen similar to ASTM C 236[3] and C 1199,[4] and AAMA 1503-88.[5]

This new method will show the impacts of expansion and contraction, shrinkage, compression-set, fabrication and design integrity, and material and component interactions in the air leakage rates of window and doors products.

This paper compares the two methods and provides some typical test data.

KEYWORDS: air leakage rates, differential air temperatures, windows, doors, weatherseals, sealants, cold temperature air leakage rates, field tests

In 1982 the ASTM E6 Committee on Building Materials and Components established a task group in the windows, doors, and curtain walls subcommittee E6.51, which was charged with the responsibility to develop a test method for measuring the air leakage rate of windows and doors under differential air temperature and pressure conditions.

The purpose for developing such a standard was the presumption that fenestration products and many of their components may change their sealing characteristics when subjected to differential air temperatures typical of building applications throughout the United States.

The starting assumptions for the task group were that air leakage measurements at or near standard atmospheric conditions (approximately 21°C (70°F) air temperature) may not be

---

[1] President and owner, ETC Laboratories (Commercial and Research Product Testing), Rochester, NY 14624.

[2] ASTM Test Method for Rate of Air Leakage Through Exterior Windows, Curtain Walls, and Doors (E 283-84).

[3] ASTM Test Method for Steady-State Thermal Performance of Building Assemblies by Means of a Guarded Hot Box (C 236-87).

[4] Test Method for Measuring the Steady-State Thermal Transmittance of Fenestration Systems Using Hot Box Methods (C 1199-91).

[5] American Architectural Manufacturers Association Voluntary Specification for Aluminum and Poly (Vinyl Chloride) (PVC) Prime Windows and Glass Doors, 1993.

accurate for product rating and certification programs and, more importantly, for energy calculation purposes.

The task group charged with the responsibility for developing the new test method was ASTM E6.51.04. The members focused their efforts on expanding the scope of the current ASTM test method, E 283, for measuring fenestration air leakage, by adding a temperature variable. The new test method was approved by the ASTM society ballot in November 1990, and given the designation E 1424.

Throughout the standard development process, the task group reasoned that the coefficient of thermal expansion of many commonly used materials in fenestration products, when subjected to elevated or depressed exterior ambient air temperatures, was enough to cause components in windows and doors to alter their dimensional relationships and thereby alter their sealing characteristics. This was based on manufacturers' material expansion coefficients date.

The following issues were before the task group throughout the eight-year development period.

1. The term "effective leakage area" (from the science of whole-building infiltration analysis) defines an area which is equivalent to the sum of all the cracks, holes, joints, etc. which the air leakage will flow through. For fenestration products, it may vary based on the dimensional relationship of sashes, panels, frames, weatherseal profile geometries and their relationship to mating surfaces at differential ambient air temperatures. Were there correlations to fenestration products?

2. Another issue was that of sealants and weatherseals (weatherstrips) and their ability to remain effective in sealing fenestration products. Physical deterioration of these products as a result of exposure to environmental degradation factors (heat, moisture, ultraviolet light, etc.), shrinkage, and compression set (permanent loss of sealing geometry due to chemical and physical changes in the polymeric structure of the plastic) causing seals and sealants to lose sealing efficiency over time. Often the time frame is relatively short, from 3 to 24 months [1].

3. Another issue was the operating style of fenestration products. It has been well documented that some fenestration designs are less prone to air leakage than others based on how they operate. Would "tight" designs remain so at elevated or depressed ambient air temperatures?

4. Another issue was frame material. Was there a difference in the way frame materials (polyvinyl chloride, steel, wood, aluminum, fiber glass, composites, or others) sealed fenestration products, either new or over time? The difference between the coefficient of expansion of materials in conjunction with the other components in a fenestration system indicated that sealing inefficiencies would occur during some point of a product's service life.

5. Another issue was the design of the fenestration system. Were some designs better than others, even though the frame materials and operating styles were the same? Had a few select designers of fenestration products been able to pinpoint typical flaws associated with "leaky" fenestration designs and compensated for them with unique or improved materials, components, designs and fabrication techniques.

6. Another issue was the fabrication process. Even though a designer had mastered the trade on the drawing board, was process engineering and quality control up to meeting the challenge of producing these designs repeatedly?

7. Another issue was installation. Typical comments from the fenestration industry, blamed window and door installers for improper installation of these products and the conse-

quent poor field performance of many fenestration systems. A lack of experience or training, carelessness, and budget constraints could lead to improper installation techniques.

It was obvious that many factors could contribute to the loss of airtightness of fenestration systems. Those that are controllable by the manufacturer are discussed in this paper.

What was needed was a means to understand and measure these variables in the laboratory. The task group developed an expanded version of the already existing method, ASTM E 283, by specifying the air leakage test be conducted in a conventional guarded or calibrated hot box, environmental chamber, insulated glass test chamber, or other such apparatuses which was capable of creating differential air temperatures across the test specimen.

The ASTM E 283 Standard, because it is referenced worldwide, is seen as a benchmark for all other air leakage methods used for building components. The successful development of this standard was partially based on the fact that the current standard was left intact. Many members of the task group and the voting community felt that a new standard for measuring fenestration air leakage should not be mandated without being evaluated by the fenestration community at large for several years. Therefore, several groups are evaluating the new test method as part of their air leakage test method development program. The most notable group is the National Fenestration Rating Council (NFRC).

The ASTM E 283 Procedure: How It Works

The existing test method for measuring fenestration air leakage is ASTM E 283. It was developed as a comparative tool for fenestration designers to determine relative differences between products under ideal laboratory settings. The products tested are typically prototypes from new design concepts. Production line units are also often analyzed with this method to determine fabrication quality. The method does not produce results which would indicate how fenestration products might be expected to perform after installation in a building envelope, over time, and under real-world temperature differentials.

In order to determine how fenestration products will perform in a building envelope, ASTM developed their Test Method for Field Measurement of Air Leakage Through Installed Exterior Windows and Doors (E 783-84). This method is similar to the E 283 method except that attention to wall cavity airflow, site temperature, pressure, humidity, and wind speed is referenced and must be accounted for.

Manufacturers and trade associations have, however, been using the E 283 method as a primary marketing and sales tool over the years for promoting the airtightness of their products. In fact, manufacturers will go to great lengths in order to reduce their measurable leakage rates of their products below the national allowable limit of (0.37 cfm/ft) (see footnote 5).

The E 283 standard is generally performed in a controlled laboratory environment, typically (21°C) (70 ± 20°F). Humidity is generally not controlled. The standard does not allow or require the user to introduce a differential test temperature across the specimen.

Test pressures that are most often referenced by specifiers, while somewhat high compared with actual field test pressures, allow for greater repeatability and accuracy of airflow measurements in volumetric measuring instruments. The method has a default pressure of (1.57 psf) (0.3 in w.g.), which is equivalent to a wind velocity of 40 km/h (25 mph). Table 1 gives the average wind velocities for each state taken from the *ASHRAE 1989 Fundamentals Handbook* [2]. These velocities are averaged from multiple sites throughout each state measured at airport locations. The state with the lowest average wind speed is California at 8

TABLE 1—*ASHRAE wind speeds.*

States	No. of Sites	Wind Speed (Knots)			
		High	Low	Avg.	mph
	15	10	5	7.6	8.7
Alaska	6	14	3	6.8	7.9
Arizona	10	6	4	5.3	6.1
Arkansas	10	9	6	8.0	9.2
California	61	9	3	4.4	5.0
Colorado	13	9	5	7.0	8.1
Connecticut	8	13	5	8.2	9.4
Delaware	2	9	9	9.0	10.4
D.C.	2	11	11	11.0	12.7
Florida	23	12	6	8.0	9.2
Georgia	17	12	4	7.9	9.1
Hawaii	4	12	5	8.0	9.2
Idaho	9	9	3	6.0	6.9
Illinois	27	11	8	9.5	10.9
Indiana	24	11	7	9.3	10.7
Iowa	16	11	9	9.8	11.2
Kansas.	16	14	8	10.7	12.3
Kentucky	10	9	6	7.8	9.0
Louisiana	11	9	7	8.4	9.7
Maine	7	11	7	9.0	
Maryland	6	10	9	9.4	10.8
Massachusetts	13	16	8	11.1	12.8
Michigan	21	12	5	8.4	9.6
Minnesota	15	12	8	9.2	10.6
Mississippi	13	8	6	6.8	7.8
Missouri	19	12	6	9.1	10.5
Montana	14	12	5	8.0	9.2
Nebraska	14	10	8	9.0	10.4
Nevada	9	10	3	6.3	7.2
New Hampshire	7	11	7	8.7	10.0
New Jersey	8	11	9	10.3	11.9
New Mexico	17	8	5	6.2	7.1
New York	35	15	4	8.4	9.6
North Carolina	17	12	6	7.8	8.9
North Dakota	8	12	7	9.6	11.0
Ohio	30	12	6	8.8	10.1
Oklahoma	16	14	10	11.2	12.8
Oregon	15	12	4	6.3	7.3
Pennsylvania	22	11	0	9.1	10.4
Rhode Island	2	11	10	10.5	12.1
South Carolina	12	8	6	7.0	8.1
South Dakota	9	11	8	9.0	10.4
Tennessee	13	10	6	8.2	9.4
Texas	48	23	7	10.9	12.6
Utah	10	6	5	5.5	6.3
Vermont	3	7	7	7.0	7.3
Virginia	11	10	6	8.0	9.2
Washington	18	15	4	7.2	8.2
W. Virginia	10	10	6	8.4	9.7
Wisconsin	16	10	8	9.0	10.4
Wyoming	11	11	5	8.6	9.9

Avg. Wind Speed	No. of States
5–6	= 1
6–7	= 3
7–8	= 6
8–9	= 5
9–10	= 14
10–11	= 13
11–12	= 3
12–13	= 6
Total	= 51

National Avg.
$N = 51$
$X = 9.6$ mph
$S = 1.8$ mph

NOTES: X = average. S = standard deviation.

km/h (5 mph) and the highest is Massachusetts at 20.6 km/h (12.8 mph). The national average is 15.7 km/h (9.8 mph). As can be seen from this chart, the 40 km/h (25 mph) test velocity is extremely high no matter where one resides in the United States.

The E 283 method allows test specifies or users to test at pressures other than (1.57 psf). The default is provided only as guidance to the specifier if they are not familiar with wind load tables, building design, or product performance capabilities.

The method requires users to install products in similar fashion as they would see in the field. For commercial storefronts or curtain wall mock-ups this is usually no problem. For residential and commercial windows and doors, however, the laboratory installations are typically in a wood buck (often pine) which is a 2 × 6 or 2 × 8 in. approximation of the rough opening in a building. Products are set up in the laboratory test apparatus in a square and plum arrangement. This set up should provide the most ideal condition for obtaining the lowest possible leakage rate, assuming all other factors are accounted for, e.g., weatherseals, sealants, and proper fit between mating components.

One final observation should be made regarding this method. That is, its popularity due to its low cost and ease of use in conducting the test. It also provides a simplistic and approximate method for comparing products under ideal laboratory settings. A typical E 283 test apparatus arrangement is illustrated in Fig. 1.

Deficiencies in the ASTM E 283 Method

The method has the following deficiencies:

1. There are no means to account for exterior air temperatures on expansion or contraction of materials and possibly the products effective leakage area.
2. There are no means to account interference mating of components manufactured from dissimilar materials which can cause thermal bowing of window and door panels.
3. There are no means to account for shrinkage on weatherseals, seals, sealants, and glazing beads.
4. There are no means to account for framing materials and dimensional stability over a variety of temperature extremes.

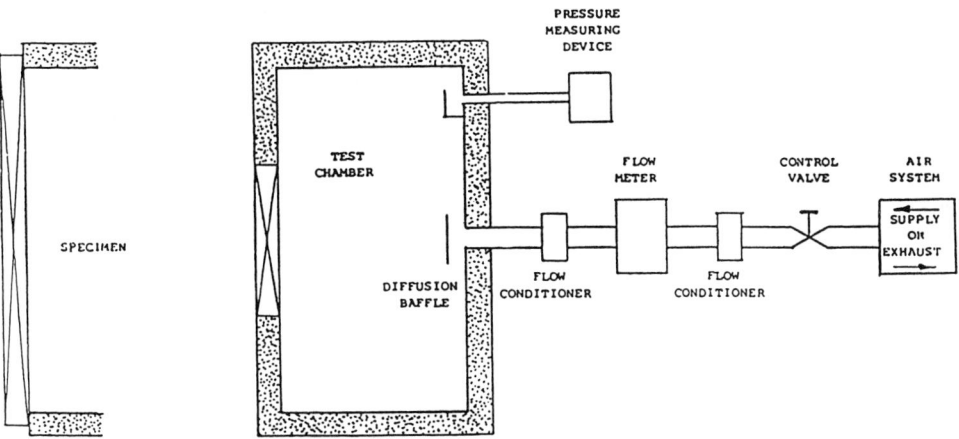

FIG. 1—*General arrangement of air leakage apparatus.*

5. There are no means to account for user, environmental and design factors on hardware functionality.
6. There are no means to account for occupant/user discomfort during periods of cold weather.
7. There are no means to account for material and component degradation due to natural and accelerating aging processes.
8. There are no means to account for installation defects.
9. There are no means to account for manufacturing quality.
10. There are no means to allow users and specifiers of the method to understand the significance of the air leakage rate number and the significance between rated products.
11. There are no provisions for understanding how products will perform in a buildings envelope after being rated in the laboratory.
12. The results are always normalized to standard atmospheric conditions of 69.9°F, 29.92 in. Hg, and 50% relative humidity (RH). The problem with this approach is that fenestration products are open for ventilation most of the times during these conditions. There really are no concerns over a product's airtightness when the outside air temperature is 70°F. There are concerns, however, for occupant comfort, energy consumption, and heating, ventilation and air conditioning (HVAC) sizing when the outside air temperature is substantially beyond standard atmospheric conditions: 40°F or lower, or, 90°F or higher.

The ASTM E 1424 Procedure

The ASTM E 1424 test method is identical to the ASTM E 283 method with one exception. The E 1424 method provides a means to apply differential air temperatures across the product concurrently with the differential air pressure. These test conditions can provide a means to simulate environmental exposures typical of winter and summer air temperatures and wind velocities. This exposure is believed by many in the fenestration community to be more realistic for testing and rating fenestration products for air tightness because it emulates actual field conditions which may occur anytime during the year, in any part of the country.

The data obtained from air leakage tests no longer have to be idealistic and limited to structural certification programs. The new method can be used to generate data which provide much greater accuracy for energy calculation programs such as those used by architects, state energy offices, NFRC, energy engineers, code officials, HVAC engineers, utilities, and others who must know how fenestration products perform after installation and under adverse weather conditions.

The test method requires that fenestration products be mounted in a buck or thermal mask which is then assembled and mounted on or installed into an apparatus which is capable of maintaining a differential air temperature across the specimen for the required test period. A generic arrangement of such an apparatus is illustrated in Fig. 2. This arrangement must allow the person performing the test to enter either side of the chamber in order to check for icing conditions which may form in critical airflow paths such as the operable crack perimeter of operable fenestration products (vertical and horizontal sliding windows and doors, swing doors, casement/awning/hopper windows, operable skylights, etc.). The apparatus should also have a means to condition the air entering the weathering compartment of the apparatus. This is especially true when the air temperature is cold. An inexpensive means to achieve this is to place a rolled-up tube inside the weathering compartment which

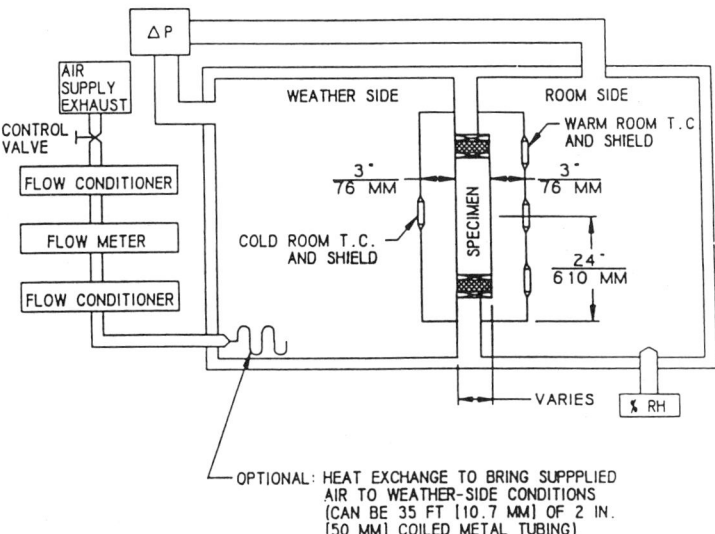

FIG. 2—*Environmental chamber schematic. Note that airflow and metering system may be set up on room side for positive or negative pressure leaking measurements.*

has incoming pressurized airflow entering through it. The longer the tube, the longer the time available to pressurize the weathering compartment with conditioned air. The air in the tube will be at the temperature of the weathering compartment. When the airflow equipment is turned on, warm laboratory air will begin to fill the tube, displacing the conditioned air in the tube. Depending on the volumetric capacity of the tube, a specific amount of conditioned air will be available to pressurize the chamber for a specific time period. This unique and inexpensive concept assumes nonmixing, piston flow through the tube.

Another means to perform the test is to evacuate air through the specimen from the warm side of the chamber. This is for infiltration and exfiltration test modes.

The test also requires that an approximate steady-state condition be reached before any air leakage measurements are recorded. The "steady-state" criteria for this test method are met when each of five consecutive surface temperature measurements, at 10-min intervals, are within 2°F of the average of all five thermocouple readings. Thermocouple placements are specified for various types of products in the specification. Air and surface temperature measurements are recorded every 10 min and are required data in the report.

Because this method is not quantifying energy transfer similar to ASTM C 236 or the ASTM Test Method for Thermal Performance of Building Assemblies by Means of a Calibrated Hot Box (C 976-82), the tolerance on air and surface temperature measurements is broader. There are no requirements for measuring any energy input into the test apparatus because the test does not quantify heat flow.

The air leakage of the product can be obtained by pressurizing or depressurizing either the room side (room air temperature compartment) of the test apparatus or the weathering side (elevated or depressed air temperature side of the compartment). When the room side is depressurized or when the weathering side is pressurized, cold or warm exterior air infiltrates the test specimen from its exterior surface. Conversely, when the room side is pressurized or the weathering side is depressurized, room temperature air exfiltrates through test specimen from its interior surface.

For either of these test modes the extraneous air leakage must be measured at each differential air temperature. Because tape is difficult to adhere on cold surfaces, a thin-gage (2 to 3 mil) plastic film may be taped on the weathering side of the specimen before the exterior side compartment air temperature is raised or lowered. Although this creates a slight insulating effect by separating the air space immediately in front of the specimen which is separate from the weathering room conditioned air space, the effect is minimal and on the order of 2°F or less. The weathering side air thermocouple is placed between the exterior of the specimen and the plastic film. In this way, the air temperature directly in front of the specimen is at the set point. Another way to achieve the extraneous leakage measurement is to tape over all cracks on the interior or exterior surface of the specimen with a high tack tape, similar to duct tape, before the temperature is lowered. This technique, which allows the specimen to be exposed directly to the exterior air, is a small improvement in the technique.

After "steady state" is obtained and the extraneous air leakage is measured, the plastic film or tape can be removed by opening up either the sash or panel and stripping the tape from the interior side of the specimen. Alternatively, the technician may enter the weathering side compartment and remove the tape, although this technique can be very uncomfortable at very low or high air temperatures. When the specimen is measured for exfiltration, the tare measurement is obtained by placing the plastic film or tape on the room side of the specimen.

One critical element in this method is the requirement for the operation of the sashes or panels before the air leakage is measured. This operation is also performed in the E 283 test method as a means to assure that the sashes or panels are operable within industry allowable limits. This operation also "settles" or aligns the weather seals so that a fair measurement is recorded, i.e., that sashes or panels are not sealed or favorably aligned to reduce leakage.

The E 1424 method, however, also assures that no icing is present during the test which may act to further seal the test specimen. The icing may result from the interior room compartment having too high a humidity level for the air and surface temperatures during the test. The test laboratory must take adequate precautions when performing this test so as not to allow this condition to occur. It is recommended that the interior humidity levels be as low as possible when performing low-temperature measurements. A table is provided in the E 1424 standard which has suggested indoor humidity levels for various air temperature conditions.

Conclusion

This new ASTM test method will provide air leakage data under real-life, installed conditions to which fenestration products are exposed. These data may provide a more accurate assessment of the energy associated with air leakage of fenestration products.

References

[1] Kehrli, D. W., "Window Air Leakage Performance as a Function of Differential Temperatures and Accelerated Environmental Aging," *Proceedings,* Thermal Performance of the Exterior Envelopes of Buildings III, ASHRAE/DOE/BTECC, Clearwater, FL, 2–5 Dec. 1985.
[2] *ASHRAE Handbook, Fundamentals,* American Society of Heating, Refrigeration and Air Conditioning Engineers, Chapter 24, "Weather Data," 1993.

DISCUSSION

Dave Jones[1] (*written discussion*)—Why would you want to pressurize in such a way as to promote the movement of warm-moist air to the cold side of the specimen give the potential for ice buildup?

David W. Kehrli (*author's closure*)—This method allows for exfiltration, room air leaving out through the specimen or, for infiltration, hot or cold exterior air entering the building through the specimen. This method may also be performed in two additional modes: the first being the pressure being applied after steady-state is achieved. And the second is the pressure applied throughout the steady-state period and the period before steady-state. In this condition, the exfiltration mode may produce ice build-up.

[1] Dupont, Richmond, VA.

David Kehrli[1]

Whole House Fenestration Energy Consumption as a Function of Variable Window Air Leakage Rates

REFERENCE: Kehrli, D., **"Whole House Fenestration Energy Consumption as a Function of Variable Window Air Leakage Rates,"** *Airflow Performance of Building Envelopes, Components, and Systems, ASTM STP 1255,* Mark P. Modera and Andrew K. Persily, Eds., American Society for Testing and Materials, Philadelphia, 1995, pp. 90–107.

ABSTRACT: Residential building energy consumption is dependent on many variables. The heat loss or gain attributable to fenestration products can be a significant portion of the whole building load. The fenestration industry is currently developing and implementing new test methods and rating procedures to more accurately account for fenestration energy transfer. One of the tools being developed by the National Fenestration Rating Council (NFRC) is a PC-based program called Residential Fenestration (RESFEN) heating and cooling load use and costs.

This paper will provide a review of the energy and cost impacts that variable air leakage rates of several types of window products can have on overall window energy usage as modeled in four typical building designs located in the United States. The analysis was performed with the RESFEN software as part of an NFRC sensitivity study on this issue.

KEYWORDS: air leakage, windows, energy consumption, National Fenestration Rating Council (NFRC), Residential Fenestration (RESFEN), computer modeling, degradation

The National Fenestration Rating Council (NFRC) was organized in 1989 for the sole purpose of developing accurate, fair, and repeatable energy rating systems for fenestration products. This organization has many committees that are developing test methods for a variety of energy transfer functions. The NFRC Air Leakage Subcommittee is charged with the responsibility for developing an accurate and repeatable air leakage test method and specification.

The NFRC rating program, when fully implemented, will have fenestration products labeled with energy performance indices such as U-factor, air leakage, solar heat gain, condensation resistance, durability, solar/optical properties, and possibly others. Most, if not all, of these indices will use newly developed ASTM test methods.

The Air Leakage Subcommittee is developing test and rating specifications based on ASTM E 283, Test Method for Measuring the Air Leakage through Exterior Windows, Curtain Walls, and Doors under Specified Pressure Differences across the specimen. This national standard has been in place for many years and is referenced by many trade associations, building codes, architects, local, state, and national government agencies, as well as a host of other countries. This test method is used for comparing relative air leakage rates of windows, doors, and curtain walls at standard atmospheric conditions: 70°F (21.1°C), 50% RH, and 29.92 in. Hg (101 KPa).

---

[1] President, ETC Laboratories, Rochester, NY 14624.

The method was designed and has been promulgated as a simple tool to compare the relative differences between products. Because the test is performed at ideal environmental conditions, accurate information as to how windows and doors will perform in buildings under seasonal and diurnal temperature swings is questionable. When fenestration energy transfer is calculated using data from this standard, the errors associated with it can be significant due to variations in air leakage rates that may occur as a result of environmental temperatures and pressures that are normally encountered in actual building installations: $-30°F$ to $120°F$ (-34.4 to $48.9°C$) and 0 mph to 50 mph (typical range for air leakage considerations).

Durability and installation factors are not accounted for in this test method, but are being investigated by the NFRC and ASTM for possible inclusion into performance standards [1]. These factors will provide the degradation in air leakage rates as simulated in this study.

Changes in fenestration product leakage rates with variations in air temperature and wind speed has been documented over the last ten years by researchers, manufacturing companies involved in producing sealing products, trade associations, and window test laboratories [2–6]. It as for these reasons that ASTM Committee E06 developed a new fenestration air leakage test method, published in 1991, which takes these variables into account. Its designation is ASTM E 1424, Test Method for Determining the Rate of Air Leakage through Exterior Windows, Curtain Walls, and Doors under Specified Pressure and Temperature Differences across the Specimen.

Production line quality issues are also a concern, but are not directly addressed by either the ASTM E 283 or the E 1424 test methods or any fenestration industry specifications.

ASTM has a field test method for fenestration air leakage designated as ASTM E 783, Test Method for Field Measurement of Air Leakage through Installed Exterior Windows and Doors. This method can be used anytime during the year, under any temperature extreme. It is this method that provides the best assessment of product leakage rates because it accounts for actual conditions of product quality, installation, aging (durability), wind loads, and ambient air temperatures.

The purpose for basing ASTM E 1424 on temperature differentials across the test specimen is to account for framing material expansion and contraction, material and component shrinkage, and thermal bowing [6,7]. Research by this author and others has illustrated that these are primary mechanisms for variations in fenestration air leakage rates.

The NFRC Issues

In order to determine which of the above methods will provide the most accurate data for energy and cost analysis, the NFRC Air Leakage Subcommittee embarked on a sensitivity study to determine how variations in window air leakage rates would affect energy and cost consumption in several typical buildings. The subcommittee's initial assumption was that air leakage was an insignificant portion of the total fenestration energy transfer as long as the leakage rates remained below 0.37 cfm/fcp. Therefore, it was not necessary to have a test method with greater accuracy than the ASTM E 283 method.

There are two unknowns to this assumption. First, what are energy and cost impacts if leakage rates increased, for any reason, above the 0.37 cfm/fcp? And second, what factors could cause the rates to increase to an unacceptable level and thereby compromise the NFRC rating system?

Another major issue was the possible requirement, mandated through the NFRC certification program, that manufacturers use a test method that is new and that may cost more to use. The fenestration manufacturing community was partly unresponsive to the possibility of having to produce all new air leakage data for a new test method. For this reason, any differences between ASTM E 283 and E 1424 must be significant to warrant any changes to the current test method.

RESFEN Overview

The RESFEN program calculates building heating and cooling energy loads and costs by interpolation of best fit regressions of thousands of DOE-2 runs. Regression analysis of a database of DOE-2 building energy simulations of single and two-story residential buildings were used in this analysis. For further detail of this software, the reader is directed to the references [8].

In order to perform the sensitivity study, and for the NFRC Air Leakage Subcommittee to be able to use these results as a basis for selecting an appropriate air leakage test method, the following assumptions were made:

1. RESFEN accurately calculates energy loads and costs. The subcommittee has made no attempts to verify the algorithms of this program. As it was developed by LBL for the NFRC, and because the NFRC has adopted this software for use in residential building analysis, it has been assumed by the NFRC Air Leakage Subcommittee and the author that its outputs are valid.
2. The DOE-2 runs, which RESFEN are based on, are accurate. Again, these results are also obtained from LBL and have not been questioned [9].
3. The standard flow equation, $Q = k\Delta p$, is accurate for this analysis, i.e., products with narrow, sharp-edged flow orifices with turbulent flow.
4. The flow coefficient in this equation, $n = \frac{1}{2}$, is correct and therefore the reference test pressure used in the DOE-2 runs of 4 Pa does not affect product air leakage [10].
5. The Sherman-Grimsrud single-cell whole house infiltration models are accurate [10].
6. Fenestration products do not change their effective leakage areas as a function of wind pressure, exterior air temperature, or weatherseal degradation.

The RESFEN software was selected by the NFRC Air Leakage Subcommittee for its operating and input simplicity and its development within the framework of the NFRC was deemed to be important for validating the results with other working committees within the NFRC.

The Sensitivity Study

Several studies are planned and therefore this first phase was simplified by adjusting some of the input variables and setting others constant.

The RESFEN software allows the user to vary the following parameters: geographic locations, fenestration air leakage rates, window size, U-factor, shading coefficient, window operating style, window area as a percentage of the overall floor area, energy costs, building heating and cooling equipment, external and internal shading devices, external wall construction, and building type and size.

The following parameters were varied in the study:

1. Air leakage rates; using both ASTM E 283 data and ASTM E 1424 data.
2. Window size and operating type.
3. Total window area as a percent of floor area.
4. House type and size.
5. Geographic location.

The following variables were set constant:

1. Energy cost: electric at $0.08 per kWh; gas at $0.50 therm.
2. Shading coefficient: 0.7.
3. U-factor: 0.5 BTU/HR*FT²*F.
4. Percentage of window area was equal for each side of the building.
5. Wall construction.
6. Heating and cooling equipment.
7. Interior loads.

These variables were adjusted for each analysis:

1. Locations: Atlanta, GA, Madison, WI, Los Angeles, CA, and Denver, CO.
2. Total window area (as a percentage of building floor areas):
    (a) 1% windows per building side = 4% total (61.1 ft² (5.7 m²) windows for the 1540
        ft² (143.1 m²) ranch house and 123 ft² (11.4 m²) of windows for the 3080 ft² (286.1
        m²) colonial house).
    (b) 2% windows per building side = 8% total (123.2 ft² (11.4 m²) ranch and 246.4 ft²
        (22.9 m²) colonial).
    (c) 4% windows per building side = 16% total (246.4 ft² (22.9 m²) ranch and 492.8
        ft² (45.8 m²) colonial).
3. House type: 1540 ft² (143.1 m²) single story ranch
                3080 ft² (286.1 m²) two story colonial.
4. Window types: casement—2 ft wide × 4 ft high (0.6 × 1.2 m)
                 double hung—2.5 ft wide × 5 ft high (0.8 × 1.5 m)
                 horizontal slider—6 ft wide × 3 ft high (1.8 × 0.9 m).
5. Air leakage rates: 0.0 scfm/fcp, 0.5 scfm/fcp, 1.0 scfm/fcp, 1.5 scfm/fcp, 2.0 scfm/
    fcp (NOTE: scfm/fcp = standard cubic feet per min per foot of operable crack
    perimeter).

For each of the four locations, the leakage rate was varied from 0.0 to 2.0 scfm/fcp, in
increments of 0.5 scfm/fcp for each window and floor ratio. This series was duplicated for
both house types and sizes, and these were duplicated for each of the three fenestration
product operating styles.

Analysis of Results

The output from the study is detailed in Table 1. The table is laid out with the geographic
location, window operating style, window dimensions (w × h), building square footage,
window square footage, air leakage rates, total window energy transfer (heating and cooling)
of all the windows in the house for one year, total energy fraction (ratio of infiltration heat
transfer to conduction heat transfer), total energy cost for one year (heating and cooling),
the cost associated with just the air leakage for heating and cooling for one year, the cost
fraction (ratio of infiltration energy cost to conduction energy cost), and the peak heating
and peak cooling loads.

The data illustrate that when the air leakage rate for any location gets above 0.50 scfm/
fcp, the cost associated with heating or cooling the air becomes significant. For this study,
significant is a 5% increase above the "no leakage" condition. A summary analysis on the
yearly cost of heating and cooling the infiltrating air is found in Table 2. The concept of
significant air leakage rates relies on a cost fraction of 5% or greater for any category in the

TABLE 1—*Continued. Cost fraction:* $F(city/ft^2/type)$, %.

	Building Floor Area (ft²)							
	Atlanta		Los Angeles		Madison		Denver	
	1540	3080	1540	3080	1540	3080	1540	3080
Casements	7.0	7.6	10.2	11	18.4	19.6	13.2	14.1
Horizontal Slider	5.4	5.9	6.8	6.5	14.3	15.3	10.5	11.1
Double Hung	6.7	7.2	8.5	10.0	17.4	18.3	12.3	13.3

NOTE: 1 ft² = 0.09 m².

analysis. Five percent was chosen because it occurs at all leakage rates above 0.5 cfm/fcp. All new windows have to be below 0.37 cfm/fcp for certification to national standards or to compete with certified products. This analysis assumes an adjustment factor for installation and therefore provides the manufacturer or building owner an additional allowance of 0.13 cfm/fcp before they fall into the significance category.

These numbers indicate that for each city, building size, and window operating style there is at least a 5% cost fraction at 0.5 cfm/fcp window leakage. The greater the cost fraction, the greater the air leakage impact is on the total energy consumption for each case. The base case (no leakage) varies with each model and can be found in Table 2.

There is little difference between the size of the building on the air cost fraction, but in each case the larger building has a higher fraction. This is due to the incremental increase in the number of windows for the larger building.

Casement windows have a higher fraction than the horizontal sliders or double hungs because there are more of them due to their smaller size as a percentage of the floor area; e.g., there are 31 2 × 4 casements at 246 ft² (22.9 m²) of overall glass versus 14 6 × 3 horizontal sliders. This rational is the same for the double hungs.

Madison has the highest cost fraction, followed by Denver, Los Angeles, and Atlanta. The Los Angeles case has no heating loads on many of the 0.5 cfm/fcp runs. This is indicated by negative energy numbers in several of the leakage and glass area categories.

The single highest air cost is $174 at 2.0 cfm/fcp for 493 ft² (45.8 m²) of glass in the 3080 ft² (286.1 m²) house with a total energy cost of $414. The no-leakage base-case is $240.

Conclusion

Air leakage can become a significant portion of any fenestration product's total energy consumption when window leakage rates exceed the 0.5 cfm/fcp. This study illustrates this example with four cities located through the United States.

The NFRC is currently analyzing the impact of fenestration framing materials, size, test pressures and temperatures, operating styles, and installation on the air leakage rate of windows and doors. These data will be used in conjunction with the results from this sensitivity study to determine if a more rigorous test method is appropriate for use by the NFRC rating system. The study will look at framing materials, wind pressures, temperature differentials, and operating styles.

Field testing by various individuals and organizations along with production-line testing indicates that fenestration products are often higher than their stated 0.37 cfm/fcp maximum allowed rating and are more likely to be on the order of 50 to 200% higher.

TABLE 2—NFRC Air Leakage Subcommittee sensitivity analysis of RESFEN air leakage, 14 Sept. 1992. Atlanta Casement (2×4).

House (ft²)	Glass (ft²)	Window Leakage (cfm/fcp) (75 pa)	Tot Energy (KBTU) (4 Pa)	Energy Fraction Q_i/Q_c (%) (4 Pa)	Tot Cost ($) (4 Pa)	Air Cost ($) (4 Pa)	Cost Fraction C_i/C_c (%) (4 Pa)	Cool Pk (kW)	Heat Pk (KBTU/H)
1540	246.4	2	14 542.2	79.4	195	42	27.8	1.6	25.3
		1.5	12 933	59.6	184	32	20.9	1.5	24.3
		1	11 323.8	39.7	174	21	13.9	1.5	23.3
		0.5	9 714.6	19.9	163	11	7.0	1.4	22.3
		0	8 105.4		153			1.4	21.4
	123.2	2	6 991.4	85.3	97	21	28.1	0.8	13.4
		1.5	6 186.8	64.0	91	16	21.1	0.8	12.9
		1	5 382.2	42.7	86	11	14.1	0.8	12.4
		0.5	4 577.5	21.3	81	5	7.0	0.7	11.9
		0	3 772.9		75			0.7	11.4
	61.6	2	3 425.7	88.6	48	11	28.3	0.4	6.9
		1.5	3 023.4	66.4	45	8	21.1	0.4	6.6
		1	2 621.1	44.3	43	5	14.1	0.4	6.4
		0.5	2 218.8	22.1	40	3	6.9	0.4	6.1
		0	1 816.5		38			0.4	5.9
3080	492.8	2	28 338.1	83.2	369	85	30.0	3.2	51.9
		1.5	25 119.6	62.4	347	64	22.5	3.1	50
		1	21 901.2	41.6	326	43	15.0	3	48
		0.5	18 682.8	20.8	305	21	7.5	2.8	46.1
		0	15 464.4		284			2.7	44.1
	246.4	2	13 573.2	90.2	182	43	30.4	1.7	31.1
		1.5	11 964	67.6	172	32	22.7	1.7	30.2
		1	10 354.8	45.1	161	21	15.2	1.6	29.2
		0.5	8 745.6	22.5	150	11	7.6	1.6	28.2
		0	7 136.4		140			1.5	27.2
	123.2	2	6 638.1	94.1	91	21	30.5	1	19.8
		1.5	5 833.5	70.6	85	16	22.9	0.9	19.3
		1	5 028.9	47.1	80	11	15.3	0.9	18.8
		0.5	4 224.3	23.5	75	5	7.6	0.9	18.3
		0	3 419.7		69			0.9	17.8

TABLE 2—Continued. Horizontal Slider (6×3).

House (ft²)	Glass (ft²)	Window Leakage (cfm/fcp) (75 pa)	Tot Energy (KBTU) (4 Pa)	Energy Fraction Q_i/Q_c (%) (4 Pa)	Tot Cost ($) (4 Pa)	Air Cost ($) (4 Pa)	Cost Fraction C_i/C_c (%) (4 Pa)	Cool Pk (kW)	Heat Pk (KBTU/H)
1540	246.4	2	13 111.8	61.8	186	33	21.6	1.5	24.4
		1.5	11 860.2	46.3	177	25	16.2	1.5	23.6
		1	10 608.6	30.9	169	17	10.8	1.4	22.9
		0.5	9 357	13.8	161	8	5.4	1.4	22.1
		0	8 105.4		153			1.4	21.4
	123.2	2	6 276.2	66.3	92	17	21.9	0.8	13
		1.5	5 650.4	49.8	88	12	16.4	0.8	12.6
		1	5 024.6	33.2	84	8	10.9	0.7	12.2
		0.5	4 398.7	16.6	80	4	5.4	0.7	11.8
		0	3 772.9		75			0.7	11.4
	61.6	2	3 068.1	68.9	46	8	21.9	0.4	6.7
		1.5	2 755.2	51.7	44	6	16.5	0.4	6.5
		1	2 442.3	34.5	42	4	10.9	0.4	6.3
		0.5	2 129.4	17.2	40	2	5.3	0.4	6.1
		0	1 816.5		38			0.4	5.9
3080	492.8	2	25 477.2	64.7	350	66	23.3	3.1	50.2
		1.5	22 974	48.6	333	50	17.5	3	48.7
		1	20 470.8	32.4	317	33	11.7	2.9	47.1
		0.5	17 967.6	16.2	300	17	5.8	2.8	45.6
		0	15 464.4		284			2.7	44.1
	246.4	2	12 142.8	70.2	173	33	23.6	1.7	30.3
		1.5	10 891.2	52.6	165	25	17.7	1.6	29.5
		1	9 639.6	35.1	156	17	11.8	1.6	28.8
		0.5	8 388	17.5	148	8	5.9	1.5	28
		0	7 136.4		140			1.5	27.2
	123.2	2	5 922.9	73.2	86	17	23.8	0.9	19.4
		1.5	5 297.1	54.9	82	12	17.9	0.9	19
		1	4 671.3	36.6	78	8	12.0	0.9	18.6
		0.5	4 045.5	18.3	70	4	5.9	0.9	18.2
		0	3 419.7		69			0.9	17.8

TABLE 1—Continued. Double Hung (2.5×5).

House (ft²)	Glass (ft²)	Window Leakage (cfm/fcp) (75 pa)	Tot Energy (KBTU) (4 Pa)	Energy Fraction Q_i/Q_c (%) (4 Pa)	Tot Cost ($) (4 Pa)	Air Cost ($) (4 Pa)	Cost Fraction C_i/C_c (%) (4 Pa)	Cool Pk (kW)	Heat Pk (KBTU/H)
1540	246.4	2	14 113.1	74.1	192	40	26.0	1.6	25
		1.5	12 611.2	55.6	182	30	19.5	1.5	24.1
		1	11 109.3	37.1	172	20	13.0	1.5	23.2
		0.5	9 607.3	18.5	162	10	6.5	1.4	22.3
		0	8 105.4		153			1.4	21.4
	123.2	2	6 776.8	79.6	95	20	26.3	0.8	13.3
		1.5	6 025.8	59.7	90	15	19.6	0.8	12.8
		1	5 274.9	39.8	85	10	13.1	0.8	12.3
		0.5	4 523.9	19.9	80	5	6.5	0.7	11.9
		0	3 772.9		75			0.7	11.4
	61.6	2	3 318.5	51.4	47	10	26.4	0.4	6.8
		1.5	2 943	34.3	45	7	19.7	0.4	6.6
		1	2 567.5	17.1	42	5	13.1	0.4	6.4
		0.5	2 192	0.0	40	3	6.7	0.4	6.1
		0	1 816.5		38			0.4	5.9
3080	492.8	2	27 479.8	77.7	363	79	28.0	3.2	51.4
		1.5	24 476	58.3	343	60	21.0	3.1	49.6
		1	21 472.1	38.8	323	40	14.0	3	47.7
		0.5	18 468.2	19.4	303	20	7.0	2.8	45.9
		0	15 464.4		284			2.7	44.1
	246.4	2	13 144.1	84.2	179	40	28.3	1.7	30.9
		1.5	11 642.2	63.1	170	30	21.2	1.7	30
		1	10 140	42.1	160	20	14.2	1.6	29.1
		0.5	8 638.3	21.0	150	10	7.1	1.6	28.1
		0	7 136.4		140			1.5	27.2
	123.2	2	6 423.5	87.8	89	20	28.5	1	19.7
		1.5	5 672.6	65.9	84	15	21.5	0.9	19.2
		1	4 291.6	43.9	79	10	14.3	0.9	18.8
		0.5	4 170.6	22.0	74	5	7.2	0.9	18.3
		0	3 419.7		69			0.9	17.8

TABLE 2—Continued. Los Angeles Casement (2×4).

House (ft²)	Glass (ft²)	Window Leakage (cfm/fcp) (75 pa)	Tot Energy (KBTU) (4 Pa)	Energy Fraction Q_i/Q_c (%) (4 Pa)	Tot Cost ($) (4 Pa)	Air Cost ($) (4 Pa)	Cost Fraction C_i/C_c (%) (4 Pa)	Cool Pk (kW)	Heat Pk (KBTU/H)
1540	246.4	2	2 991.7	328.0	49	9	20.9	2.2	8
		1.5	2 418.5	246.0	47	6	15.8	2.1	7.5
		1	1 845.4	164.0	45	4	10.3	2	7
		0.5	1 272.2	82.0	43	2	5.2	2	6.6
		0	699		41			1.9	6.1
	123.2	2	1 009.9	839.9	19	4	28.6	1.1	4.1
		1.5	723.3	629.9	18	3	21.8	1.1	3.8
		1	436.7	419.9	17	2	14.3	1	3.6
		0.5	150.1	210.0	16	1	6.8	1	3.4
		0	− 136.5		15			1	3.1
	61.6	2	383.4	302.1	8	2	37.3	0.6	2.1
		1.5	240.2	226.6	8	2	27.1	0.5	1.9
		1	96.9	151.1	7	1	18.6	0.5	1.8
		0.5	− 46.4	75.5	7	1	10.2	0.5	1.7
		0	− 189.7		6			0.5	1.6
3080	492.8	2	5 140.7	825.8	94	17	22.2	4.3	13.8
		1.5	3 994.4	619.3	89	13	16.6	4.2	12.8
		1	2 848	412.9	85	9	11.1	4	11.8
		0.5	1 701.7	206.4	81	4	5.5	3.9	10.9
		0	555.3		77			3.8	9.9
	246.4	2	1 405.8	258.5	35	9	32.6	2.3	6.5
		1.5	823.6	192.9	33	6	24.5	2.2	6
		1	259.4	129.2	30	4	16.5	2.2	6.5
		0.5	− 313.8	64.6	28	2	8.0	2.1	5
		0	− 886.9		26			2	4.6
	123.2	2	412.4	156.2	14	4	43.0	1.2	2.7
		1.5	125.8	117.1	13	3	32.0	1.2	2.5
		1	− 160.8	78.1	12	2	22.0	1.2	2.2
		0.5	− 447.4	39.0	11	1	11.0	1.1	2
		0	− 733.9		10			1.1	1.8

TABLE 2—Continued. Horizontal Slider (6×3).

House (ft²)	Glass (ft²)	Window Leakage (cfm/fcp) (75 pa)	Tot Energy (KBTU) (4 Pa)	Energy Fraction Q_i/Q_c (%) (4 Pa)	Tot Cost ($) (4 Pa)	Air Cost ($) (4 Pa)	Cost Fraction C_i/C_c (%) (4 Pa)	Cool Pk (kW)	Heat Pk (KBTU/H)
1540	246.4	2	2 482.2	255.1	47	7	16.3	2.1	7.6
		1.5	2 036.4	191.3	46	5	12.3	2.1	7.2
		1	1 590.6	127.6	44	3	8.1	2	6.8
		0.5	1 144.8	63.8	42	2	4.2	2	6.5
		0	699		41			1.9	6.1
	123.2	2	755.1	653.2	18	3	22.4	1.1	3.9
		1.5	532.2	489.9	17	3	17.0	1.1	3.7
		1	309.3	326.6	16	2	10.9	1	3.5
		0.5	86.4	163.3	16	1	5.4	1	3.3
		0	− 136.5		15			1	3.1
	61.6	2	256.1	235.0	8	2	28.8	0.5	2
		1.5	144.6	176.2	7	1	22.0	0.5	1.9
		1	33.2	117.5	7	1	15.3	0.5	1.8
		0.5	− 78.3	58.7	6	0	6.8	0.5	1.7
		0	− 189.7		6			0.5	1.6
3080	492.8	2	4 121.7	642.2	90	13	17.2	4.2	12.9
		1.5	3 230.1	481.7	87	10	12.9	4.1	12.2
		1	2 338.5	321.1	83	7	8.6	4	11.4
		0.5	1 446.9	160.6	80	3	4.3	3.9	10.7
		0	555.3		77			3.8	9.9
	246.4	2	896.3	201.1	33	7	25.3	2.2	6
		1.5	450.5	150.8	31	5	19.2	2.2	5.7
		1	4.7	100.5	29	3	12.6	2.1	5.3
		0.5	− 441.1	50.3	28	2	6.5	2.1	4.9
		0	− 886.9		26			2	4.6
	123.2	2	157.7	121.5	13	3	33.0	1.2	2.5
		1.5	− 65.2	91.1	13	3	25.0	1.2	2.3
		1	− 288.1	60.7	12	2	17.0	1.2	2.1
		0.5	− 511	30.4	11	1	9.0	1.1	2
		0	− 733.9		10			1.1	1.8

TABLE 2—Continued. Double Hung (2.5×5).

House (ft²)	Glass (ft²)	Window Leakage (cfm/fcp) (75 pa)	Tot Energy (KBTU) (4 Pa)	Energy Fraction Q_i/Q_c (%) (4 Pa)	Tot Cost ($) (4 Pa)	Air Cost ($) (4 Pa)	Cost Fraction C_i/C_c (%) (4 Pa)	Cool Pk (kW)	Heat Pk (KBTU/H)
1540	246.4	2	2 838.9	306.1	49	8	20.2	2.2	7.9
		1.5	2 303.9	229.6	47	6	14.5	2.1	7.4
		1	1 768.9	153.1	45	4	9.9	2	7
		0.5	1 234	76.5	43	2	4.9	2	6.5
		0	699		41			1.9	6.1
	123.2	2	933.5	783.9	19	4	26.5	1.1	4
		1.5	666	587.9	18	3	20.4	1.1	3.8
		1	398.5	391.9	17	2	13.6	1	3.6
		0.5	131	196.0	16	1	6.8	1	3.3
		0	- 136.5		15			1	3.1
	61.6	2	345.2	282.0	8	2	33.9	0.6	2
		1.5	211.5	211.5	7	2	25.4	0.5	1.9
		1	77.8	141.0	7	1	16.9	0.5	1.8
		0.5	-56	70.5	6	1	8.5	0.5	1.7
		0	- 189.7		6			0.5	1.6
3080	492.8	2	4 835	770.7	92	16	20.6	4.2	13.5
		1.5	3 765.1	578.0	89	12	15.5	4.1	12.6
		1	2 695.2	385.4	85	8	10.3	4	11.7
		0.5	1 625.2	192.7	81	4	5.1	3.9	10.8
		0	555.3		77			3.8	9.9
	246.4	2	1 252.9	241.3	34	8	30.3	2.3	6.3
		1.5	718	181.0	32	6	23.0	2.2	5.9
		1	183	120.6	30	4	15.3	2.1	5.0
		0.5	- 352	60.3	28	2	7.7	2.1	5
		0	- 886.9		26			2	4.6
	123.2	2	336	145.8	14	4	40.0	1.2	2.7
		1.5	68.5	109.3	13	3	30.0	1.2	2.4
		1	-199	72.9	12	2	20.0	1.2	2.2
		0.5	- 466.5	36.4	11	1	10.0	1.1	2
		0	- 733.9		10			1.1	1.8

TABLE 2—Continued. Madison Casement (2×4).

House (ft²)	Glass (ft²)	Window Leakage (cfm/fcp) (75 pa)	Tot Energy (KBTU) (4 Pa)	Energy Fraction Q_i/Q_c (%) (4 Pa)	Tot Cost ($) (4 Pa)	Air Cost ($) (4 Pa)	Cost Fraction C_i/C_c (%) (4 Pa)	Cool Pk (kW)	Heat Pk (KBTU/H)
	246.4	2	28 615.1	128.8	213	87	69.3	1.9	22.8
		1.5	24 587.5	95.6	191	65	51.9	1.8	20.6
		1	20 559.8	64.4	169	44	34.7	1.7	18.3
		0.5	16 532.2	32.2	148	22	17.3	1.6	16.1
		0	12 504.6		126			1.6	13.9
	123.2	2	13 783	140.6	104	44	72.7	1	11.3
		1.5	11 769.2	105.5	93	33	54.5	0.9	10.2
		1	9 755.4	70.3	82	22	36.3	0.9	9
		0.5	7 741.6	35.2	71	11	18.2	0.8	7.9
		0	5 727.8		60			0.8	6.8
	61.6	2	6 760.4	147.4	51	22	74.4	0.5	5.6
		1.5	5 753.5	110.5	46	16	55.6	0.5	5.1
		1	4 746.6	73.7	40	11	37.2	0.5	4.5
		0.5	3 739.7	36.9	35	5	18.4	0.4	3.9
		0	2 732.7		29			0.4	3.4
	492.8	2	56 829.2	130.9	414	174	72.7	3.8	46.2
		1.5	49 773.9	98.2	370	131	54.5	3.6	41.7
		1	40 718.7	65.5	327	87	36.4	3.4	37.2
		0.5	32 663.5	32.7	283	44	18.2	3.3	32.7
		0	24 608.2		240			3.1	28.3
	246.4	2	27 360.2	143.2	201	87	76.3	2.1	21.7
		1.5	23 332.6	107.4	179	65	57.2	2	19.5
		1	19 305	71.6	158	44	38.2	1.9	17.3
		0.5	15 277.4	35.8	136	22	19.1	1.8	15
		0	11 249.8		114			1.7	12.8
	123.2	2	13 416.8	150.2	99	44	78.0	1.2	9.7
		1.5	11 403	112.7	88	33	58.8	1.1	8.6
		1	9 389.2	75.1	77	22	39.2	1.1	7.5
		0.5	7 375.4	37.6	67	11	19.6	1	6.4
		0	5 361.6		56			1	5.3

TABLE 2—Continued. Horizontal Slider (6×3).

House (ft²)	Glass (ft²)	Window Leakage (cfm/fcp) (75 pa)	Tot Energy (KBTU) (4 Pa)	Energy Fraction Q_i/Q_c (%) (4 Pa)	Tot Cost ($) (4 Pa)	Air Cost ($) (4 Pa)	Cost Fraction C_i/C_c (%) (4 Pa)	Cool Pk (kW)	Heat Pk (KBTU/H)
	246.4	2	25 035	100.2	194	68	53.9	1.8	20.8
		1.5	21 902.4	75.2	177	51	40.4	7.2	77.4
		1	18 769.8	50.1	160	34	27.0	1.7	17.3
		0.5	15 637.2	25.1	143	17	13.5	1.6	15.6
		0	12 504.6		126			1.6	13.9
	123.2	2	11 992.9	109.4	94	34	56.5	0.9	10.3
		1.5	10 426.7	82.0	85	25	42.3	0.9	9.4
		1	8 860.4	54.7	77	17	28.3	0.9	8.5
		0.5	7 294.1	27.3	69	9	14.2	0.8	7.7
		0	5 727.8		60			0.8	6.8
	61.6	2	5 865.3	114.6	46	17	57.7	0.5	5.1
		1.5	5 082.2	86.0	42	13	43.3	0.5	4.7
		1	4 299	57.3	38	8	29.0	0.4	4.2
		0.5	3 515.9	28.7	34	4	14.3	0.4	3.8
		0	2 732.7		29			0.4	3.4
	492.8	2	49 669	101.8	375	134	55.6	3.6	42.2
		1.5	43 403.8	76.4	341	100	41.6	3.5	38.7
		1	37 138.6	50.9	307	66	27.5	3.4	35.2
		0.5	30 873.4	25.5	274	33	13.5	3.2	31.8
		0	24 607		241				
	246.4	2	23 780.1	111.4	182	68	59.3	2	19.8
		1.5	20 647.5	83.5	165	51	44.5	1.9	18
		1	17 514.9	55.7	148	34	29.7	1.8	16.3
		0.5	14 382.3	27.8	131	17	14.9	1.8	14.5
		0	11 249.8		114			1.7	12.8
	123.2	2	11 626.7	116.9	90	34	61.0	1.1	8.7
		1.5	10 060.4	87.6	81	25	45.7	1.1	7.9
		1	8 494.1	58.4	73	17	30.6	1	7
		0.5	6 927.8	29.2	64	8	15.3	1	6.1
		0	5 361.6		56			1	5.3

TABLE 2—Continued. Double Hung (2.5×5).

House (ft²)	Glass (ft²)	Window Leakage (cfm/fcp) (75 pa)	Tot Energy (KBTU) (4 Pa)	Energy Fraction Q_i/Q_c (%) (4 Pa)	Tot Cost ($) (4 Pa)	Air Cost ($) (4 Pa)	Cost Fraction C_i/C_c (%) (4 Pa)	Cool Pk (kW)	Heat Pk (KBTU/H)
246.4	246.4	2	27 541	120.2	207	81	64.7	1.9	22.2
		1.5	23 781.9	90.2	187	61	48.5	1.8	20.1
		1	20 022.8	60.1	166	41	32.4	1.7	18
		0.5	16 263.7	30.1	146	20	16.2	1.6	15.9
		0	12 504.6		126			1.6	13.9
	123.2	2	13 246	131.3	101	41	67.8	1	11
		1.5	11 366.4	98.4	91	31	50.8	0.9	9.9
		1	9 486.9	65.6	80	20	33.8	0.9	8.9
		0.5	7 687.3	32.8	70	10	17.0	0.8	7.9
		0	5 727.8		60			0.8	6.8
	61.6	2	6 491.9	137.6	50	20	69.3	0.5	5.5
		1.5	5 552.1	103.2	45	15	51.9	0.5	4.9
		1	4 612.3	68.8	40	10	34.8	0.4	4.4
		0.5	3 672.5	34.4	34	5	17.4	0.4	3.9
		0	2 732.7		29			0.4	3.4
492.8	492.8	2	54 681.1	122.2	402	163	67.8	3.8	45
		1.5	47 162.9	91.7	362	122	50.9	3.6	40.8
		1	39 644.7	61.1	321	81	33.9	3.4	36.6
		0.5	32 126.5	30.6	280	41	16.9	3.2	32.4
		0	24 608.2		240			3.1	28.3
123.2	246.4	2	26 286.2	133.7	195	81	71.3	2	21.1
		1.5	22 527.1	100.2	175	61	53.5	1.9	19.1
		1	18 768	66.8	155	41	35.7	1.9	17
		0.5	15 008.9	33.4	134	20	17.8	1.8	14.9
		0	11 249.8		114			1.7	12.8
	123.2	2	12 879.8	140.2	96	41	73.2	1.1	9.4
		1.5	11 000.2	105.2	86	30	54.9	1.1	8.4
		1	9 120.7	70.1	76	20	36.7	1.1	7.3
		0.5	7 241.1	35.1	66	10	18.3	1	6.3
		0	5 361.6		56			1	5.3

TABLE 2—Continued. Denver Casement (2×4).

House (ft²)	Glass (ft²)	Window Leakage (cfm/fcp) (75 pa)	Tot Energy (KBTU) (4 Pa)	Energy Fraction Q_i/Q_c (%) (4 Pa)	Tot Cost ($) (4 Pa)	Air Cost ($) (4 Pa)	Cost Fraction C_i/C_c (%) (4 Pa)	Cool Pk (kW)	Heat Pk (KBTU/H)
1540	246.4	2	14 676.8	171.8	156	49	45.7	1.8	25.4
		1.5	12 357.8	128.8	143	37	34.3	1.8	24.2
		1	10 038.7	85.9	131	24	22.9	1.8	23
		0.5	7 719.7	42.9	119	12	11.4	1.7	21.7
		0	5 400.6		107			1.7	20.5
	123.2	2	6 562.4	241.0	73	24	50.6	0.9	13.4
		1.5	5 402.8	180.8	67	18	38.0	0.9	12.7
		1	4 243.4	120.5	60	12	25.3	0.9	12.1
		0.5	3 083.8	60.3	54	6	12.7	0.9	11.5
		0	1 924.3		48			0.9	10.9
	61.6	2	3 087.2	301.9	35	12	53.5	0.5	6.8
		1.5	2 507.4	226.4	32	9	39.9	0.5	6.5
		1	1 927.7	151.0	29	6	26.8	0.5	6.2
		0.5	1 347.9	75.5	26	3	13.2	0.4	5.9
		0	768.1		23			0.4	5.6
3080	492.8	2	29 119.4	175.6	303	98	47.5	3.7	52.8
		1.5	24 481.3	131.7	279	73	35.6	3.6	50.3
		1	19 843.2	87.8	254	49	23.8	3.5	47.8
		0.5	15 205	43.9	230	24	11.9	3.5	45.4
		0	10 566.9		205			3.4	42.9
	246.4	2	12 923.2	254.4	141	49	53.0	1.9	30.4
		1.5	10 604.1	190.8	129	37	39.8	1.9	29.1
		1	8 285.1	127.2	116	24	26.5	1.9	27.9
		0.5	5 966	63.6	104	12	13.3	1.8	26.7
		0	3 647		92			1.8	25.4
	123.2	2	6 053.1	327.8	68	24	56.4	1	18.3
		1.5	4 893.5	245.9	62	18	42.3	1	17.7
		1	3 734	163.9	56	12	28.2	1	17.1
		0.5	2 574.5	82.0	49	6	14.1	1	16.5
		0	1 414.9		43			1	15.8

TABLE 2—Continued. Horizontal Slider (6×3).

House (ft²)	Glass (ft²)	Window Leakage (cfm/fcp) (75 pa)	Tot Energy (KBTU) (4 Pa)	Energy Fraction Q_i/Q_c (%) (4 Pa)	Tot Cost ($) (4 Pa)	Air Cost ($) (4 Pa)	Cost Fraction C_i/C_c (%) (4 Pa)	Cool Pk (kW)	Heat Pk (KBTU/H)
1540	246.4	2	12 615.4	133.6	145	38	35.6	1.8	24.3
		1.5	10 811.7	100.2	135	28	26.7	1.8	23.4
		1	9 008	66.8	126	19	17.8	1.8	22.4
		0.5	7 204.3	33.4	116	10	8.9	1.7	21.5
		0	5 400.6		107			1.7	20.5
	123.2	2	5 531.7	187.5	67	19	39.2	0.9	12.8
		1.5	4 629.8	140.6	62	14	29.5	0.9	12.3
		1	3 728	93.7	58	9	19.5	0.9	11.8
		0.5	2 826.1	46.9	53	6	9.8	0.9	11.4
		0	1 924.3		48			0.9	10.9
	61.6	2	2 571.8	100.0	32	9	41.7	0.5	6.6
		1.5	2 120.09	176.0	30	7	31.1	0.5	6.3
		1	1 670	117.4	28	5	20.6	0.5	6.1
		0.5	1 219.1	58.7	25	2	10.5	0.4	5.8
		0	768.1		23			0.4	5.6
3080	492.8	2	24 996.6	136.6	281	76	37.0	3.6	50.6
		1.5	21 389.2	102.4	262	57	27.7	3.5	48.7
		1	17 781.8	68.3	243	38	18.5	3.5	46.7
		0.5	14 174.4	34.1	224	19	9.3	3.4	44.8
		0	10 566.9		205			3.4	42.9
	246.4	2	10 861.8	197.8	130	38	41.2	1.9	29.3
		1.5	9 058.11	148.4	120	28	30.9	1.9	28.3
		1	7 254.4	98.9	111	19	20.5	1.9	27.4
		0.5	5 450.7	49.5	102	10	10.3	1.8	26.4
		0	3 647		92			1.8	25.4
	123.2	2	5 022.4	255.0	62	19	43.9	1	17.8
		1.5	4 120.5	191.2	58	14	32.8	1	17.3
		1	3 218.7	127.5	53	10	21.9	1	16.8
		0.5	2 316.8	63.7	48	5	11.1	1	16.3
		0	1 414.9		43			1	15.8

TABLE 2—Continued. Double Hung (2.5×5).

House (ft²)	Glass (ft²)	Window Leakage (cfm/fcp) (75 pa)	Tot Energy (KBTU) (4 Pa)	Energy Fraction Q_i/Q_c (%) (4 Pa)	Tot Cost ($) (4 Pa)	Air Cost ($) (4 Pa)	Cost Fraction C_i/C_c (%) (4 Pa)	Cool Pk (kW)	Heat Pk (KBTU/H)
1540	246.4	2	14 058.4	160.3	152	46	42.7	1.8	25.1
		1.5	11 894	120.2	141	34	32.1	1.8	24
		1	9 729.5	80.2	130	23	21.4	1.8	22.8
		0.5	7 565.1	40.1	118	11	10.7	1.7	21.7
		0	5 400.6		107			1.7	20.5
	123.2	2	6 253.2	225.0	71	23	47.1	0.9	13.2
		1.5	5 170.9	168.7	65	17	35.3	0.9	12.6
		1	4 088.7	112.5	60	11	23.4	0.9	12
		0.5	3 006.5	56.2	54	6	11.8	0.9	11.5
		0	1 924.3		48			0.9	10.9
	61.6	2	2 932.6	281.8	34	11	50.0	0.5	6.7
		1.5	2 391.5	211.4	31	9	37.3	0.5	6.5
		1	1 850.4	140.9	29	6	25.0	0.5	6.2
		0.5	1 309.2	70.4	26	3	12.3	0.4	5.9
		0	768.1		23			0.4	5.6
3080	492.8	2	27 882	163.9	297	91	44.4	3.6	52.1
		1.5	23 553.6	122.9	274	68	33.3	3.6	49.8
		1	19 224.7	81.9	251	46	22.2	3.5	47.5
		0.5	14 895.8	41.0	228	23	11.1	3.5	45.2
		0	10 566.9		205			3.4	42.9
	246.4	2	12 304.8	237.4	138	46	49.5	1.9	30
		1.5	10 140.3	178.0	126	34	37.1	1.9	28.9
		1	7 975.9	118.7	115	23	24.7	1.9	27.7
		0.5	5 811.4	59.3	103	11	12.4	1.8	26.6
		0	3 647		92			1.8	25.4
	123.2	2	5 743.8	306.0	66	23	52.7	1	18.2
		1.5	4 661.6	229.5	60	17	39.5	1	17.6
		1	3 579.4	153.0	55	11	26.3	1	17
		0.5	2 497.2	76.5	49	6	13.2	1	16.4
		0	1 414.9		43			1	15.8

NOTE: 1 ft² = 0.09 m².

If air leakage rates are prone to dramatically vary based on test method or other variables, it will be critical that the NFRC develop a rating system that accounts for these variables in order to accurately predict fenestration energy consumption.

A second phase of this study is being prepared that will analyze the significant differences between varying fenestration air leakage rates, U-factors, and solar heat gain for all the cities in RESFEN.

References

[1] Ontario Hydro, *Residential Window Installation Guideline,* Interim Version 1.0, April 1991.

[2] Cahlers Du Centre Scientifique et Technique Du Batiment No. 132, "Variations Windows Air Permeability According to Outside Temperature," September 1972.

[3] Kehrli, D. W., "Window Air Leakage Performance as a Function of Differential Temperature and Accelerated Aging," Thermal Performance of the Exterior Envelopes of Buildings III—ASHRAE/DOE/BTECC, 5 December 1985, Clearwater, FL.

[4] Kehrli, D. W., "Window and Door Air Leakage Tests Predict Real World Performance," *Fenestration Magazine,* December 1989.

[5] Kehrli, D. W., "Fenestration Air Tightness Limitations: Serviceability/Durability," Symposium on Air Infiltration, Ventilation and Moisture Transfer, BTECC, December 1986.

[6] Sasaki, J. R., "Cold-Weather Performance a Hinged Exterior Door," Research paper 623, NRC Canada, Building Research Division, September 1974.

[7] Weidt, J. and Weidt, J., "Air Leakage of Newly Installed Residential Windows," LBL, Energy and Environment Division, 1980 LBL-11111, uc-94d, EEB-w-80-12, w-68.

[8] Regents of the University of California, Building Technologies Program, Lawrence Berkeley Labs, *RESFEN 1.3,* NFRC, 1993.

[9] American Society of Heating, Refrigeration and Air Conditioning Engineers, *1993 Fundamentals, Chapter 24, Weather Data.* ASHRAE, Atlanta, GA, 1993.

[10] American Society of Heating, Refrigeration and Air Conditioning Engineers, *1993 Fundamentals, Chapter 24, Infiltration and Ventilation.* ASHRAE, Atlanta, GA, 1993.

Michael J. Louis[1] and Peter E. Nelson[1]

Extraneous Air Leakage from Window Perimeters

REFERENCE: Louis, M. J. and Nelson, P. E., **"Extraneous Air Leakage from Window Perimeters,"** *Airflow Performance of Building Envelopes, Components, and Systems, ASTM STP 1255,* Mark P. Modera and Andrew K. Persily, Eds., American Society for Testing and Materials, Philadelphia, 1995, pp. 108–122.

ABSTRACT: The window industry has ratings for air leakage through window assemblies. These tests exclude air leakage from the joint between window and wall assemblies or from the sides of the windows. Such leakage is discounted by present window testing standards and labeled "extraneous air leakage." Air leakage to the window perimeter can significantly effect the overall performance of the window/wall assembly.

A test methodology is presented for quantifying the portion of air leakage that is not measured by current industry standards, but that nonetheless is real and significant air leakage through some types of installed window assemblies.

Design guidelines are proposed to minimize air leakage through the window frame and around the window perimeter. This requires careful selection and detailing of the window and wall system to insure that the air barrier effectively seals to the window frame. These details must be carefully coordinated with the design of the flashing and cavity water-proofing, which have distinct and sometimes conflicting purposes. In three examples, the proper location for air barriers in different types of wall construction is identified along with methods to seal the air barreir to the window.

KEYWORDS: air leakage, condensation, drafts, extraneous air leakage, thermal bypass, window surround

Windows are rated for air infiltration according to guidelines established by the American Architectural Manufactures Association (AAMA) [1] and the National Wood Window and Door Association (NWWDA) [2]. These organizations promulgate standards for various classes of windows that limit the air and water leakage through the window unit. They evaluate window performance based on the length of weatherstripped joints or the area of the fixed glazing within the perimeter of the unit in accordance with the ASTM Standard Test Method for Field Measurement of Air Leakage Through Installed Exterior Windows and Doors (E 783-91). Windows are tested to this ASTM standard by measuring the air leakage from the outside of the assembly to the inside of the assembly. Any air leakage from the wall cavity through the sides of the window is excluded from the air leakage measurement. Such leakage is measured only as an unqualified portion of the test chamber leakage and is discounted as "extraneous air leakage."

This "extraneous air leakage" can adversely affect the overall performance of the wall/window system by allowing air from the cavities inside the wall to leak into the building through openings in the window frame, by passing around the perimeter of the window

---

[1] Senior engineer and senior associate, respectively, Simpson Gumpertz & Heger Inc., Consulting Engineers, 297 Broadway, Arlington, MA 02174.

frame, or through the interior trim. Thus, even though the window element may be "airtight" within accepted industry standards, the installed window assembly, including its perimeter joints, may be highly air permeable, leading to excessive drafts, condensation, and energy loss. The purpose of this paper is to provide a test methodology for quantifying that portion of the air leakage that is not measured by current industry standards and to show some methods for minimizing air leakage through the window frame and around the window perimeter.

Diagnosis

Excessive air leakage can cause the following problems:

- Condensation—Condensation formation on the sides of window frames occurs when cool exterior air comes in contact with warmer interior surfaces, causing air vapor in the moist interior air to condense onto the cooler frame surface.
- Energy loss—Air in the cavity between the window and wall construction can bypass thermal breaks in metal frames, causing interior surfaces of framing and trim to cool, lose heat, and contribute to convective airflow in the room.
- Drafts—Air in the cavity between the window and wall construction will flow through small cracks between window frames and interior trim and through openings in the frame due to pressure gradient across the wall.

Proper evaluation of a window assembly for air leakage must start with a thorough understanding of the function and performance of the window as well as of the surrounding wall areas and the wall/window interface. In many cases, air leakage results from the way in which the window is integrated into the surrounding wall construction and not from any window-related defects. We have seen remedial window projects that have failed to solve both air and water leakage problems because of misguided investigations which focused on windows alone and failed to consider the conditions of the window surround, that is, perimeter joints and adjacent wall construction, which were the primary sources of both air and water leaks.

To provide an accurate diagnosis of window problems a thorough investigation, emphasizing the following, should be conducted:

- Review documents of the existing construction (architectural, shop drawings, and as-built drawings, if available), reports of complaints (drafts, condensation, etc.), and specification and manufacturer's literature.
- Survey interior and exterior wall conditions for cracks, failed sealants, or other defects that can allow air leaks to bypass exterior weather barriers. Determine the location of the air barriers in the exterior wall assembly.
- Probe internal conditions by selective disassembly of wall and window components.
- Test the existing window units, perimeter joints, and adjoining wall surfaces to isolate air leakage paths.

Proposed Testing Procedure

ASTM Test Method E 783 is applicable for measuring air leakage only through window and door assemblies. Measurements and calculations derived by this test method discount the air leakage between the window/door frame and the adjacent construction as an unqualified portion of the chamber leakage, or extraneous air leakage. The test method discounts that portion of extraneous air leakage without regard for the effect of this air leakage on the overall performance of the wall/window system.

The following test procedure is a modification of ASTM E 783-91, designed to allow segregation of "extraneous air leakage" into the following two components: (1) equipment/test chamber air leakage, and (2) window surround air leakage. To quantify the air leakage through the surrounding construction, apply the following test setup and sequence in lieu of test steps in paragraphs 11.1, 11.2.1, and 11.2.2 of ASTM E 783-91, which describe guidelines for application of the test chamber and the procedure to test for air leakage.

Step 1—Seal off the high-pressure side of the test window (Tare A) using a continuous sheet of polyethylene film. Seal the perimeter of the polyethylene film to the outermost perimeter of the test window; make sure to exclude the perimeter joint and adjacent wall construction from this seal. Also, seal off the low-pressure side of the test window using a continuous sheet of polyethylene film (Tare B). Cover the test window, including perimeter joints, at the window frame and trim. Attach Tare B to wall finishes adjacent to perimeter joint. The test setup for Steps 1 through 7 below is shown in Fig. 1.

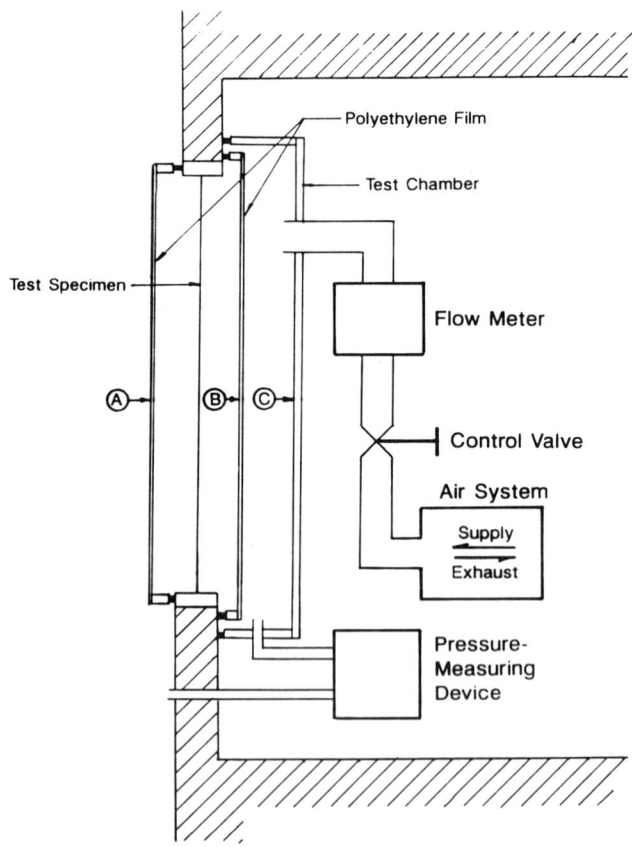

FIG. 1—*Schematic general air leakage test setup. Notes:* Ⓐ *polyethylene film for Tare A;* Ⓑ *polyethylene film for Tare B;* Ⓒ *polyethylene film for test chamber.*

Step 2—Install the test chamber (Tare C) at the low-pressure side of the test area just outboard of the polyethylene film (Tare B) installed in Step 1, that is, also sealed to the wall finish. Test chamber shall be fitted with a door that can be opened, shut, and resealed during the test without disturbing the perimeter chamber seal and without altering the quality of the chamber seal. If the chamber construction utilizes a tape seal on the door, verify that the door can be opened and resealed without variation of the chamber leakage measurement. Fit the test chamber with a suitable pressure measuring device, airflow meter, and an adjustable airflow fan.

Step 3—Adjust the airflow from the test chamber to provide the test pressure differential established for the test. Measure and record the metered airflow. Designate this measurement Q_e.

Step 4—Open the chamber door and remove the polyethylene film (Tare B) and all tape and sealant from the low-pressure side of the test window without disturbing any other seals between the test chamber, the test area, and the surrounding construction.

Step 5—Close and reseal the chamber door.

Step 6—Adjust the airflow out of the test chamber to provide the test pressure differential established for the test. Measure and record the metered airflow. Designate this measurement Q_{e1}.

Step 7—Remove the polyethylene film and all tape and sealant from the high-pressure side of the test window (Tare A) without disturbing any other seals between the window, the surrounding construction, and within the window itself.

Step 8—Continue with the calibration and test procedures in paragraphs 12 and 13 of ASTM E 783 to determine the metered airflow through the window (Q_t).

Step 9—Calculate the air leakage through the test window and through the window surround as outlined below, where:

Q_s = air leakage through test window, ft³/min (m³/s),
Q_e, Q_{e1}, Q_t = metered airflows from tests, ft³/min (m³/s),
Q_{sur} = air leakage through specimen surround, ft³/min (m³/s),
B = barometric pressure at test site corrected for temperature, in. Hg (Pa),
r = relative humidity at test site (dimensionless),
p = vapor pressure of dry air at dry bulb temperature t, in. Hg (Pa),
t = dry bulb temperature, °F (°C),
W = density of air at test site, lb/ft³ (kg/m³), and
W_s = density of air at reference standard conditions, W_s = 0.075 lb/ft³ (1.202 kg/m³).

Express the air leakage through the test window (Q_s) as follows:

$$Q_S = Q_t - Q_{e1}$$

Express the air leakage through the window surround (Q_{sur}) as follows:

$$Q_{sur} = Q_{e1} - Q_e$$

Convert the measured air leakage through the test specimen (Q_s) and the specimen surround (Q_{sur}) to air leakage at reference standard conditions (Q_{st} and Q_{ssur}, respectively) as follows:

$$Q_{st} = Q_s(W/W_s)^{1/2} \qquad Q_{ssur} = Q_{sur}(W/W_s)^{1/2}$$

$$W = 1.326(B - 0.378rp)/(t + 460), \text{ lb/ft}^3, \text{ for metric conversion}$$

$$W = 3.485 \times 10^{-3} (B - 0.378rp)/(t + 273), \text{ kg/m}^3$$

Step 10—Calculate the rate of air leakage for the test specimen as outlined in paragraph 14.3 of ASTM E 783.

Step 11—Compare the volume of air leakage through the test specimen (Q_{st}) with the volume of air leakage through the specimen surround (Q_{ssur}).

The window/curtain wall industry has no standards for acceptable levels of air leakage around or through the sides of its window/curtain wall frames. Until such standards are developed, the primary usefulness of this test method will be limited to establishing the relative contribution to air leakage of the manufactured window unit compared with its installed condition.

Further Consideration

The investigator should consider that the quantitative value for surround air leakage derived from this test method may include a component of interior air leakage depending on the nature of the wall construction. Consideration must be given to quantifying the contribution of interior versus exterior air leakage sources, if any. Examples to look for which may contribute to airflow from interior sources include interior openings linked to exterior wall cavities. Such air leakage from internal sources may not occur during normal service and exposure of the window/door system being evaluated.

The relative contribution of interior versus exterior air leakage can be evaluated using methods such as tracer gases, temperature and air velocity probes, or by isolating internal sources to the exterior wall cavity. Interior air leakage sources may be eliminated from the test area by creating negative pressure in the room outside the test area. This paper does not address such procedures for determining the extent of contribution of this airflow and its sources. Our general experience is that with most common wall constructions air leakage is primarily from exterior sources.

Case Studies

The following examples depict various types of wall construction. In these examples we identify the locations of air barriers and show methods for reducing air leakage around the sides of the windows and subsequently into the building by integrating the air barriers with the window perimeter. Two of these examples include field data using the testing technique described above. These field data show how failure to evaluate the components of the error that can occur with the extraneous air leakage can mislead those investigating complaints of excessive air leakage at windows.

Example 1: Wood Frame Wall

Problem—In a mid-Atlantic state we investigated complaints of poor window performance in a residential housing community. The windows in question were installed in wood-framed walls that were insulated with glass fiber batts. The typical windows had continuous integral perimeter flanges for anchoring the window to the outboard face of the surrounding construction. The air barrier, which was an asphaltic building paper applied over the sheathing, was located at Point X in Fig. 2. Continuity of the air barrier was maintained at the window perimeter by compression of the window flanges to the air barrier. Homeowners complained of condensation problems on the window frames and cold air drafts in the vicinity of the windows during cold weather. The complaints were consistent throughout the homes with no difference with changes in building exposure. The most persistent complaints occurred at multi-window openings where two or more windows were joined, side-by-side, within a single wall opening, as shown on the right-hand side of Fig. 2. A discontinuity in the air seal occurred where the integral side flanges of the windows were removed to "mull" the windows together. Because the vertical mullion was sealed to the outside face of both windows, it was positioned outboard of the air barrier and allowed air to enter the cavity between the windows from above and below and then into the building through poorly sealed interior window trim.

Tables 1a and 1b give the results of a series of tests performed on these windows using the test method described herein. The tables list the results for perimeter, extraneous (equipment and chamber), and window air leakage for single window openings (test locations 1, 2) and 3), and double- and triple-mulled window openings (test locations 4 and 5). Comparing the results for perimeter air leakage versus air leakage through the windows shows that in four of the five test areas perimeter air leakage was about two times, or more, than that of the air leakage through the window. Also, four of the five test areas show that perimeter air leakage exceeds the allowable air leakage for the windows. In test locations 4 and 5 the magnitude of window perimeter air leakage caused unpleasant drafts inboard of the windows, and condensation formation on the interior surface of the window frames during cold weather. In test locations 1, 2, and 3, the enclosed perimeter cavity shown on the left-

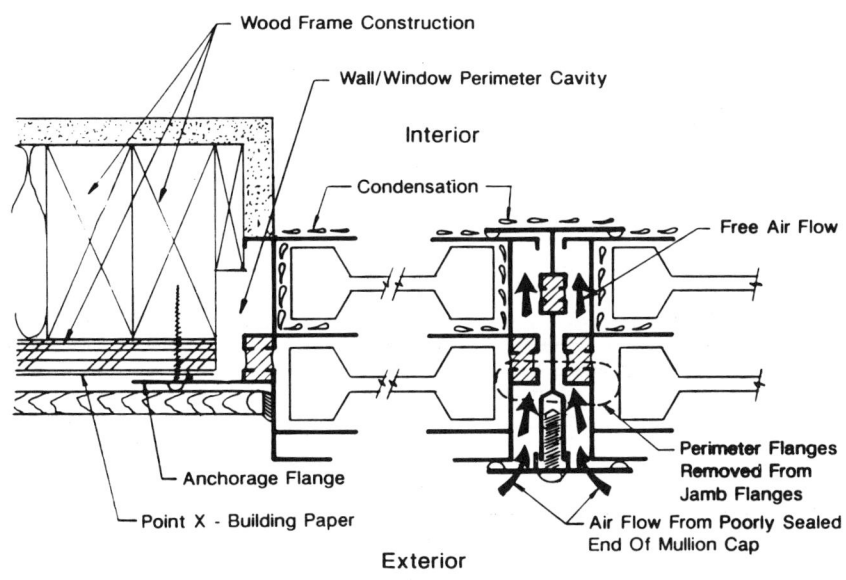

FIG. 2—*Residential wall/window assembly schematic. Cold air enters to sides of windows from top and bottom of intermediate mullion, causing drafts and condensation formation on interior surfaces.*

TABLE 1a—*Air leakage measurements in accordance with proposed modifications to ASTM E 783.*

Test Location[a]	Measured Air Leakage, ft³/min (m³/min)			Air Leakage, ft³/min (m³/min)			
	Test 1 Tare[b] @ A,B&C (Q_e)	Test 2 Tare[b] @ A&C (Q_{e1})	Test 3 Tare[b] @ C (Q_t)	Extraneous (Q_e)	Surround ($Q_{e1} - Q_e$)	Window ($Q_t - Q_{e1}$)	Allowable for Windows (AAMA)[c]
1	5.4 (0.15)	10.0 (0.28)	16.0 (0.45)	5.4 (0.15)	4.6 (0.13)	6.0 (0.17)	4.0 (0.11)
2	19.0 (0.54)	23.5 (0.67)	25.0 (0.71)	19.0 (0.54)	4.5 (0.13)	1.5 (0.04)	5.1 (0.14)
3	11.0 (0.31)	17.5 (0.50)	20.0 (0.57)	11.0 (0.31)	6.5 (0.19)	2.5 (0.07)	4.3 (0.12)
4	8.7 (0.25)	19.0 (0.54)	19.5 (0.55)	8.7 (0.25)	10.3 (0.29)	0.5 (0.01)	8.7 (0.25)
5	11.5 (0.33)	30.0 (0.85)	39.5 (1.12)	11.5 (0.33)	18.5 (0.52)	9.5 (0.27)	12.0 (0.34)

[a] Test locations 1, 2, and 3 are of single window openings, test locations 4 and 5 are of a double and triple window openings; respectively, that is, the windows are mulled together at their jamb(s).

[b] "Tare" refers to a polyethylene film placed at one or more locations on the test specimen to isolate areas for testing: A = exterior mounted film; B = interior mounted film; C = interior mounted test chamber.

[c] See Ref *1*.

hand side of Fig. 2 allowed air to circulate by convection to cool the interior surface of the window frame and trim. Tests 4 and 5 demonstrated air leakage through the mullions. Under extreme cooling conditions convection caused condensation formation on the interior surface of the window frames.

Solution—To correct these problems, the air barrier had to be connected to the intermediate mullions at their top and bottom to stop free flow of air (see Fig. 3). Break-metal closure pieces were mounted and sealed at the top and bottom of the intermediate mullion to make an easy transition in this location. The perimeter cavity between the window and wall was insulated to resist airflow and help maintain the temperature of the inside window frame and trim near that of the inside air, that is, to reduce drafts, cold spots, and the potential for condensation.

Example 2: Masonry Wall Construction with Cavity

Problem—In a northeastern state we investigated complaints of poor window performance, namely, air and water leakage and poor energy efficiency. The pertinent aspects relating to the air leakage problems were as follows:

TABLE 1b—*Air leakage measurements in accordance with ASTM E 783.*

Test Location[a]	Measured Air Leakage, ft³/min (m³/min)		Air Leakage, ft³/min (m³/min)	
	Test 1 Tare[b] @ A&C (Q_{e1})	Test 2 Tare[b] @ C (Q_t)	Extraneous (Q_{e1})	Window ($Q_t - Q_{e1}$)
1	10.0 (0.28)	16.0 (0.45)	10.0 (0.28)	6.0 (0.17)
2	23.5 (0.67)	25.0 (0.71)	23.5 (0.67)	1.5 (0.04)
3	17.5 (0.50)	20.0 (0.57)	17.5 (0.50)	2.5 (0.07)
4	19.0 (0.54)	19.5 (0.55)	19.0 (0.54)	0.5 (0.01)
5	30.0 (0.85)	39.5 (1.12)	30.0 (0.85)	9.5 (0.27)

NOTE: See footnotes a and b in Table 1a.

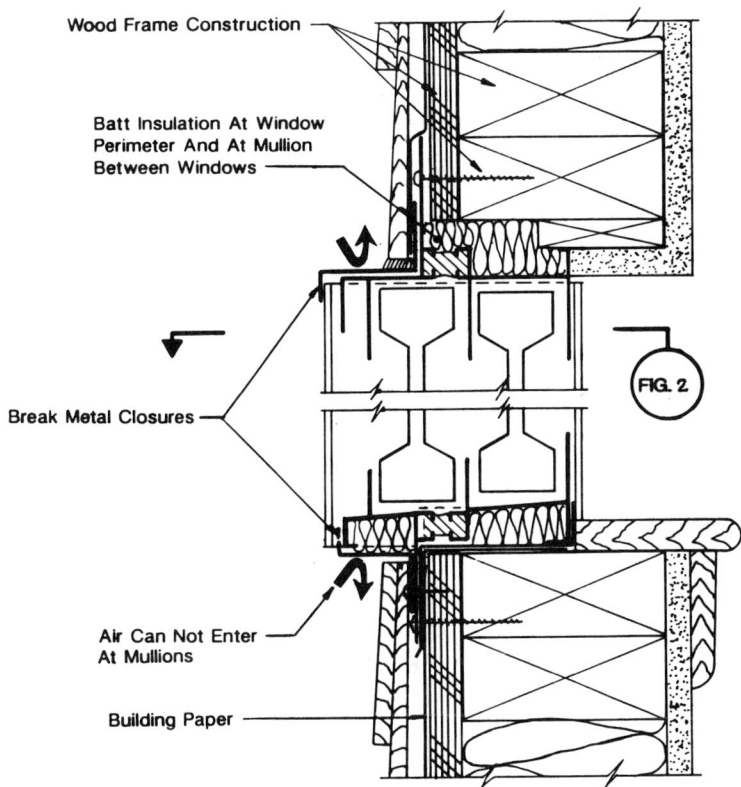

Wood Frame Construction

Batt Insulation At Window
Perimeter And At Mullion
Between Windows

FIG. 2

Break Metal Closures

Air Can Not Enter
At Mullions

Building Paper

VERTICAL SECTION AT INTERMEDIATE MULLION

FIG. 3—*Retrofit assembly. Break metal closures applied at top and bottom of intermediate mullion prevent air from flowing to voids at window perimeter. Insulating the perimeter voids helps to reduce convection and drafts.*

- The windows were installed in masonry walls consisting of, from outside to inside, brick veneer or painted concrete masonry unit, a ½-in. (12.7 mm) air space, a cementitious parge coat, and a concrete masonry unit backup.
- A single perimeter sealant joint was used as the sole means for sealing the window to the adjacent wall (Fig. 4). Although easy to design and build, the system did not provide reliable protection from water or air leakage. In this example, the air barrier is located at the outboard face of the backup masonry wall (Point X in Fig. 4). The cementitious parge coat on the backup masonry made a reasonably effective air barrier since it was generally sound. Continuity of the air barrier at Point X was broken at the window perimeter due to the window-to-wall air seal (perimeter sealant joint) being located outboard of the air barrier. Air circulating within the wall cavity had a clear path to the sides of the window, bypassing the exterior air seal. This airflow path caused unpleasant drafts at the window/trim interface and condensation on the window frame and interior trim.

Solution—To correct the airflow problem at Point X, the air barrier was made continuous at the sides of the window opening using sealants and insulation to block the air path, Fig. 5.

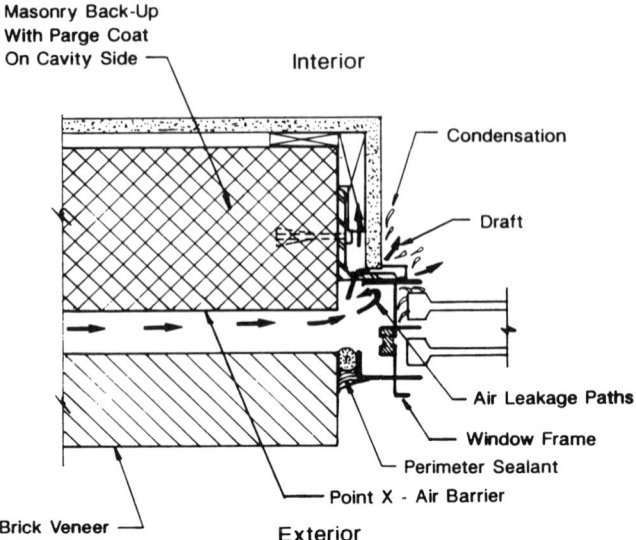

FIG. 4—*Masonry wall with cavity. Cold air in wall cavity flows to window perimeter, causing drafts and condensation formation on interior surfaces.*

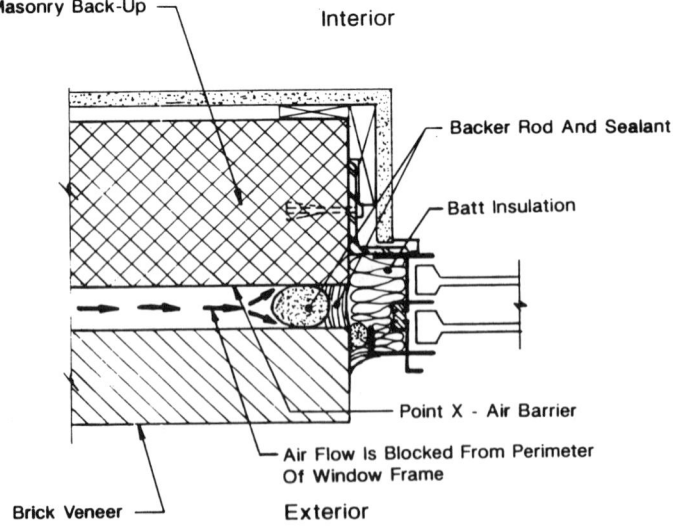

FIG. 5—*Retrofit assembly (Option 1). Cavity seal prevents air from flowing to voids at window perimeter. Insulation helps reduce convection and drafts.*

An alternative method (Fig. 6) uses elastomeric flashing in conjunction with metal jamb flashings and sealants to provide the necessary seal between the existing air barrier (Point X) and the window frame. These seals are easily integrated to head and sill flashings to provide a continuous air seal around the window perimeter.

Example 3: Double Wythe Walls with Dual Windows

Problem—We investigated a 75-plus-year-old historic structure in a Northeastern state consisting of a solid masonry wall system with wood double-hung windows in punched openings. The building had been rehabilitated and converted to condominiums. The original wood double-hung windows were left in place and a new window and insulated wall system with metal studs on the inboard side of the wall was installed to improve energy efficiency while maintaining the original exterior building appearance. We got involved with the project after the rehabilitation was complete, as a result of tenant complaints of severe cold drafts during cold weather. The drafts were reported to be strong enough to cause interior window treatments to flutter, even though the inside windows would be closed and locked. The location of the air barrier in this system was at Point X in Fig. 7. As is common with many dual window systems, occupants adjust the exterior windows seasonally or just leave them "ajar" and operate the inside windows day-to-day. Thus, an air barrier at Point Y is not practical since the air seal may be bypassed due to improper window operation. As shown in Fig. 7, continuity of the air barrier at Point X was broken at several locations around the perimeter of both windows. In addition, the sashes of the exterior windows were not well weatherstripped and allowed considerable airflow around them. This system had a variety of air leakage paths as well as air leakage into the insulated space between the original and remedial wall systems.

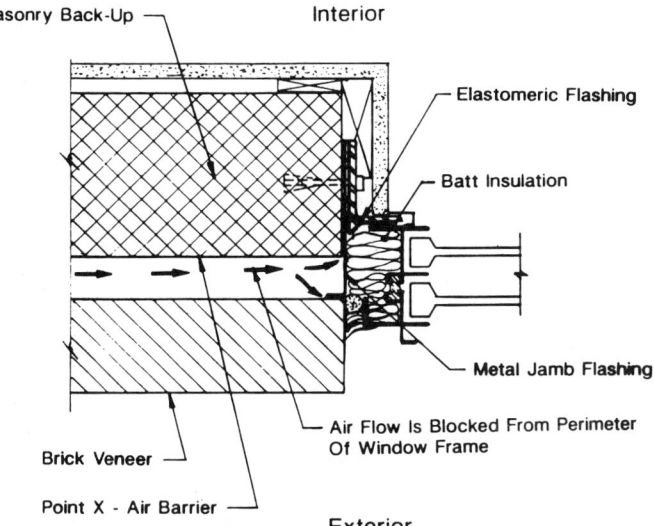

FIG. 6—*Retrofit assembly (Option 2). Composite flashing and insulation provide equivalent protection from air leakage compared with Option 1 (Fig. 5). This flashing also serves as an effective and durable barrier to water leakage.*

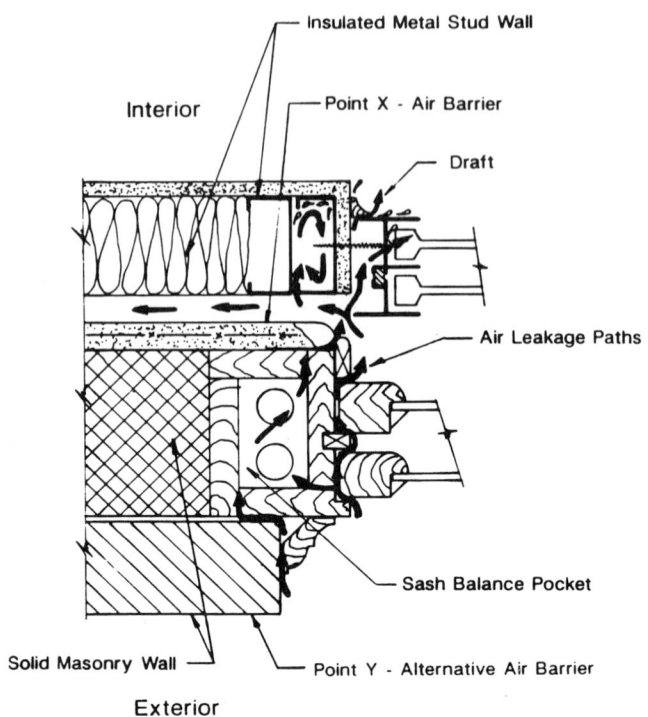

FIG. 7—*Dual wall/window assembly. Cold air bypasses exterior window assembly at numerous locations.*

The window testing procedure outlined at the beginning of this paper was modified slightly to provide information regarding each individual window unit, as well as for the sash balance pocket for the original wood double-hung windows, that is, the void space between the window frame and the masonry wall used to hide counterweights for assisting the window operation, and the window surround. The modified test setup for this window/wall assembly is shown in Fig. 8.

Tables 2a and 2b show the results of tests performed on this window assembly. Comparison of the results for perimeter, extraneous, and window air leakage shows that significant air leakage occurred from the window perimeter which exceeded the allowable air leakage for the inside window units. Perimeter air leakage paths effectively bypassed the interior window system, causing drafts and condensation to occur at the interior window/trim interface. Air leakage between the wall systems caused drafts to occur at switch plates for lights, at utility outlets, and at HVAC (heating, ventilating, and air conditioning) duct penetrations in the wall; these leaks promoted condensation on finishes, and within insulation and wall framing, reduced the thermal value of the wall insulation, and caused corrosion of the metal wall studs away from the window opening. (Water leakage caused corrosion of the wall studs in the vicinity of the wall opening.) In addition, the test data showed that the new interior windows were extremely drafty themselves and required retrofit of gaskets and seals.

Solution—To correct the airflow problems from the window perimeters, the air barrier had to be connected to the sides of the window opening, and the cavity between the wall systems had to be sealed. Figure 9 shows the use of elastomeric flashing to seal between the wood

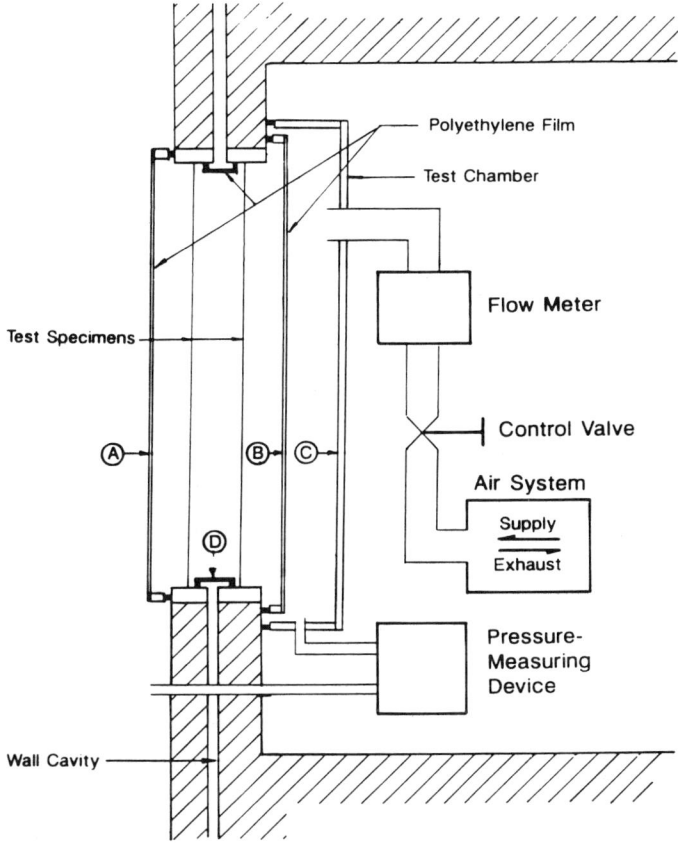

FIG. 8—*Special air leakage test setup. Notes:* Ⓐ *polyethylene film for Tare A;* Ⓑ *polyethylene film for Tare B;* Ⓒ *polyethylene film for test chamber;* Ⓓ *polyethylene film for Tare D.*

frame of the exterior window assembly and the outboard face of the interior window. This procedure assumes that the interior window will serve as the primary barrier to resist air leakage. The condition at the sill should be properly flashed to prevent water penetration between the new and existing windows.

Design Considerations

To integrate a window system into a wall opening to minimize perimeter air leakage, the designer should do the following:

1. Evaluate the continuity of the wall's air barrier seal to the window frame. Identify potential sources of air leaks through wall cavities, framing members, hardware, receptors, exterior trim, weep holes and sill flashings. Determine whether the air barrier can be bypassed at any of these conditions.
2. Locate the air barrier within the wall construction. Usually there are many possible locations for air barriers in a wall assembly. The best location to effect a seal to the

TABLE 2a—Air leakage measurements in accordance with proposed modifications to ASTM E 783.

| Test Location | Measured Air Leakage, ft³/min (m³/min) | | | | | Air Leakage, ft³/min (m³/min) | | | | | |
| | | | | | | Surround | | | Window | | |
	Test 1 Tare[a] @ A,B,C&D $(Q_{ea})^c$	Test 2 Tare[a] @ A,C&D $(Q_{e1a})^c$	Test 3 Tare[a] @ C&D Outside Open, Inside Closed $(Q_{ta})^c$	Test 4 Tare[a] @ C&D Outside Closed, Inside Open $(Q_{tb})^c$	Test 5 Tare[a] A&C (Q_{e1})	Extraneous (Q_{ea})	Window Frame (WF) $Q_{e1a} - Q_{ea}$	Cavity $Q_{e1} - Q_{ea} - WF$	Inside Unit $Q_{ta} - Q_{ea}$	Outside Unit $Q_{tb} - Q_{ea}$	Allowable for Inside Unit $(AAMA)^d$
Y	8.7 (0.25)	27.5 (0.78)	55.3 (1.56)	46.6 (1.32)	51.9 (1.47)	8.7 (0.25)	18.8 (0.53)	24.4 (0.69)	46.6 (1.31)	37.9 (1.07)	10.9 (0.31)
Z	12.8 (0.36)	55.5 (1.57)	27.7 (0.78)	56.6 (1.60)	67.2 (1.90)	12.8 (0.36)	42.7 (1.21)	11.7 (0.33)	14.9 (0.42)	43.8 (1.24)	9.7 (0.27)

[a] Tare refers to a polyethylene film placed at one or more locations on the test specimens to isolate areas for testing: A = exterior mounted film; B = interior mounted film; C = film mounted between windows (see Fig. 4b); D = film mounted in interior mounted test chamber.
[b] Projected values are beyond the capacity of the test equipment used; equivalent to extraneous air leakage plus inside unit or outside unit air leakage, as applicable.
[c] Subscript "a" or "b" indicates modified test version.
[d] See Ref 1.

TABLE 2b—Air leakage measurements in accordance with ASTM E 783.

| Test Location | Measured Air Leakage, ft³/min (m³/min) | | | Air Leakage, ft³/min (m³/min) | | |
| | | | | Window | | |
	Test 1 Tare[a] @ A&C (Q_{e1})	Test 2 Tare[a] @ C Outside Open, Inside Closed $(Q_{ta})^c$	Test 3 Tare[a] C Outside Closed, Inside Open $(Q_{tb})^c$	Extraneous (Q_{e1})	Inside Unit $(Q_{ta} - Q_{e1})$	Outside Unit $(Q_{tb} - Q_{e1})$
Y	51.9 (1.47)	98.5^b (2.78)	89.8^b (2.54)	51.9 (1.47)	46.6 (1.31)	37.9 (1.07)
Z	67.2 (1.90)	82.1^b (2.32)	111.0^b (3.14)	67.2 (1.90)	14.9 (0.42)	43.8 (1.24)

NOTE: See footnotes a, b, and c in Table 2a.

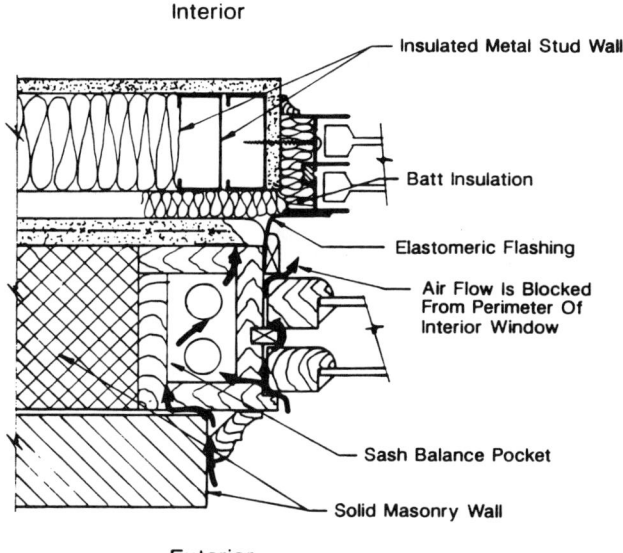

Interior

Exterior

FIG. 9—*Retrofit assembly. Provide a continuous seal between the interior and exterior windows in conjunction with insulating the voids between the units to correct the airflow defects and to improve the performance of the dual window assembly.*

sides of the window unit depends on many factors, including window style, window hardware, the nature of the air barrier material, accessibility for joining the window system to the air barrier, and compatibility of construction materials.

3. Select an air barrier that has durability commensurate with the durability of the window if these air seals are to be enclosed within the wall construction, because they will be difficult or impractical to maintain.

Conclusions

The window and door industries have come a long way in improving the air infiltration of their products and in providing verifiable means to assess performance. However, the industry needs to develop guidelines and standards for establishing acceptable levels of air leakage around fenestration products to help maintain overall wall construction quality at a level consistent with the performance of the windows, and to help prevent air leakage around the window perimeter from significantly reducing the overall performance of the window/ wall assembly. The testing procedures and guidelines presented herein are a start to the development of such a standard and can be used in the interim by designers to address this problem on a job-to-job basis.

References

[1] "Voluntary Specifications for Aluminum Prime Windows & Sliding Glass Doors," AAMA Publication 101-88, Architectural Aluminum Manufacturers Association, Chicago, 1983.
[2] "Wood Windows," NWWDA Publication 1.5. 2-87, National Wood Window & Door Association, Des Plaines, IL,1987.

Discussion

Stephen N. Flanders[1] (*written discussion*)—You have developed a means for determining where in the envelope system the leak occurs. Can you also determine whether the air is coming from indoors or outdoors?

P. E. Nelson (*authors' closure*)—The source of the air leakage from the walls surrounding the window could be from the indoors or outdoors. It would depend on the air resistance of the wall layers. An additional test chamber(s) could be constructed around the window test to evaluate air leakage through the wall layers.

---

[1] USA CRREL, Hanover, NH 03755.

Gary Proskiw[1]

Air Leakage Characteristics of Various Rough-Opening Sealing Methods for Windows and Doors

REFERENCE: Proskiw, G., "**Air Leakage Characteristics of Various Rough-Opening Sealing Methods for Windows and Doors**," *Airflow Performance of Building Envelopes, Components, and Systems, ASTM STP 1255,* Mark P. Modera and Andrew K. Persily, Eds., American Society for Testing and Materials, Philadelphia, 1995, pp. 123–134.

ABSTRACT: The air leakage characteristics of eight methods of sealing, or otherwise treating, the rough openings (R/O) found around window and door frames were evaluated under laboratory conditions. The eight methods studied were:

1. no treatment (empty),
2. conventional (fiber glass),
3. densely packed, fiber glass,
4. backer rod,
5. casing tape,
6. poly-return,
7. poly-wrap, and
8. foamed-in-place urethane.

The untreated R/O (Method 1) displayed the greatest leakage while the second largest occurred using the conventional practice of packing fiber glass into the R/O space (Method 2). In contrast, Methods 5, 6, 7 and 8 were able to reduce R/O leakage to negligible levels.

To relate these results to the overall building leakage, an estimate was made of the percentage of the total building leakage which would occur through the R/Os in a typical 97 m^2 (1040 ft^2) bungalow. Two levels of total house airtightness were assumed: 1.5 air changes per hour at 50 Pa (ac/h_{50}) representing tight construction and 5.0 ac/h_{50} (representing loose construction). For the tight house with Method 1 (no treatment), the R/O leakage accounted for 39% of the total house leakage; with Method 2 (conventional), this figure dropped to 14%. With each of Methods 5 to 8, the contribution of R/O leakage to total house leakage was less than 1%. For the loose house and Method 1, the R/O leakage was 12% of the total; with Method 2, it dropped to 4% and with Methods 4 to 8, it was less than 1%.

The incremental builder cost of each technique (i.e., exclusive of mark-up for overhead, profit, etc.) was estimated assuming Method 2 as conventional practice with a base cost of zero. Incremental costs ranged from a low of $18 per house for the urethane approach (Method 8) to $84 per house for the poly-wrap technique (Method 7).

KEYWORDS: residential airtightness, air leakage, rough-opening air leakage, window and door air leakage

Excessive air leakage through the building envelope can accelerate moisture damage to the structure, increase energy consumption, and contribute to the transmission of outdoor noise into the house. From a building science perspective, measures which reduce leakage

---

[1] Proskiw Engineering Ltd., Winnipeg, Manitoba R3H 0H1, Canada.

are desirable, particularly if they can be incorporated at reasonable cost and without excessive disruption to the normal construction process.

Since the introduction of the blower door in the 1970's and the establishment of standardized testing procedures,[2,3] the airtightness of thousands of houses has been measured and documented [1]. However, one limitation of the blower door test is its inability to quantify the leakage of individual building components. Component leakage can be measured under laboratory or field (in situ) conditions using ASTM Test Methods E 283 and E 783 respectively.[4,5] Some typical values for the effective leakage areas of a number of residential building components are given in the *ASHRAE Handbook of Fundamentals* [2].

One component of the building envelope of particular interest is the rough-opening (R/O) space found around window and door frames. Rough-opening leakage is often very noticeable to building occupants although many people (including some within the industry) attribute it to defects in the window or door units themselves and do not fully appreciate its significance. Historically, various methods have been used to seal this space, although detailed information on the performance of alternative methods is limited.

The objectives of this study were first to measure the air leakage rates of various methods of sealing the rough openings around doors and windows in residential wood frame construction, and second to estimate the incremental costs of these methods.

Test Program

Methodology

Prior to the laboratory testing program described in this paper, a series of field trials was conducted to measure R/O leakage of windows installed in actual houses. However, this proved to be highly problematic because of the difficulties created by extraneous leakage into the wall system and nonuniform crack geometries around the windows. For these reasons, the field tests were abandoned.

A laboratory testing program was then established to measure the air leakage characteristics of eight commonly used (or proposed) methods of sealing, or otherwise treating, the rough-opening space around windows and doors. The program was restricted to methods applicable to wood frame residential construction. An experimental apparatus, shown in Fig. 1, was constructed which consisted of a well-sealed chamber with one side built to replicate a typical 38 × 140 (2 × 6 in.) frame wall with plywood sheathing, building paper, and horizontal lap wood siding on the exterior and drywall on the interior. A standard, non-operating 0.97 m × 0.67 m (38 × 26½ in.) wood window was installed in the wall section. The window was secured into the R/O using screws through the brick mold. A polymethylmethacrylate (Plexiglas) panel was installed on the opposite side of the chamber to permit access and viewing.

Rough-opening leakage was defined as the air movement which occurred along the shaded pathway shown in Fig. 2, i.e., between the brick mold and sheathing, into the space between the framing and window and then around the casing into the living space. For the testing

---

[2] Canadian General Standards Board (CAN/CGSB) Standard 149.10-M86, Determination of the Airtightness of Building Envelopes by the Fan Depressurization Method (1986).
[3] ASTM Standard Test Method for Determining Air Leakage Rate by Fan Pressurization (E 779-87).
[4] ASTM Standard Test Method for Rate of Air Leakage Through Exterior Windows, Curtain Walls, and Doors (E 283-84).
[5] ASTM Standard Test Method for Field Measurement of Air Leakage Through Installed Exterior Windows and Doors (E 783-84).

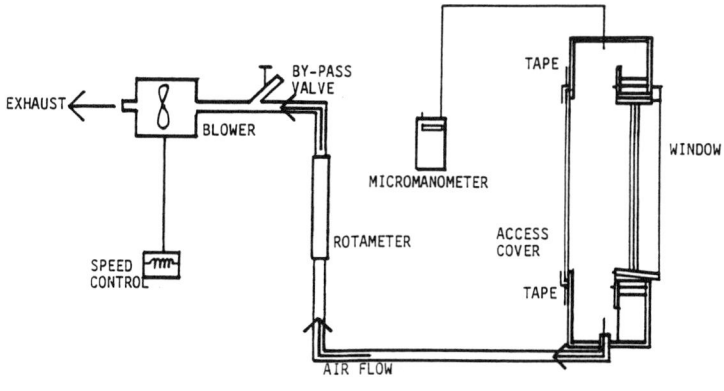

EXPERIMENTAL ARRANGEMENT

FIG. 1—*Experimental arrangement.*

program, the sides of the framing/window passageway were sealed to minimize extraneous flow into the wall cavity.

For testing purposes, the various sealing methods were applied using a level of care which was considered typical of normal residential construction; that is, no attempt was made to produce an "ideal" seal. In most cases, two different installers were used for the replicate tests (described below). To maintain equivalent crack geometries—a critical factor for comparative studies of this type—the R/O dimensions were controlled by installing the window against wooden spacer blocks to give uniform crack widths around the frame (approximately 13 mm (½ in.) on the sides and top and 32 mm (1¼ in.) on the bottom). The interior drywall was cut flush with the framing members to provide uniform conditions.

Extraneous chamber leakage (i.e., leakage other than that through the window rough opening) was evaluated by first measuring the total chamber leakage (i.e., R/O plus extraneous).

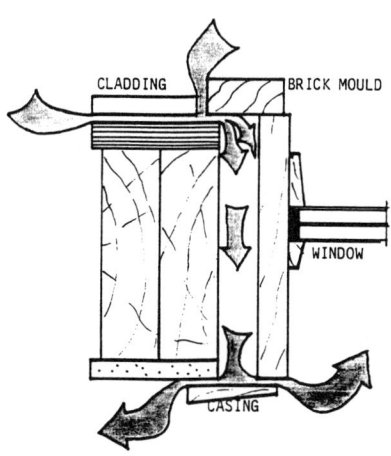

ROUGH-OPENING LEAKAGE

FIG. 2—*Rough-opening leakage.*

The window casing was then taped to eliminate R/O leakage and the test repeated. By subtracting the "taped" leakage from the "total" leakage, the extraneous leakage was measured.

Five complete, replicate tests were conducted for each of the eight sealing methods. For each test, the R/O was sealed using the method under evaluation, the interior casing installed, and the chamber closed. Once the test was completed, the casing and R/O seal were removed and the process repeated for the next test. In some instances, the window was also removed and reinstalled for the next test.

Airflow rates were measured using rotameters with ranges of 0.04 L/s to 0.39 L/s and 0.47 L/s to 4.7 L/s. Test chamber pressure differentials were measured using a digital micromanometer. All hose connections were sealed, tested, and checked.

Air leakage rates were measured at 10 to 15 Pa increments between 10 and 90 Pa. Analysis of the test data was performed using a procedure similar to that described in CAN/CGSB 149.10-M86 in which a regression curve, of the form shown in Eq 1, was generated to describe the leakage:

$$Q = C \cdot \Delta p^n \tag{1}$$

where

Q = airflow rate (L/s),
C = flow coefficient (L/s $\cdot$ Pa^n),
Δp = pressure differential (Pa), and
n = flow exponent (dimensionless).

Note that the CGSB procedure assumes a discharge coefficient of 0.611. Once C and n were determined, leakage rates (Q) could be calculated for any value of Δp as discussed later in the "Results" section. Air leakage rates were corrected to the reference conditions specified in CAN/CGSB 149.10-M86 (20°C and 101.325 kPa).

Leakage rates were also expressed using the equivalent leakage area at 10 Pa (ELA$_{10}$) defined as:

$$\mathrm{ELA}_{10} = 1.157 \cdot (\rho_r)^{0.5} \cdot C_r \cdot 10^{n-0.5} \tag{2}$$

where

ELA_{10} = equivalent leakage area at 10 Pa (m^2),
ρ_r = air density at reference conditions (kg/m^3), and
C_r = flow coefficient at reference conditions (L/s $\cdot$ Pa^n).

One advantage of using the ELA$_{10}$ is that the leakage areas of individual components can be added together directly to estimate the overall building leakage.

Sealing Methods Evaluated

Eight methods of sealing rough openings were evaluated (see Fig. 3); several of these are in common use while others have fairly limited application.

1. No Treatment—Although not an advocated method, this represents the worst-case scenario which might occur due to faulty workmanship or which may exist in older houses. The only significant restrictions to airflow are those provided by the interior casing and the exterior brick mold.

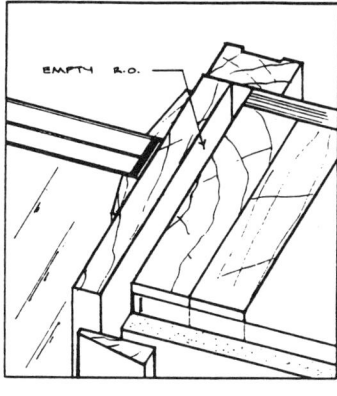

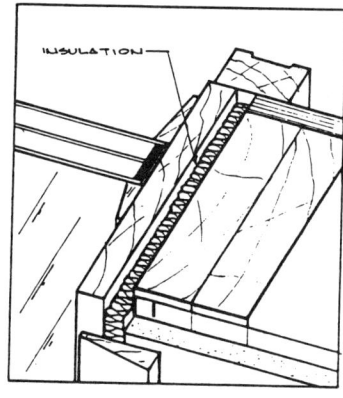

1. NO TREATMENT

2. CONVENTIONAL (FIBREGLASS)

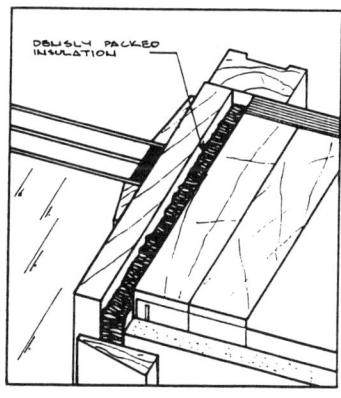

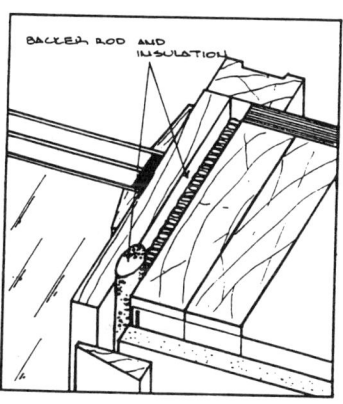

3. DENSELY PACKED FIBREGLASS

4. BACKER ROD

FIG. 3—*Sealing methods evaluated.*

2. Conventional—This method, which consists of packing pieces of fiber glass batt insulation into the R/O space, is the most common in use for residential construction. It is inexpensive and easy to apply but is often criticized since fiber glass insulation is not intended as an air barrier material.

3. Densely Packed Fiber Glass—This technique consists of packing extra insulation into the R/O to further reduce air leakage and has been suggested as a low-cost improvement over conventional practice. Care is needed with this technique to prevent inward bowing of window and door frames, which can occur if excessive insulation is forced into the R/O space. For the tests, the insulation density which was used was equal to twice that employed with the conventional method.

4. Backer Rod—The backer rod technique was first suggested as part of the Airtight Drywall Approach (ADA) and uses suitably sized, round polyethylene gaskets which are forced into the R/O space. It is now recognized that this material is susceptible to creep under

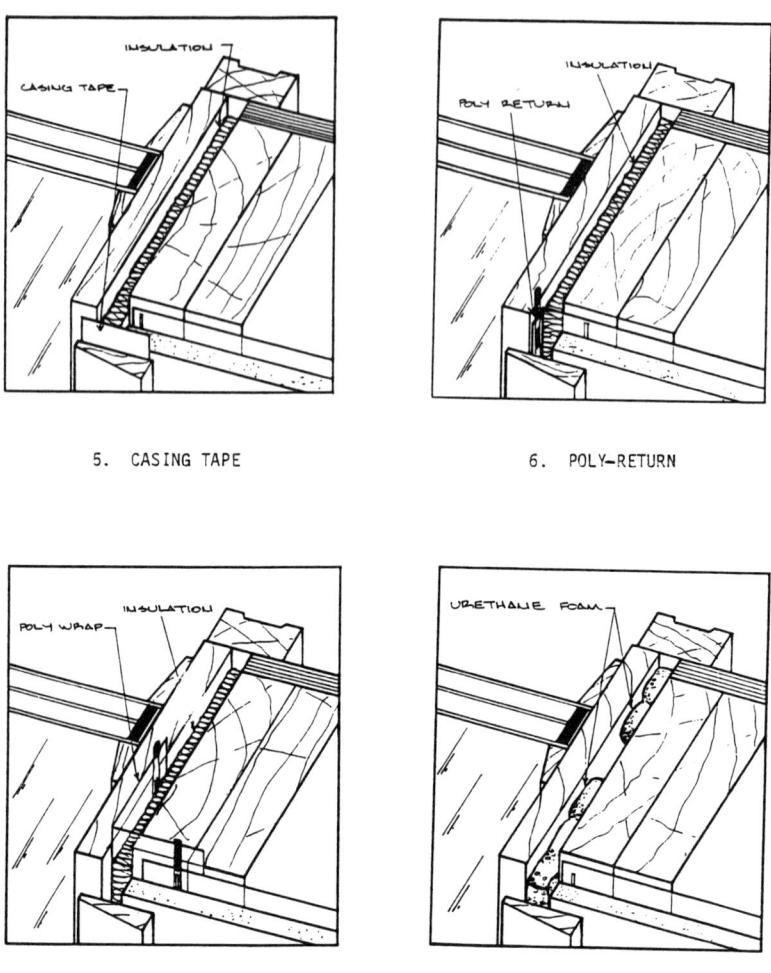

5. CASING TAPE   6. POLY-RETURN

7. POLY-WRAP   8. FOAMED-IN-PLACE URETHANE

FIG. 3—*Continued.*

compressive loads and may not have the necessary longevity for such applications and, for this reason, is no longer recommended. However, it was included in the test program to provide an example of what could be achieved, albeit with a better product.

5. Casing Tape—This novel method has also been suggested as a low-cost alternative to conventional practice. After fiber glass has been packed into the R/O space and the drywall installed, a strip of contractor's sheathing tape is applied across the R/O gap from the drywall to the window frame. This product is normally used for taping joints in exterior insulated sheathing. The casing is then applied over the tape.

6. Poly-Return—Although not commonly used, this method has been suggested for use with polyethylene air barrier systems. After fiber glass has been installed, the wall's polyethylene air/vapor barrier is returned to the window or door frame, caulked and stapled in place.

7. Poly-Wrap—The poly-wrap technique is commonly used in airtight construction and was developed more than a decade ago for double-wall houses. A polyethylene apron is caulked and stapled around the window frame; after the window and fiber glass have been installed, the apron is folded back and sealed to the wall polyethylene.

8. Foamed-in-Place Urethane—With this method, the R/O space is partially filled with foamed-in-place urethane. For the test program, two rows of foam were used. One- and two-component foams are available, as are both expanding and nonexpanding types. Several application methods can be used ranging from simple spray cans to spray guns, which permit accurate control of the foam injection rate. The latter method, with low-expansion foam, was used for this study.

All eight R/O sealing methods can be used with conventional 38 × 89 (2 × 4 in.) or 38 × 140 (2 × 6 in.) frame walls and for frame walls with exterior insulated sheathing. If double-wall construction is used, however, then two of the techniques, the casing tape and poly-return methods (5 and 6), can be used only when the windows are mounted on the inside of the rough opening. Likewise, interior strapped walls should use only the backer rod, casing tape, poly-return and urethane methods (4, 5, 6 and 8) if the joints in the R/O spaces, created where the strapping meets the wall framing, are first sealed.

Incremental Cost Data

Development of the Cost Data

The incremental costs of six of the eight rough-opening sealing methods were estimated for a standard 97 m² (1040 ft²) bungalow with seven main floor windows, two basement windows, and two doors. The combined window and door R/O crack length for this example house was calculated as 53.7 m (176 ft). Incremental costs were estimated relative to those which would be incurred if the conventional practice (i.e., Method 2) had been employed. Costs were calculated assuming the house was constructed by a large tract builder with efficient subtrades familiar with the various air leakage control techniques. A labor rate of $15.00/hour (Canadian) was used with materials costs based on 1990 prices. No provision was made for builder overhead, profit, or taxes since these vary widely between builders; i.e., the data represent direct costs to the builder.

Since actual costs are highly dependent upon labor rates, inflation, local market characteristics, etc., the cost data are primarily intended for comparison of the incremental costs of the measures relative to each other.

Results

Rough-Opening Air Leakage

The measured air leakage rates, expressed in "litres per second per metre of rough-opening crack length" at pressure differentials of 50 and 75 Pa are summarized in Table 1. The former pressure differential is referenced in CAN/CGSB 149.10-M86 while 75 Pa is used by ASTM E 283 and E 783. The ELA_{10} values are also shown in Table 1 and in Figs. 4 and 5.

Mean Air Leakage Rates and ELA_{10}'s

The measured air leakage rates of the eight sealing methods were found to vary significantly. Method 1 (no treatment) displayed the greatest leakage, 1.38 L/s · m @ 50 Pa with an ELA_{10} of 1.81 cm²/m. Method 2 (conventional) had a leakage rate about one-third this

TABLE 1—*Measured R/O air leakage rates and ELA$_{10}$'s.*

METHOD	AIR LEAKAGE (l/s•m)		ELA$_{10}$/m (cm^2/m)	METHOD	AIR LEAKAGE (l/s•m)		ELA$_{10}$/m (cm^2/m)
	50 Pa	75 Pa			50 Pa	75 Pa	
1. No Treatment				**5. Casing Tape**			
Test A	1.35	1.80	1.74	Test A	0.0043	0.0054	0.0063
Test B	1.44	1.91	1.88	Test B	0.0059	0.0076	0.0094
Test C	1.32	1.75	1.70	Test C	0.0033	0.0045	0.0031
Test D	1.38	1.83	1.78	Test D	0.0009	0.0034	0.0000
Test E	1.41	1.85	1.95	Test E	0.0033	0.0046	0.0031
MEAN	1.38	1.83	1.81	MEAN	0.0035	0.0051	0.0044
S	0.0486	0.0585	0.102	S	0.0018	0.0016	0.0036
S/X (%)	3.52	3.20	5.65	S/X (%)	51.4	30.8	81.8
2. Conventional				**6. Poly-Return**			
Test A	0.442	0.652	0.374	Test A	0.0117	0.0162	0.0126
Test B	0.534	0.749	0.557	Test B	0.0243	0.0317	0.0314
Test C	0.555	0.796	0.528	Test C	0.0182	0.0238	0.0252
Test D	0.497	0.713	0.475	Test D	0.0658	0.0881	0.0818
Test E	0.514	0.749	0.459	Test E	0.0140	0.0173	0.0220
MEAN	0.508	0.732	0.479	MEAN	0.0268	0.0354	0.0346
S	0.0428	0.0536	0.0704	S	0.0223	0.0301	0.0272
S/X (%)	8.42	7.38	14.7	S/X (%)	83.3	85.0	78.6
3. Dense F/G				**7. Poly-Wrap**			
Test A	0.147	0.219	0.120	Test A	0.0022	0.0029	0.0031
Test B	0.375	0.545	0.337	Test B	0.0006	0.0007	0.0000
Test C	0.305	0.448	0.261	Test C	0.0076	0.0078	0.0189
Test D	0.452	0.667	0.381	Test D	0.0046	0.0047	0.0094
Test E	0.441	0.651	0.374	Test E	0.0061	0.0068	0.0126
MEAN	0.344	0.506	0.294	MEAN	0.0042	0.0046	0.0088
S	0.125	0.183	0.109	S	0.0028	0.0029	0.0075
S/X (%)	36.3	36.2	36.9	S/X (%)	67.3	62.9	85.2
4. Backer Rod				**8. Urethane**			
Test A	0.0246	0.0329	0.0314	Test A	0.0000	0.0000	0.0000
Test B	0.117	0.163	0.123	Test B	0.0000	0.0000	0.0000
Test C	0.0685	0.0955	0.0723	Test C	0.0000	0.0000	0.0000
Test D	0.0258	0.0342	0.0346	Test D	0.0000	0.0000	0.0000
Test E	0.116	0.162	0.126	Test E	0.0431	0.0470	0.104
MEAN	0.0704	0.0974	0.0773	MEAN	0.0086	0.0094	0.0208
S	0.0457	0.0643	0.0457	S	0.0193	0.0210	0.0464
S/X (%)	64.9	66.0	59.1	S/X (%)	224	224	224

NOTES:
1. For each method (1 through 8), the results of the five replicate tests (A through E) are shown.
2. S = Standard Deviation.
3. S/X = Coefficient of Variation.

value, i.e., 0.508 L/s · m @ 50 Pa with an ELA$_{10}$ of 0.479 cm^2/m. The densely packed fiber glass (Method 3) produced a slight reduction relative to the conventional method while the backer rod approach (Method 4) showed a pronounced improvement. Methods 5 to 8 (casing tape, poly-return, poly-wrap and urethane foam) all displayed leakage rates which were essentially negligible relative to conventional practice.

Method 1 can, of course, be ignored on the grounds that it does not represent good building practice. However, it is frequently encountered in older houses built in the earlier part of this century. Even if Method 2 is considered the norm, the performance of some of the alternative sealing methods is still quite surprising. Note, for example, that the mean leakage rate for the conventional method was 10 to 100 times that of Methods 5, 6, 7 or 8.

Variation in Leakage Among Replicate Tests

The variation in leakage among the replicate tests provides an indication of the reproducibility of the results—an important factor from a quality control perspective. Table 1 contains

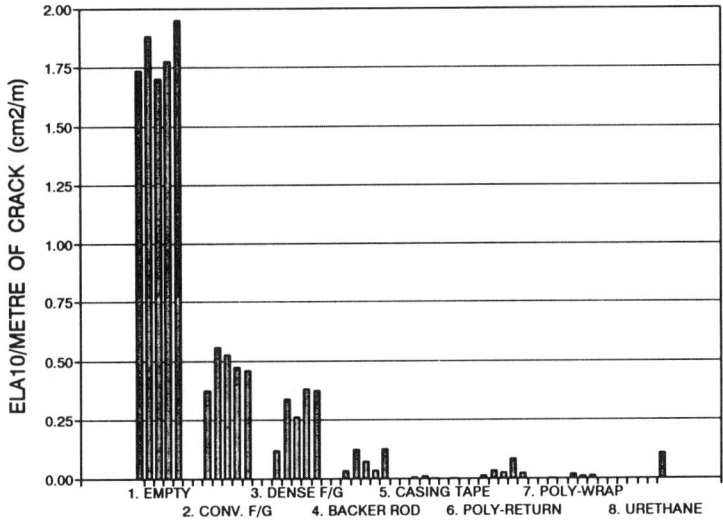

FIG. 4—*Rough-opening* ELA_{10}.

the standard deviation and coefficient of variation for the eight sets of tests. The coefficient of variation is the standard deviation divided by the mean and is expressed as a percentage.

In general, the measured air leakage rates were reasonably consistent. The largest standard deviation occurred with Method 3 (densely packed fiber glass). This may have been caused by the difficulty of installing the material in a consistent fashion. Method 8 (foamed-in-place urethane) displayed the largest coefficient of variation because the leakage (i.e., the denominator in the definition) was so small.

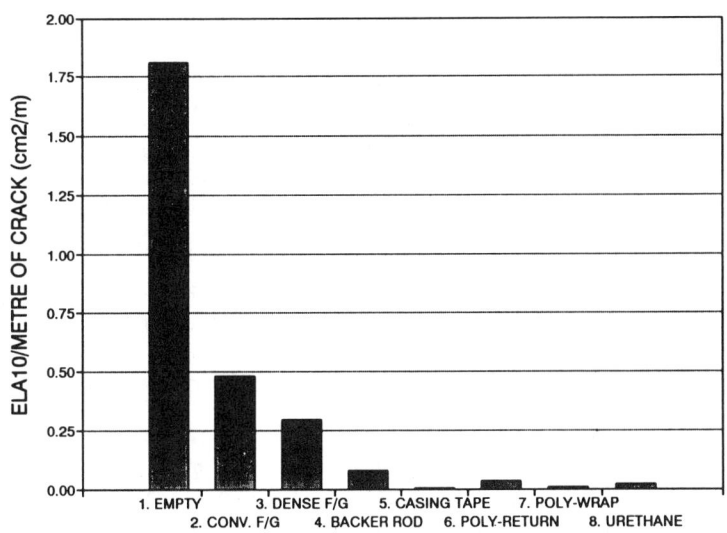

FIG. 5—*Mean rough-opening* ELA_{10}.

None of the individual test results showed dramatically higher leakage rates relative to others in the group, indicating that there were no "catastrophic failures" for individual tests. Since each method was applied with normal care, the results suggest that consistent air leakage performance can be achieved for each sealing method.

Additional Comments

Frame Materials—Methods 6 and 7 (poly-return and poly-wrap) are best suited to wood windows and doors because of the need to staple into the frame, although tape could be used for vinyl, metal, or fiber glass frames.

Ease of Installation—Method 6 (poly-return) was found to be difficult to use and somewhat vulnerable to damage during installation of the drywall. Methods 6 and 7 (poly-return and poly-wrap) involved caulking, which resulted in a significant cleanup time. Methods 5 and 8 (casing tape and urethane) were both judged as easy to apply. Some care was required with Method 5 (casing tape) to insure none of the fiber glass was left protruding under the tape to form small leakage passageways.

The performance of Method 8 (urethane) was found to depend heavily on the type of applicator used to install the foam. For the test program, a professional-style gun was used with a one-component, low-expansion foam supplied in 1 kg (2.2 lb) cans. The cost of the gun was approximately $150 and manufacturer's instructions were carefully followed. Learning time was minimal. Using the proper equipment and product, the foamed-in-place urethane was found to be particularly quick and easy to apply.

House Cladding System—The test apparatus was constructed with an air-permeable wall cladding system (wood siding); thus the results apply strictly only to windows and doors installed in walls with equivalent types of cladding systems such as brick, wood, vinyl, or metal siding. Air leakage rates for low-permeability claddings, such as stucco, would be expected to be lower than the values measured.

Significance of Rough-Opening Leakage

The significance of rough-opening leakage can be placed into perspective by comparing it with the total house leakage. To illustrate this point, the measured R/O leakage rates from Table 1 were applied to the 97 m² (1040 ft²) bungalow described earlier. Two different levels of overall house airtightness were assumed: 1.5 ac/h$_{50}$ (air changes per hour at 50 Pa), which is often regarded as an unofficial demarcation between tight and loose construction, and 5 ac/h$_{50}$, which is more typical of older, conventional structures. The results, along with the incremental construction costs, are summarized in Table 2.

For the methods normally used, or advocated, for residential construction (i.e., 2 to 8), the R/O leakage in the tight house (1.5 ac/hr$_{50}$) ranged from less than 1% to 14% of the total house leakage, whereas with Method 1 (no treatment), 39% of the house leakage occurred through the R/Os. In the loose structure, the corresponding range was from less than 1% to 12%.

These results show that rough-opening leakage can represent a significant portion of the total house leakage if conventional sealing methods are used. Further, they clearly show that it is possible to reduce the initial R/O leakage to negligible levels using relatively simple and inexpensive techniques, particularly the foamed-in-place urethane technique (Method 8).

TABLE 2—*Impact of R/O leakage relative to total house leakage.*

SEALING METHOD	INCREMENTAL BUILDER COST PER HOUSE[1]	R/O LEAKAGE (AC/H_{50})	PERCENTAGE OF TOTAL HOUSE LEAKAGE	
			TIGHT HOUSE, $AC/H_{50} = 1.5$	LOOSE HOUSE, $AC/H_{50} = 5.0$
1. No treatment	not estimated	0.59	39%	12%
2. Conventional	$ 0	0.22	14%	4%
3. Densely packed fibreglass	not estimated	0.15	10%	3%
4. Backer rod	$31	0.030	2%	<1%
5. Casing Tape	$22	0.002	<1%	<1%
6. Poly-return	$44	0.011	<1%	<1%
7. Poly-wrap	$84	0.002	<1%	<1%
8. Foamed-in-place urethane	$18	0.004	<1%	<1%

1. Relative to Method 2.

Conclusions

1. Significant differences were found in the air leakage characteristics of the eight methods of sealing, or treating, the rough-opening cracks around windows and doors.
2. The maximum air leakage occurred with the rough-opening unsealed (Method 1) while the second largest was produced with the conventional practice of packing fiber glass into the R/O space (Method 2). In contrast, leakage rates were found to be negligible for Methods 5 to 8 (casing tape, poly-return, poly-wrap and foamed-in-place urethane), although significant cost differences were noted for the various methods.
3. The test results showed that substantial reductions in R/O leakage can be achieved using relatively simple and inexpensive sealing techniques such as foamed-in-place urethane (Method 8).
4. An analysis was conducted to estimate the impact of rough-opening leakage relative to the total house leakage for a typical 97 m² (1040 ft²) bungalow. Two different levels of whole-house airtightness were assumed: 1.5 ac/hr$_{50}$ (tight construction) and 5.0 ac/hr$_{50}$ (loose construction). For the tight house, with unsealed R/Os (Method 1), 39% of the total house leakage took place through the rough openings. Using the conventional sealing technique (Method 2), the R/O leakage dropped to 14%. However, if any of Methods 5, 6, 7 or 8 were used, the R/O leakage was reduced to less than 1% of the total house leakage. For the loose house with unsealed R/Os (Method 1), R/O leakage was 12% of the total, while with Method 2 the figure was 4%. With Methods 4, 5, 6, 7 or 8 the R/O leakage was reduced to less than 1% of the total.

Acknowledgments

This work was conducted as part of the Flair Homes Energy Demo/CHBA Flair Mark XIV Project. Support was provided by Energy, Mines and Resources Canada under the Energy Demo Program and by Manitoba Energy and Mines under the Manitoba/Canada Conservation and Renewable Energy Demonstration Agreement (CREDA).

References

[1] *AIRBASE Bibliographic Database,* International Energy Agency Air Infiltration and Ventilation Centre, Coventry, U.K., 1992.
[2] *ASHRAE Handbook of Fundamentals,* Chapter 23. American Society of Heating, Refrigerating and Air Conditioning Engineers, 1993.

Envelope Heat and Mass Transfer

Anton TenWolde,[1] Charles Carll,[1] and Vyto Malinauskas[1]

Airflows and Moisture Conditions in Walls of Manufactured Homes

REFERENCE: TenWolde, A., Carll, C., and Malinauskas, V., **"Airflows and Moisture Conditions in Walls of Manufactured Homes,"** *Airflow Performance of Building Envelopes, Components, and Systems, ASTM STP 1255,* Mark P. Modera and Andrew K. Persily, Eds., American Society for Testing and Materials, Philadelphia, 1995, pp. 137–155.

ABSTRACT: Unintentional airflow in exterior framed walls frequently causes diminished thermal performance. Furthermore, even very small infiltrative or exfiltrative airflows with little or no effect on the thermal performance are capable of voiding the moisture protection provided by a vapor retarder. Combined with high indoor or outdoor humidity, such walls may accumulate excessive amounts of moisture. However, more data were needed on airflows in and through wall assemblies and the effect of airflows on moisture conditions in walls. Twenty wood frame walls of ten different designs, characteristic of walls of manufactured homes, were constructed and installed in a test building near Madison, Wisconsin. In-place pressurization and depressurization tests were performed to separately determine airtightness of the interior and exterior membranes of the wall. This paper describes the methodology, equipment, and results of these measurements. Air pressure differences across these walls were monitored during an entire winter season, along with temperatures and moisture conditions in the walls. This paper describes the instrumentation and methodology used in these measurements. The measurements showed that providing wall cavity ventilation may cause pressurization of the wall cavity and can lead to a significant increase in air leakage and heat loss without reliably providing the intended protection from moisture accumulation in the cavity. Although several of the walls experienced condensation and mold growth, the moisture conditions in the walls showed no clear correlation with airtightness or with the amount of air flowing through the wall. Indoor humidity correlated strongly with wall cavity moisture conditions, suggesting that indoor humidity control is the most effective strategy to prevent excessive moisture accumulation in the exterior building envelope during winter.

KEYWORDS: airflows, buildings, condensation, humidity, manufactured housing, moisture, ventilation, walls

This paper concerns airflows in exterior framed walls of manufactured houses during the heating season in a northern U.S. climate and the measured influence of the airflow on moisture accumulation within the walls. Air leakage through building envelopes has long been recognized as reducing thermal performance of the envelopes. Air leakage is also recognized as a mechanism by which moisture can accumulate in building envelopes. The influence of air leakage on condensation within building envelopes was recognized as early as 1950 [1] and was documented in published research literature by the early 1960s [2].

---

[1] Research Physicist, Research Forest Products Technologist, and General Engineer, respectively, USDA Forest Service, Forest Products Laboratory, One Gifford Pinchot Drive, Madison, WI 53705-2398.

The Forest Products Laboratory is maintained in cooperation with the University of Wisconsin. This paper was written and prepared by U.S. Government employees on official time, and it is therefore in the public domain and not subject to copyright.

Recent field studies by Platts [*3*], Tsongas [*4*], and Tsongas and Nelson [*5*] indicated that high sheathing moisture contents were often associated with air leakage through wall cavities. Simulation modeling by Burch and Thomas [*6*] indicated that circulation of indoor air through a wall cavity at an assumed rate of one air cavity change per hour resulted in significant moisture accumulation within the wall in northern U.S. climates. Simulation modeling by Ojanen and Kumaran [*7*] indicated the importance of exfiltration through wall cavities as a mechanism for moisture accumulation within them, particularly in very cold climates.

Providing cavity ventilation is currently an approved construction method in wood frame walls of manufactured homes in the United States [*8*]. The rationale for wall cavity ventilation is that circulation of outside air through the cavity promotes drying. TenWolde and Carll [*9*] showed that cavity ventilation may lead to increased air leakage and presented calculations showing that air leakage through a wall cavity can either increase or decrease wall moisture levels, depending on airflow direction. TenWolde [*10*] also showed that in cold climates the detrimental wetting effect of exfiltration is not offset by the drying effect of an equal amount of infiltration. Sherwood [*11*] found neither a consistent positive nor negative influence of cavity ventilation. By applying simulation modeling to wall cavities of known leakage, TenWolde and Carll [*9*] concluded that cavity ventilation is not a reliable strategy for avoiding condensation within walls. Analysis with a more detailed mathematical model led to the same conclusion [*12*].

To confirm these conclusions with measured data, we conducted a field study in which airflow and moisture conditions in walls were monitored frequently and repetitively. The purposes of this study were to

(*a*) quantify air leakage characteristics of walls representative of walls of manufactured homes in the United States,

(*b*) identify the relationship between airflows and moisture accumulation in these walls in a northern U.S. climate,

(*c*) identify the influence of wall construction details on air leakage and moisture accumulation, and

(*d*) identify the role of indoor relative humidity (RH) on moisture accumulation in walls and how the influence of indoor RH interacts with the other factors investigated.

Description

Test Building

The test walls were built as modules and installed in a 15.2-by-2.4-m (50 by 8 ft) building at a test site near Madison, Wisconsin (Fig. 1). The long dimension of the building was oriented in an east-west direction. The building was partitioned into three interior rooms: two 6.1-m-long (20 ft) rooms at opposite ends of the building and a central 3.0-m-long (10 ft) room. We installed the wall modules on the north side of the two end rooms. The center room contained control and data-gathering equipment. All three rooms were maintained at 21°C (70°F) through the heating season with fan-equipped portable electric heaters on thermostatically switched circuits. The east room was maintained at 35% RH and the west room at 45% RH through the heating season using steam vaporizers controlled by humidistats. During spring and fall, higher humidities and somewhat higher temperatures occurred because the building had no air-conditioning or dehumidification equipment.

FIG. 1—*Test building near Madison, Wisconsin, with wall modules installed.*

Test Walls

The test walls were 122-cm-wide (4 ft) modules, each containing two stud bays 36.8 cm (14.5 in.) wide by 213 cm (7 ft) tall to represent stud spaces with studs on 40.6-cm (16 in.) centers (Fig. 2). The two stud spaces in each wall module were separated by multiple studs and two narrow cavities. Wall framing was standard 38-by-140-mm (nominal 2 by 6 in.) softwood dimension lumber. We built two replicates of each module; the first was installed in the east room (35% RH), the second in the west room (45% RH). Thus, the building contained 10 wall modules, with a total of 20 test wall cavities. Of the 20 test wall cavities, 16 included a rectangular electrical box (designed for installation in gypsum board) with a standard duplex outlet (or receptacle). We performed air leakage tests on these 16 test wall cavities and monitored pressure differentials across the membranes of the cavities during the heating season.

By sealing the joint between the framing and the gypsum board, we constructed the wall modules such that all air exchange between the inside room and the wall cavity would be through the electrical box or around its edges. This was necessary to obtain useful results from the pressurization tests and thus an accurate correlation of pressure differences and air exchange between the room and the cavity.

Wall modules were fabricated from dry lumber (9.5% to 11% moisture content). Joints in the framing lumber were caulked with acrylic-latex caulk to prevent lateral air movement. A caulk seal was also applied between the sheathing and the center stud of the wall module for the same purpose. The framing lumber did not contain defects that would allow lateral air movement; any small knots and checks were filled with latex wood filler. To prevent moisture diffusion exchange between test wall cavities, we painted the back face of the framing members and both faces of the center stud with an alkyd vapor retarder paint pigmented with aluminum flakes (Fig. 2).

We attached thermocouples and moisture sensors to the cavity side of the exterior sheathing and on the back side of the siding where sheathing and siding were separate entities. Wall cavities were filled with glass fiber insulation with a rated heat flow resistance of 3.3 m^2 K/W (19 hft^2 °F/Btu) and an asphalt-coated kraft paper facing. We stapled the flange of

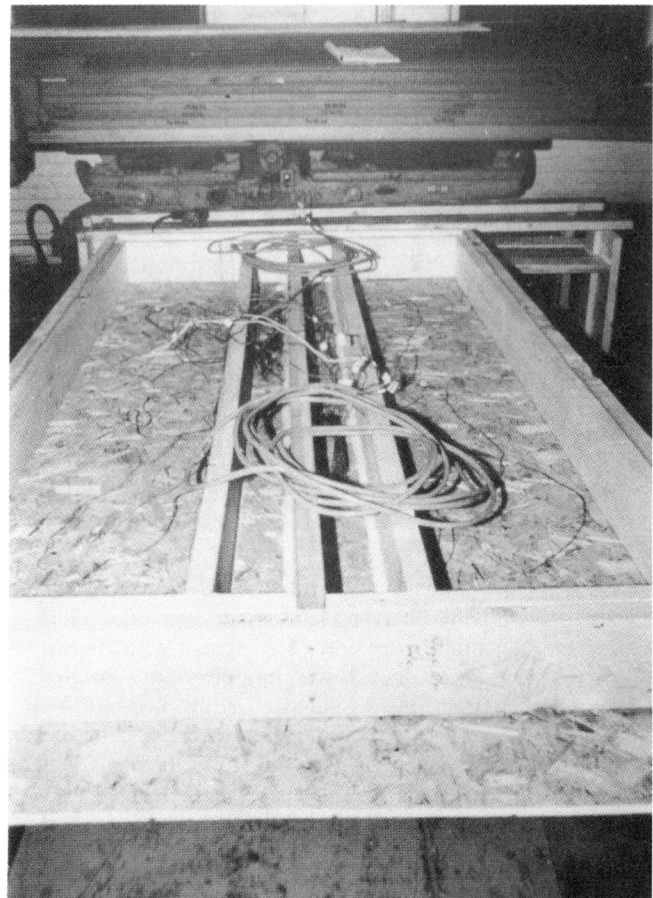

FIG. 2—*Wall module under construction.*

the kraft paper to the broad sides of the studs, as is commonly done in manufactured home construction. We left the insulation flange unstapled at the bottom 61 cm (24 in.) of the stud bay so an electrical box could be installed in the gypsum board without tearing the insulation facing. In order to simulate penetration of the facing by electrical cable, we cut a 25-mm-diameter (1 in.) hole in the insulation facing where the electrical box would be located.

In 16 wall test sections, we installed electrical boxes in the gypsum board before attaching it to the framing. The boxes were plastic and were secured to the gypsum board with built-in clamping ears. We centered them horizontally in the stud space with the lower edge of the box 36 cm (14 in.) from the floor. On 8 of the 16 receptacles, we installed foam gaskets underneath the receptacle cover plates and attached the cover plates snugly.

Acrylic-latex caulk was applied to the inside surfaces of all studs and plates. The gypsum board was 8 mm (5/16 in.) thick with vinyl finish, and we secured the gypsum board to the studs and plates in the bed of caulk with drywall screws. According to measurements carried out by the National Institute of Standards and Technology [13], the vinyl finish has a vapor permeance of about 3×10^{-11} s/m (0.5 perms) and therefore qualifies as a vapor retarder.

Shortly after installation of the wall modules in the test building, we sealed around the wall module interior perimeters and covered the exterior vertical gaps between modules with weathertight batten strips.

The wall constructions are outlined in Table 1. Where wall cavities were ventilated, spacers approximately 6 mm (¼ in.) thick were attached to the top and bottom plates at intervals of approximately 203 mm (8 in.) to provide passageways between the plates and the sheathing (Fig. 3). In these walls, we provided a horizontal vent slot in the sheathing above the stud space and an air passageway between that vent slot and the vented soffit of the test building (Fig. 4). In wall modules 4 and 5, the sheathing and siding were separate entities, with ventilation between them. In the two #4 wall modules, ventilation occurred through the vertical ribs of the steel siding. In the two #5 wall modules, the vent space was afforded by spacers approximately 6 mm (¼ in.) thick between the sheathing and the siding.

Measurements and Instrumentation

Air Pressurization Tests

During August of 1990, we performed air leakage tests with the walls installed in the test building. The tests were generally done before 10:00 a.m. because after 10:00, wind pressures usually became measurable. Ambient temperature at time of testing was between 20°C and 23°C (68°F and 74°F).

The test procedure was derived from ASTM Standard Method for Field Measurement of Air Leakage Through Installed Exterior Windows and Doors (E 783-84). We used a blower to pressurize or depressurize a specially constructed box with an open face designed to fit against the interior gypsum board over the electrical outlet (Fig. 5). To ensure an airtight seal, the edge of the open face of the box had a foam gasket, and we used a system of drawbars and turnbuckles to draw the box and its gasket snugly against the gypsum board. Leakage associated with the box and the box-to-gypsum-board seal was too small to measure at the pressures involved in our wall leakage tests. Pressure taps to the wall cavity and to the outside were at mid-height of the wall. They were copper tubes, held in place with silicone caulk, which also provided an airtight seal around their periphery. The tap into the wall cavity extended approximately 25 mm (1 in.) into the cavity, extending through the paper facing of the insulation.

TABLE 1—*Wall construction details.*

Wall	Siding	Sheathing	Cavity Ventilation	Receptacle	Gasket
1LA/1HA	waferboard	...	no	no	...
1LB/1HB	waferboard	...	no	yes	no
2LA/2HA	waferboard	...	yes	no	...
2LB/2HB	waferboard	...	yes	yes	no
3LA/3HA	waferboard	...	no	yes	yes
3LB/3HB	waferboard	...	yes	yes	yes
4LA/4HA	steel	Foamcore[a]	yes	yes	yes
4LB/4HB	steel	Foamcore	no	yes	yes
5LA/5HA[b]	waferboard	fiberboard[c]	no	yes	no
5LB/5HB[b]	waferboard	Foamcore	no	yes	no

[a] Foamcore is 5-mm-thick (0.2 in.) weatherbarrier composed of foam between kraft paper facings.

[b] Ventilation strips [approximately 6 mm (0.25 in.) thick] between sheathing and siding.

[c] Wood fiberboard, sheathing quality.

FIG. 3—*Vent spacers attached to* (a) *top plate and* (b) *bottom plate.*

The following were important features of our test procedure:

(*a*) Air pressure differentials across the interior and exterior wall membranes were measured simultaneously and individually.

(*b*) The pressure differences were measured at a sequence of different airflow rates through the wall.

(*c*) The pressure differences ranged from 0 to 150 Pa (0 to 0.6 in. of water) across the tighter membrane.

(*d*) Data points were collected with the box pressurized and with the box depressurized, thereby permitting us to identify the pressure/flow relationships in both infiltrative and exfiltrative modes.

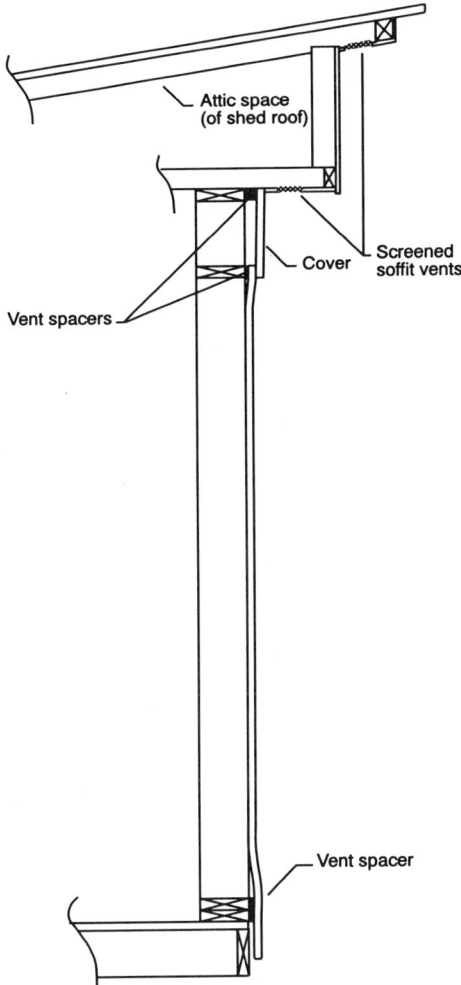

FIG. 4—*End elevation of ventilated wall as installed in test building.*

We fitted the results for each wall membrane to the following equation:

$$q = K \, \Delta p^b \tag{1}$$

where

q = airflow,
K = coefficient,
Δp = air pressure difference, and
b = exponent.

We calculated effective leakage area (ELA) of individual membranes and of whole walls

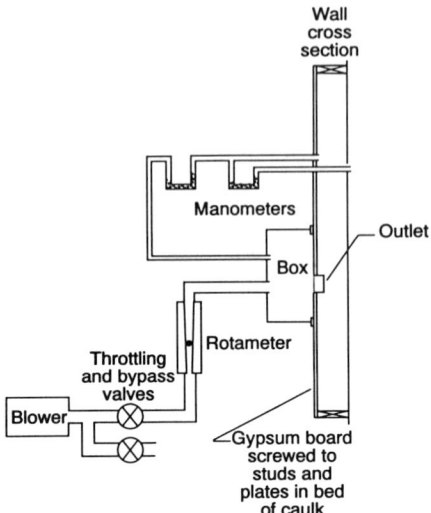

FIG. 5—*Apparatus used to measure airtightness of walls with electrical outlets (pressurization mode shown).*

at a reference pressure of 4 Pa (0.016 in. of water) using the procedure described in the *ASHRAE Handbook of Fundamentals* [14] and using a discharge coefficient value of 1:

$$L_e = q_r(\rho/2\,\Delta p_r)^{1/2} \tag{2}$$

where

L_e = effective leakage area, m²,
q_r = airflow L/s at reference pressure difference, Pa,
ρ = air density, kg/m³, and
Δp_r = reference pressure difference = 4 Pa (0.016 in. of water).

The total pressure difference across the wall is the sum of the pressure differences across each membrane:

$$\Delta p_{wall} = \Delta p_{outlet} + \Delta p_{ext.membrane} \tag{3}$$

Because of the careful caulking of the junction between the gypsum board and the wood framing, we may assume that the airflow through the outlet equals the flow through the exterior membrane:

$$q_{wall} = q_{outlet} = q_{ext.membrane} \tag{4}$$

Equation 1 can be written as

$$\Delta p = (q/K)^{1/b} \tag{5}$$

and Eq 3 can then be written as

$$\Delta p_{\text{wall}} = (q/K_{\text{outlet}})1/b_{\text{outlet}} + (q/K_{\text{ext}})1/b_{\text{ext}} \qquad (6)$$

The value of q that yielded a value of 4 Pa (0.016 in. of water) in Eq 6 was determined iteratively. This value was then used in Eq 2 to calculate ELA for the total wall.

Monitoring During Heating Season

We used a microcomputer to collect data during the heating season (November to May). These data included air pressure differentials across wall membranes and temperature and RH on the cavity side of the sheathing at the center top, middle, and bottom and on the back of the siding where sheathing and siding were separate. In walls with an electrical outlet, additional RH data were collected on the back of the sheathing immediately across from the outlet. The locations of the sensors are shown in Fig. 6. We calculated airflows across membranes from pressure differential data using the pressure/flow relationships we obtained from the air pressurization tests.

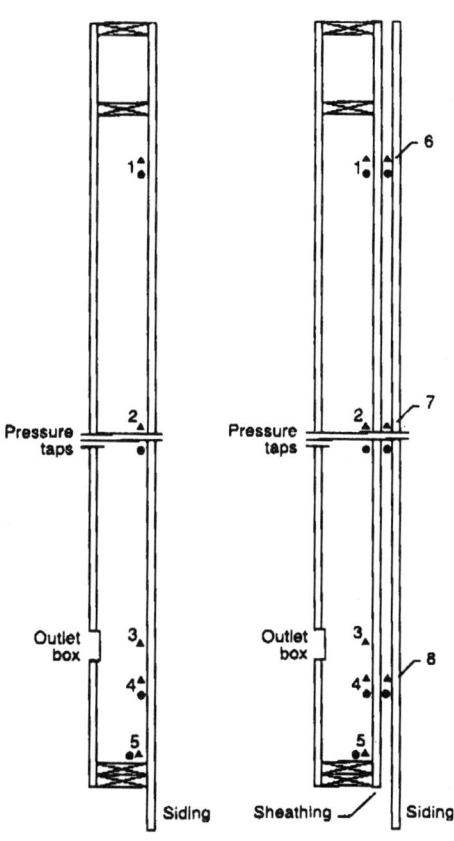

▲ Relative humidity sensor
● Thermocouple

FIG. 6—Locations of pressure taps, thermocouples, and relative humidity sensors in wall with combination sheathing-siding (left) and separate sheathing and siding (right).

We monitored temperature and RH hourly. Temperature was monitored with thermocouples. Relative humidity on the back surface of sheathing and siding was monitored with small wood electric resistance sensors as described by Duff [15] and TenWolde and Courville [16]. The sensors were connected to individual signal conditioners, and each sensor/signal conditioner pair was calibrated before installation. This type of sensor has an estimated error of approximately ±10% RH [17]. However, we probably did not attain that level of accuracy in our measurements, because the monitoring took place more than 20 months after calibration and installation of the sensors (see Discussion).

Pressure differentials across wall membranes were measured every half hour with a single precision pressure transducer that had a range of −125 to +125 Pa (−0.5 to +0.5 in. of water). The resolution of our data acquisition system was 1 Pa (0.004 in. of water). All pressures were measured at half-height. To measure pressure differentials across 36 membranes, we used an automatic switching valve controlled by the microcomputer. We used the same taps we had installed for air leakage tests for monitoring pressure differentials during the heating season. In each wall, we selected the more airtight of the two membranes for detailed sampling of pressure differential. Detailed sampling consisted of 72 s of 4-s averages (eighteen 4-s averages) every half hour. We sampled the other membrane once (one 4-s average) per half hour. Table 2 gives the schedule for air pressure differential sampling.

The monitoring took place during the period November 1991 through May 1992. The walls with instrumentation were installed in December 1989, and an indoor temperature of 21°C (70°F) and RHs of 35% and between 45% and 50% were maintained in the two rooms from mid-January through April of 1990 and during the 1990–91 heating season. However, difficulties with our instrumentation and data acquisition software prevented us from collecting data until the 1991–92 heating season.

Results

Air Pressurization Tests

Results of the air pressurization tests are presented in Table 3. For all outlets and nonventilated exterior membranes, the exponent value in Eq 1 was substantially less than 1. However, ventilated exterior membranes showed an essentially linear relationship between airflow and pressure differential.

TABLE 2—*Differential pressure measurements.*

	Number of 4-Second Samples per Half-Hour	
Wall	Across Gypsum Board	Across Sheathing/Siding
1LA/1HA	none	1
1LB/1HB	1	18
2LA/2HA	none	1
2LB/2HB	18	1
3LA/3HA	1	18
3LB/3HB	18	1
4LA/4HA	18	1
4LB/4HB	18	1
5LA/5HA	18	1
5LB/5HB	18	1

TABLE 3—*Coefficient and exponent values for pressure/flow relationship[a] and effective leakage areas (ELAs).*

Membrane	Airflow Direction[b]	Wall	K	b	ELA ($\times 10^{-6}$ m^2)[c]	Wall	K	b	ELA ($\times 10^{-6}$ m^2)[c]
Inner	ex.	1LB	0.11	0.64	98	1HB	0.06	0.74	62
	in.		0.16	0.53	130		0.06	0.73	59
Outer	ex.		0.03	0.84	33		0.01	0.95	16
	in.		0.04	0.73	45		0.01	0.90	19
Both	ex.				29				14
	in.				41				16
Inner	ex.	2LB	0.09	0.61	78	2HB	0.10	0.59	87
	in.		0.10	0.58	83		0.08	0.63	75
Outer	ex.		0.31	1.00	470		not available		
	in.		0.36	0.99	550		not available		
Both	ex.				71				86
	in.				76				75
Inner	ex.	3LA	0.05	0.76	51	3HA	0.06	0.74	62
	in.		0.07	0.63	65		0.06	0.71	63
Outer	ex.		0.01	0.82	14		0.03	0.78	31
	in.		0:01	0.84	14		0.02	0.81	29
Both	ex.				13				24
	in.				13				23
Inner	ex.	3LB	0.09	0.57	74	3HB	0.13	0.59	110
	in.		0.09	0.57	72		0.14	0.57	120
Outer	ex.		0.34	1.0	520		0.47	0.99	710
	in.		0.68	0.78	760		0.75	0.61	670
Both	ex.				69				100
	in.				70				110
Inner	ex.	4LA	0.07	0.56	62	4HA	0.04	0.67	41
	in.		0.09	0.52	68		0.02	0.81	25
Outer	ex.		0.37	1.0	560		0.35	1.0	540
	in.		0.66	0.83	790		0.42	0.99	640
Both	ex.				58				39
	in.				66				24
Inner	ex.	4LB	0.08	0.52	64	4HB	0.06	0.55	50
	in.		0.10	0.48	70		0.06	0.57	46
Outer	ex.		0.03	0.72	34		0.05	0.76	52
	in.		0.05	0.64	45		0.06	0.70	62
Both	ex.				29				32
	in.				37				34
Inner	ex.	5LA	0.13	0.57	110	5HA	0.13	0.55	110
	in.		0.13	0.56	110		0.12	0.57	100
Outer	ex.		0.03	0.95	36		0.03	0.97	45
	in.		0.02	0.96	35		0.05	0.83	64
Both	ex.				32				38
	in.				31				49
Inner	ex.	5LB	0.11	0.59	95	5HB	0.06	0.98	91
	in.		0.13	0.53	110		0.10	0.73	100
Outer	ex.		0.02	0.68	18		0.01	0.65	13
	in.		0.02	0.63	21		0.03	0.50	22
Both	ex.				18				12
	in.				20				21

[a] $q = K \Delta p^b$ (Eq 1), where q = airflow in L/s and Δp = pressure difference in Pa.

[b] ex. = exfiltration; in. = infiltration.

[c] 1 m^2 = 10.8 ft^2.

Effective leakage areas of ventilated exterior membranes were at least an order of magnitude greater than ELAs of nonventilated exterior membranes. As a general rule, nonventilated exterior membranes showed the lowest ELA values, followed by gasketed outlets, then by ungasketed outlets, and finally by vented exterior membranes, which showed the greatest ELAs. Gasketed outlets on average showed a little less air leakage [average ELA

65.0 10^{-6} m² (7.0 10^{-4} ft²)] than did ungasketed outlets [average ELA 93.4 10^{-6} m² (1.0 10^{-3} ft²)], but the variation between walls was large. The outlet showing greatest leakage was one with a gasket (wall 3HB). Although holes in the gypsum board for the electrical boxes were cut carefully by hand, they were not all identical in size. We believe this was the major source of variation in ELA values for outlets and this variation was not always overcome with gaskets.

Air Pressures and Airflows

Measured pressure differences were highly variable with location and time. Figure 7a,b shows representative data for walls 3HA and 3HB, respectively, during the period January 12 through 14, 1992. The cavity of wall 3HA was not ventilated; consequently, the pressures across the gypsum board were very small because the sheathing serves as the primary air barrier. Pressures across the gypsum board of the ventilated wall 3HB were considerably larger. On January 12, the wind as recorded at the Madison airport weather station was from the south to west at 4 to 7 m/s (9 to 16 mph) and all air pressure differences were exfiltrative (positive). On January 14, the wind had shifted to the north to northwest at 1.5 to 7.5 m/s (3 to 17 mph) and pressures across the gypsum board were infiltrative (negative) in both walls. However, at the same time, pressures across the exterior sheathing/siding of the ventilated wall 3HB (Fig. 7b) were exfiltrative, and the exterior of the nonventilated wall 3HA (Fig. 7a) experienced alternate small infiltrative and exfiltrative pressures. This indicates that the air pressure in the ventilated cavity (3HB) was considerably higher than either the inside or the outside air pressure. Adding pressure differences across both membranes to obtain an approximate pressure difference across the total wall showed that the total pressures across 3HA and 3HB were similar during the entire period. This indicates that the measurements were reliable and that the ventilated cavity was indeed pressurized with winds from the north.

Figure 8a,b provides more evidence of the difference in behavior of the ventilated and nonventilated walls. Only walls with electrical outlets are included in Fig. 8a, b. With winds ranging from south to west on January 12, 1992 (Fig. 8a), all walls experienced exfiltrative pressures. By comparing walls 1B with 2B (ventilated), 3A with 3B (ventilated), and 4B with 4A (ventilated), the difference between ventilated and nonventilated walls of otherwise similar construction can be seen. Significantly larger daily average exfiltrative pressures occurred across the gypsum board in ventilated walls than in similar nonventilated walls. This means that, at a given exfiltrative air pressure differential across a wall, significantly more indoor air flows into the wall cavity if the wall is ventilated.

Figure 8b shows that with a northerly wind on January 14, 1992, ventilated walls experienced cavity pressurization; daily average pressures across the gypsum board and the exterior were in opposite directions in ventilated walls. The only exception was ventilated wall 4HA, probably because it was located on the upwind corner. Walls 5HA and 5LA showed a small pressurization of the cavity as well, perhaps because of positive pressures in the vented space between siding and sheathing and the relative leakiness of the wood fiberboard sheathing. Ventilated cavities on the windward side were probably pressurized because of high pressure at the eaves where the top ventilation opening was located (Fig. 4). Most likely, outside air flowed into those cavities at the top and out through the electrical outlet and bottom vent. Because of these unexpected airflow patterns, which were very different from the flow pattern during the pressurization tests, we were unable to calculate airflows through the exterior membrane. Therefore, we only report flows across the gypsum board through the electrical outlet.

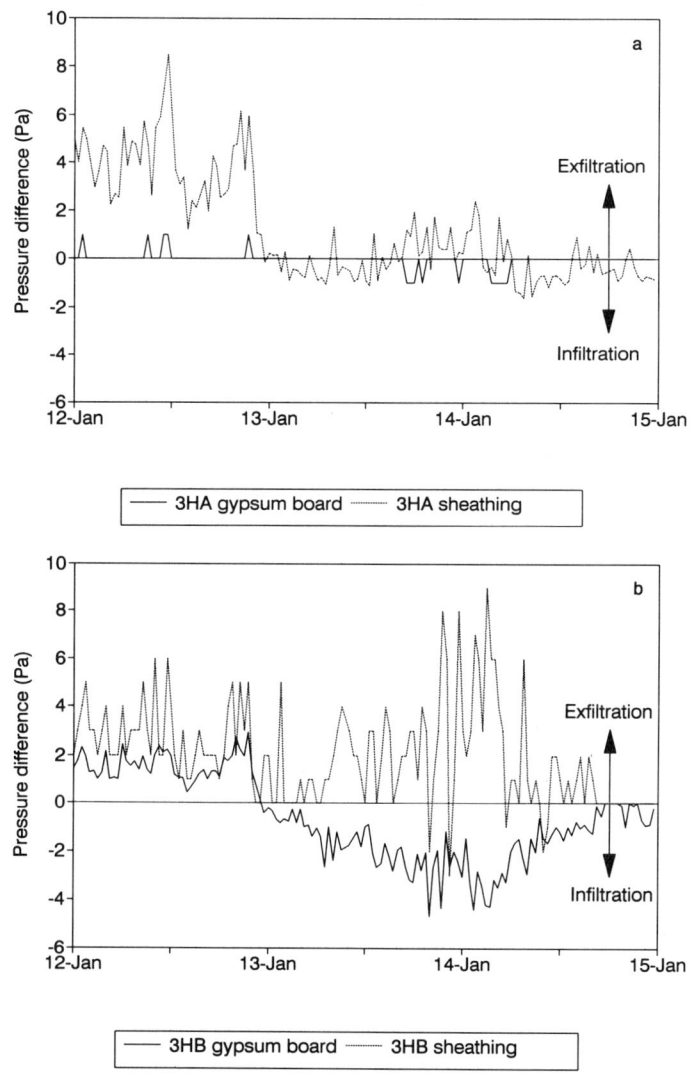

FIG. 7—*Measured air pressure differences across gypsum board and exterior sheathing of* (a) *wall 3HA and* (b) *wall 3HB, January 12–14, 1992.*

Figure 9 shows the calculated total amount of air that flowed through the electrical outlets in each direction during January 1992. Infiltration and exfiltration are shown separately because of their unequal drying and wetting potential. Clearly, ventilated walls (identified with a V in Fig. 9) generally had very large flows, with more infiltration than exfiltration; the flows in nonventilated walls were considerably smaller and more equal in both directions. Only the flows through the outlet of ventilated wall 4HA were relatively modest because of the unusual airtightness of the outlet; the measured average ELA for the outlet in wall 4HA was 33 10^{-6} m^2 (3.6 10^{-4} ft^2) compared with an average of 79 10^{-6} m^2 (8.5 10^{-4} ft^2) for

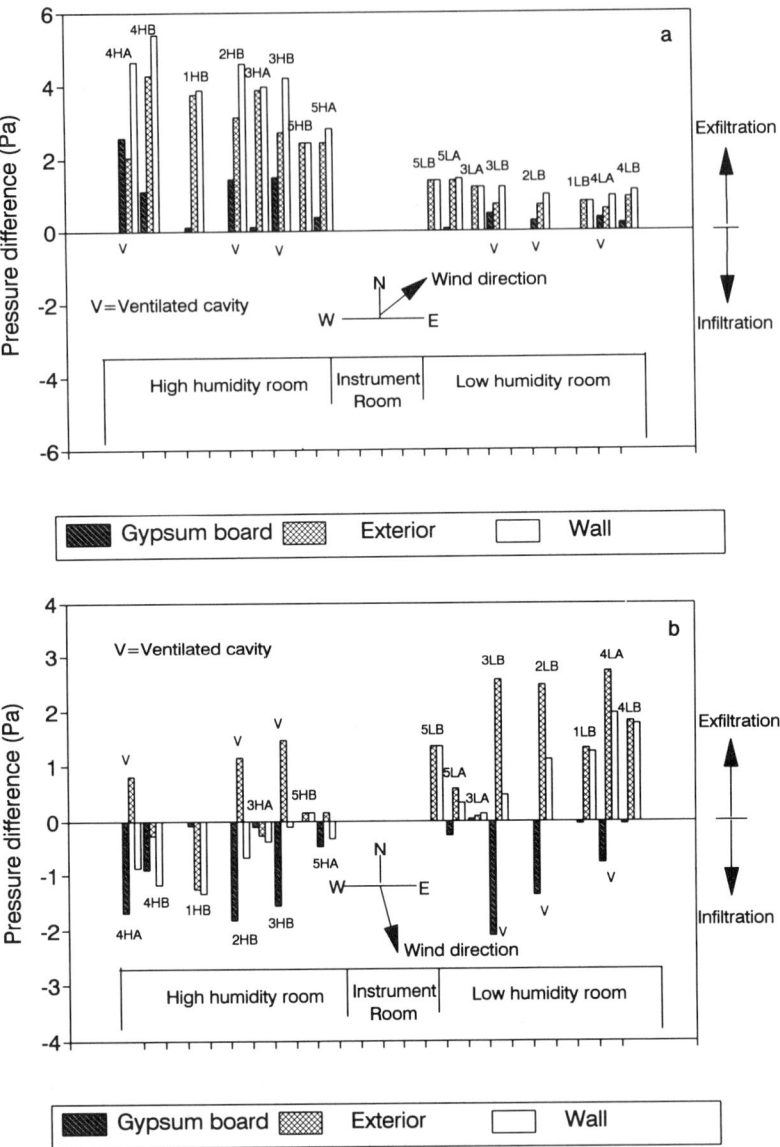

FIG. 8—*Measured daily average air pressure differences across gypsum board, exterior membrane, and both membranes of all walls with electrical outlet: (a) January 12, 1992, (b) January 14, 1992.*

all outlets. The prevailing wind direction during January 1992 was west to northwest. The flow patterns shown in Fig. 9 are typical for the period November 1991 through March 1992. In April 1992, the dominant flow in all walls was exfiltrative because of a shift of the prevailing winds towards the south. Between November 1, 1991 and April 30, 1992 total flow (sum of infiltration and exfiltration) was, on average, 16 times greater through ventilated walls than through nonventilated walls of otherwise similar construction.

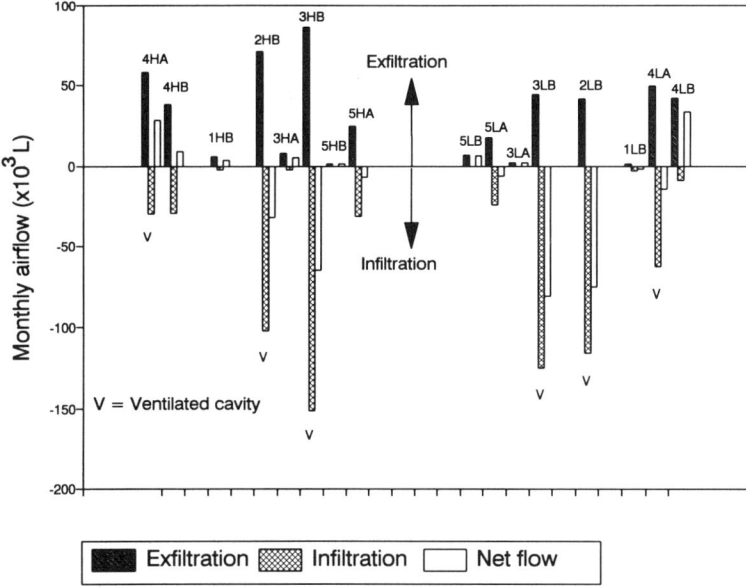

FIG. 9—*Total monthly exfiltrative, infiltrative, and net airflows through the electrical outlets during January 1992.*

Temperature and Humidity Conditions

Despite the large variation in air leakage, we found little difference in sheathing temperatures in walls of similar construction. Furthermore, temperature variation at different heights within the same wall was generally less than 1°C (1.8°F).

We used three indicators to evaluate humidity and moisture performance of the walls: average measured RH (sheathing and siding), sensor-indicated condensation, and wall inspection results. In contrast to the measured temperatures, measured humidity conditions on the back of the sheathing showed large variations. Because the sensors were in the wall for almost two years before data collection began, error in the measurements is likely to be considerable. In addition, condensation and probable sensor failures (leads coming unglued) added to the uncertainty. Therefore, we reported only average sheathing and siding humidities and eliminated data from any failed sensor or sensors obviously out of calibration. Sensor-indicated condensation occurred when the RH sensor read 100%. Several sensors recorded 100% RH for only short periods, but other sensors recorded saturation conditions for extended periods. After the 1991–92 heating season, we removed the gypsum board and insulation from the walls and inspected the sheathing and siding for signs of water stain, fungal stain, mold, or corrosion.

Table 4 lists the average RH conditions during November 1991 through March 1992. Although many walls showed elevated humidity conditions, they all dried out by May 1992. Table 4 also lists the walls where sensors indicated saturation conditions during the winter months and shows wall inspection results.

The most serious problems observed at wall cavity inspection were superficial mold growth or staining. We at no time observed problems with paint finish on the waferboard siding. Evidently, although elevated humidity and condensation conditions were reached in some test walls, none experienced serious problems as a result.

TABLE 4—*Humidity and moisture conditions on sheathing and siding.*

Wall	Average RH Nov.–Feb. (%)[a]	Sensor-Indicated Condensation Nov.–Feb. by Position #[b]	Results of Inspection
1LA	82	1[c]	no stains or mold
1LB	80	1, 3, 4[c]	no stains or mold
2LA(v)	62		no stains or mold
2LB(v)	61		no stains or mold
3LA	86		no stains or mold
3LB(v)	53	4[c]	no stains or mold
4LA(v)	66/53[d]		no stains or mold
4LB	53/57[d]		no stains or mold
5LA	<40/74[d]		no stains or mold
5LB	61/62[d]		no stains or mold
1HA	90	2[c], 4	fungal stains
1HB	83	1[c], 4	fungal stains
2HA(v)	77	1[c]	no stains or mold
2HB(v)	81		no stains or mold
3HA	86	2[c]	slight fungal stain
3HB(v)	86	1, 3[c]	very slight fungal stain
4HA(v)	46/94[d]	4, 6, 7, 8	signs of high RH on siding
4HB	88/62[d]	1, 6[c]	slight mold on sheathing
5HA	72/64[d]		no stains or mold
5HB	44/78[d]		no stains or mold

[a] Average of all readings, unless sensor was judged to be defunct or obviously out of calibration.
[b] Readings of 100% RH.
[c] One or more occasional readings of 100% RH of short duration.
[d] Average RH sheathing/average RH siding.

Short periods of condensation were recorded in walls 1LA, 1LB, 3LB, 2HA, and 2HB. Sustained condensation was recorded in walls 1HA, 1HB, 3HB, 4HA, and 4HB. The wall cavities where sensor readings indicated sustained condensation showed traces of mold or staining. No other wall showed mold or staining. Thus, the wood RH sensors adequately indicated condensation.

All walls with sustained condensation and mold or staining were in the high-humidity room (45% RH). High indoor humidity clearly tended to result in condensation and mold in walls. Walls 5HA and 5HB, constructed with sheathing and ventilated siding, were the only walls that consistently remained dry in the high-humidity room.

Cavity ventilation promoted drying in walls without outlets (1LA compared with 2LA, 1HA compared with 2HA). In walls with outlets (and thus an opportunity for air leakage between the room and the wall cavity), the benefit of cavity ventilation was not as clear as in walls without outlets. In walls with outlets in the 35% RH room, cavity ventilation either promoted drying or had no effect on wall moisture conditions (1LB compared with 2LB, 3LA compared with 3LB, 4LB compared with 4LA). In walls with outlets in the 45% RH room, cavity ventilation had no apparent influence on wall moisture conditions (1HB compared with 2HB, 3HA compared with 3HB, 4HB compared with 4HA). Apparently, if there was an opportunity for air leakage between the room and the wall cavity and if the room was maintained at high humidity, cavity ventilation was not an effective moisture control strategy. The presence of mold and evidence of sustained condensation in at least two ventilated walls (3HB and 4HA) further indicates that cavity ventilation was not always successful at providing dry conditions in wall cavities.

Air pressure measurements indicated that during the heating season, the predominant direction of pressure differential across the gypsum board membrane of ventilated walls was infiltrative. Had these walls been located on the leeward side of the building, exfiltration would have predominated and wetting would likely have occurred.

Heat Loss

To illustrate the effect of cavity ventilation on the thermal performance of a wall, we calculated the heat flow by conduction and air movement in walls 3HA (nonventilated) and 3HB (ventilated) during January 1992. Because air convection may have significantly changed conductive heat loss, we made our calculations with two extreme assumptions: (*a*) air convection was localized and did not significantly change conductive heat loss in the remainder of the wall, and (*b*) air convection was homogeneous through the cavity and wall. With localized air leakage [assumption (*a*)] convective and conductive heat flows can be considered parallel and can be added for practical purposes. Berlad and others [*18*] provided a method to calculate heat flow in the case of homogeneous airflow through the wall [assumption (*b*)]. We further assumed that the total R-value of the wall was 3.3 m² K/W (19 hft²°F/Btu). The calculation of heat flow was based on hourly indoor and outdoor temperatures and measured airflows.

The average calculated total heat flow through wall 3HA (nonventilated) during January ranged from 6.0 W (20.5 Btu/h) for homogeneous airflow to 6.1 W (20.7 Btu/h) for localized airflow. The total heat flow through wall 3HB ranged from 7.7 W (26.2 Btu/h) to 8.8 W (30.1 Btu/h), depending on the airflow distribution. This constitutes a 28% to 33% increase compared with the nonventilated wall. Increased air leakage caused by the vents in the sheathing/siding probably degraded the R-value of wall 3HB from 3.3 m² K/W (19 hft²°F/Btu) to an equivalent of about 2.3 m² K/W (13 hft²°F/Btu) to 2.5 m² K/W (14 hft²°F/Btu). With typical construction and localized leakage sites in the walls, we expect the airflows in the wall to be more localized rather than homogeneously distributed. Thus, the higher heat loss values probably apply.

We did not consider any ventilation benefits from infiltration or any credits for latent heat recovery. From the perspective of moisture control, walls should be constructed as airtight as possible, unless the airflow through the wall is controlled in magnitude and direction. Exterior vents do not provide such control.

Discussion

To cover the full range of the differential pressure gage, we were limited to a resolution of 1 Pa (0.004 in. of water). This may have led us to underestimate low-pressure flows through electrical outlets, especially in nonventilated walls. However, any pressures more than 0.5 Pa (0.002 in. of water) were rounded off to 1 Pa (0.004 in. of water), providing some compensation. The original intent was to calculate airflows based on pressure measurements across the tightest membrane, because those pressures could be measured with the greatest accuracy. After noticing the pressurization of some cavities with northerly winds, we realized that it would be better to base the flow calculation on pressures across the outlet, forcing us to use the single 4-s pressure samples for some walls (Table 2).

Another possible source of error was that air pressures were not measured simultaneously but in sequence. However, the relatively good agreement between total pressure differences across adjacent walls suggests that this is an allowable measurement technique.

Because data acquisition did not start until almost two years after the walls with moisture sensors were installed, we assume that many sensors drifted considerably since the time of calibration. Therefore, we believe that the previously reported 10% RH accuracy does not apply to our measurements, and we have treated our measured RH values with circumspection and only reported average values. The consistent differences between average RH conditions in walls in the west room and those in walls in the east room indicated that average RH may be used as an indicator of moisture performance. These sensors also accurately signaled sustained condensation by essentially short circuiting.

The inconsistent effect of cavity ventilation on moisture conditions in test walls can be explained. Cavity ventilation will result in drier walls during winter if the dominant airflow through the wall is infiltrative, or if air leakage through the wall is insignificant [9]. The latter condition occurred in walls 2LA and 2HA; these walls did not contain electrical outlets and were effectively isolated from stud bays with outlets. However, in manufactured homes, holes are typically drilled through the wall studs for electrical cable. Thus, air exchange between wall cavities without outlets and the indoors can probably occur through outlets in adjacent cavities. The effect of cavity ventilation in most stud bays in manufactured homes therefore likely depends on the prevailing direction of air leakage (infiltrative or exfiltrative). If there are leakage paths between the interior and the wall cavity, the wetting effect of exfiltration will increase with increasing indoor humidity, whereas the drying effect of infiltration will be of smaller magnitude and essentially independent of indoor RH [10]. This phenomenon explains why cavity ventilation was more effective at promoting drying in walls in the 35% RH room than in walls in the 45% RH room.

Conclusions

We conclude the following from our measurements, comparisons, and calculations:

1. Providing wall cavity ventilation can lead to a significant increase in air leakage and heat loss, especially when there is no effective air barrier on the interior side of the cavity (e.g., gypsum board with standard electrical outlet or switch).
2. Providing cavity ventilation may pressurize the wall cavity of walls on the windward side of a building, producing air movement from the cavity simultaneously to both the inside room and the exterior. Exfiltration is likely to occur in walls facing in other directions.
3. Providing cavity ventilation had an inconsistent but generally explainable influence on moisture conditions in wall cavities. The evidence suggests that providing cavity ventilation is not a reliable strategy if air leakage between the interior and the wall cavity is possible and indoor humidity is high.
4. Moisture conditions in the walls were clearly related to indoor relative humidity; all walls exposed to high indoor humidity had more moisture accumulation than did similar walls exposed to low indoor humidity. This indicates that indoor humidity control is the most important and effective strategy to prevent excessive moisture accumulation in the exterior building envelope during winter.

Acknowledgments

The authors thank Wayne E. Bewick for his expert assistance in constructing and installing wall modules, David K. Simpson for developing data collection software and for invaluable help in detecting and correcting instrumentation malfunctions, and Joseph F. Murphy for developing software used in data processing. The authors also acknowledge the U.S. Department of Housing and Urban Development for financial support for this research.

References

[1] "Moisture in Canadian House Construction: Problems, Research and Practice from 1975 to 1991," Canada Mortgage and Housing Corporation, Research Division Draft Report, Ottawa, 1992.

[2] Wilson, A. G., "Air Leakage in Buildings," *Canadian Building Digest 23,* Division of Building Research, National Research Council (Canada), 1961.

[3] Platts, R. E., "Wet Walls: Apparent Incidence of Excessive Condensation in House Envelope Construction in Canada," *Proceedings,* Air Infiltration, Ventilation and Moisture Transfer Workshop, Building Thermal Envelope Coordinating Council, Washington, DC, 1987, pp. 82–90.

[4] Tsongas, G., "The Northwest Wall Moisture Study: A Field Study of Excess Moisture in Walls and Moisture Problems and Damage in New Northwest Homes," Bonneville Power Administration, Portland, OR, 1990.

[5] Tsongas, G. A. and Nelson, G. D., "A Field Test for Correlation of Air Leakage and High Moisture Content Sites in Tightly Built Walls," *ASHRAE Transactions,* Vol. 97, Part 1, 1991.

[6] Burch, D. M. and Thomas, W. C., "An Analysis of Moisture Accumulation in a Wood Frame Wall Subjected to Winter Climate," NISTIR 4674, U.S. Department of Commerce, National Institute of Standards and Technology, Gaithersburg, MD, 1991.

[7] Ojanen, T. and Kumaran, M. K., "Air Exfiltration and Moisture Accumulation in Residential Wall Cavities," *Proceedings,* Fifth ASHRAE/DOE/BTECC Conference on Thermal Performance of the Exterior Envelopes of Buildings, Clearwater Beach, FL, 7–10 Dec. 1992.

[8] *Code of Federal Regulations,* Title 24, Part 3280, Manufactured Home Construction and Safety Standards, 4-1-90 ed., U.S. Department of Housing and Urban Development, 1990.

[9] TenWolde, A. and Carll, C., "The Effect of Cavity Ventilation on Moisture in Walls and Roofs," *Proceedings,* Fifth ASHRAE/DOE/BTECC Conference on Thermal Performance of the Exterior Envelopes of Buildings, Clearwater Beach, FL, 7–10 Dec. 1992.

[10] TenWolde, A., "Steady-State One-Dimensional Water Vapor Movement by Diffusion and Convection in a Multilayered Wall," *ASHRAE Transactions,* Vol. 91, Part 1, 1985.

[11] Sherwood, G. E., "Condensation Potential in High Thermal Performance Walls—Cold Weather Climate," Research Paper FPL 433, U.S. Department of Agriculture, Forest Service, Forest Products Laboratory, Madison, WI, 1983.

[12] Burch, D. M. and TenWolde, A., "Controlling Moisture in the Walls of Manufactured Housing," *ASHRAE Transactions,* Vol. 99, Part 2, 1993.

[13] Burch, D. M., Thomas, W. C., and Fanney, A. H., "Water Vapor Permeability Measurements of Common Building Materials," *ASHRAE Transactions,* Vol. 98, Part 2, 1992.

[14] *ASHRAE Handbook of Fundamentals,* American Society of Heating, Refrigerating and Air-Conditioning Engineers, Atlanta, GA, 1989.

[15] Duff, J. E., "A Probe for Accurate Determination of Moisture Content of Wood Products in Use," Research Note FPL-0142, U.S. Department of Agriculture, Forest Service, Forest Products Laboratory, Madison, WI, 1966.

[16] TenWolde, A. and Courville, G. E., "Instrumentation for Measuring Moisture in Buildings," *ASHRAE Transactions,* Vol. 91, Part 2, 1985.

[17] TenWolde, A. and Mei, H. T., "Moisture Movement in Walls in Warm Humid Climate," *Proceedings,* Third ASHRAE/DOE/BTECC Conference on Thermal Performance of Exterior Envelopes of Buildings, Clearwater Beach, FL, 2–5 Dec. 1985.

[18] Berlad, A. L., et al., "Air Intrusion Effects on the Performance of Permeable Insulation Systems," *Thermal Insulation Performance, ASTM STP 718,* American Society for Testing and Materials, Philadelpha, PA, 1980.

Doug Burch[1]

An Analysis of Moisture Accumulation in the Roof Cavities of Manufactured Housing

REFERENCE: Burch, D., **"An Analysis of Moisture Accumulation in the Roof Cavities of Manufactured Housing,"** *Airflow Performance of Building Envelopes, Components, and Systems, ASTM STP 1255,* Mark P. Modera and Andrew K. Persily, Eds., American Society for Testing and Materials, Philadelphia, 1995, pp. 156–177.

ABSTRACT: A detailed computer analysis is conducted to investigate whether moisture problems occur in the roof cavity of manufactured homes constructed in compliance with the current Department of Housing and Urban Development (HUD) Standards for manufactured housing. The current HUD Standards require a ceiling vapor retarder, but do not require outdoor ventilation of the roof cavity. In cold climates, the analysis revealed that moisture accumulates at lower roof surface and poses a risk of material degradation.

The analysis found the following combination of passive measures to be effective in preventing detrimental winter moisture accumulation at lower surface of the roof: 1) providing a ceiling vapor retarder, 2) sealing penetrations and openings in the ceiling construction, and 3) providing natural ventilation openings in the roof cavity.

In addition, the performance of a roof cavity exposed to a hot and humid climate is investigated. The analysis revealed that outdoor ventilation of the roof cavity causes the monthly mean relative humidity at the upper surface of the vapor retarder to exceed 80%. This condition is conducive to mold and mildew growth.

KEYWORDS: attic ventilation, HUD Manufactured Home Construction and Safety Standards, manufactured housing, mobile homes, moisture control guidelines, moisture in attics

Nomenclature

C	Specific heat, $J/kg \cdot °C$
D_γ	Diffusivity for moisture gradient, m^2/s
D_T	Diffusivity for temperature gradient, $m^2/°C \cdot s$
$f(\phi)$	Sorption isotherm function
k	Thermal conductivity of porous material, $W/m \cdot °C$
P	Pressure, Pa
t	Time, s
T	Temperature, °C
y	Distance from inside surface of wall, m
α	Solar absorptance
γ	Moisture content on dry basis, kg/kg
μ	Water-vapor permeability, $kg/s \cdot m^2 \cdot Pa$
ρ	Density, kg/m^3
ϕ	Relative humidity

[1] Heat Transfer Group, Building Environment Division, NIST Building and Fire Research Laboratory, Gaithersburg, MD 20899-0001.

Subscripts

d = Dry property
g = Saturated state
T = Temperature gradient
v = Vapor property
w = Moist or water property
γ = Moisture content gradient

Introduction

During the winter, the occupant activities in manufactured housing release moisture to the indoor air. The airtightness of mobile homes tends to be considerably better than that of site-built homes. This causes the indoor absolute humidity to be considerably higher than that of the outdoor air. The vapor pressure difference across the ceiling construction causes moisture to be transferred into the roof cavity by diffusion. In addition, the ceiling construction usually contains air leakage sites associated with lighting fixtures and other elements. The stack effect causes moist indoor air to exfiltrate through the ceiling construction and accumulate at the roof sheathing.

The moisture content of roof sheathing has not been studied in manufactured housing. However, Harrje et al. measured the moisture content of roof sheathing of a conventional house in Princeton, NJ [1]. Harrje found that the north-sloping roof sheathing absorbed water vapor during winter periods and reached a high moisture content of 20% during midwinter. When the outdoor temperature rose in the spring, the moisture content of the roof sheathing dried out and decreased to a low value of 5%.

Relative to the above discussion, the maximum amount of moisture that can be stored in roof sheathing is denoted by "fiber saturation." Above fiber saturation, liquid water appears in the pore structure of the material. Fiber saturation is generally regarded as the maximum amount of moisture that can be taken on without degradation.

Higher roof sheathing moisture contents are likely to occur in manufactured houses compared to conventional houses. Manufactured houses tend to have higher indoor relative humidity compared to conventional houses because they have smaller volumes and lower rates of natural infiltration. In addition, the Department of Housing and Urban Development (HUD) Manufactured Home Construction and Safety Standards[2] do not currently require ventilation openings in the roof cavity [2]. As a result, many manufactured houses are constructed without ventilation openings.

A few field surveys documenting moisture problems in the roof cavity of manufactured housing are reported in the literature. For example, Zieman and Waldman conducted a field survey of 49 manufactured houses located in different parts of the United States which had unresolved moisture problems [3]. Twenty-nine percent of the houses surveyed had roof cavity condensation problems, evidenced by stains at the interior surface of ceiling boards. In addition, Lee [4] surveyed 65 manufactured houses in Alberta, Canada and reported that condensation was a problem in the roof cavity, although Canadian mobile homes are not required to be constructed in compliance with the HUD Standards.

Discussion of Model

A detailed computer model, called MOIST, has been developed at the National Institute of Standards and Technology (NIST) that predicts the moisture content and temperature

---

[2] For the sake of brevity, the HUD Manufactured Home Construction and Safety Standards will henceforth be referred to as the HUD Standards.

versus time for the construction layers of a building envelope [5]. The computer model is available from NIST.

Theory

Within each layer of a roof cavity, moisture transfer is governed by the following one-dimensional conservation of mass equation

$$\frac{\partial}{\partial y}\left(D_\gamma(\gamma, T)\frac{\partial \gamma}{\partial y}\right) + \frac{\partial}{\partial y}\left(D_T(\gamma, T)\frac{\partial T}{\partial y}\right) = \frac{\partial \gamma}{\partial t} \qquad (1)$$

The selection of moisture content γ and temperature T as potentials has the advantage that the same mathematical formulation represents both diffusion transfer and capillary transfer. This formulation is equivalent to using water-vapor pressure as the moisture transfer potential in the diffusion regime and suction pressure in the capillary flow regime with a single required diffusivity.

Heat transfer is governed by the one-dimensional conservation of energy equation

$$\frac{\partial}{\partial y}\left(k(\gamma, T)\frac{\partial T}{\partial y}\right) = \rho(C_d + \gamma C_w)\frac{\partial T}{\partial t} \qquad (2)$$

Latent transport of heat is included at the boundaries of the layers. The other components of enthalpy transport by moisture movement are generally small and are therefore neglected in the analysis. The term $(C_d + \gamma C_w)$ includes the effect of energy storage in both the dry material and accumulated moisture.

In the above two governing equations, strong couplings exist between heat and moisture transfer. Both the diffusivity for the moisture gradient D_γ and the diffusivity for the temperature gradient D_T are strong functions of moisture content and temperature. The thermal conductivity k can also be a function of moisture content and temperature, but for the present analysis it is assumed to be constant.

When the moisture content of a material is below fiber saturation, the diffusivity for the moisture gradient D_γ and the diffusivity for the temperature gradient D_T are calculated by the relations

$$D_\gamma = \frac{\mu(\phi)P_{vg}(T)}{\rho_d \dfrac{\partial f(\phi)}{\partial \phi}} \quad \text{and} \quad D_T = \frac{\mu(\phi)\phi \dfrac{\partial P_{vg}(T)}{\partial T}}{\rho_d} \qquad (3)$$

The above equations may be derived by introducing the sorption isotherm function $f(\phi)$ and applying the chain rule to Fick's steady-state diffusion equation with the gradient of the water-vapor pressure as the driving-force potential.

When the moisture content of a material is above fiber saturation, a liquid diffusivity D_γ is used in Eq 1. It is calculated using procedures given in Burch and Thomas [5]. The diffusivity for the temperature gradient D_T is calculated using the second relation of Eq 3.

The model also has a provision for including nonstorage layers (e.g., an air space, glass-fiber insulation, a vapor retarder, etc.) that may be sandwiched between two storage layers. In a nonstorage layer, the storage of heat and moisture is neglected, and the transfer of heat and moisture is assumed to be steady state. A nonstorage layer may be convectively coupled to indoor and outdoor air.

Solution Procedure

Equations 1 and 2 were recast into finite-difference equations using a uniform nodal spacing within each layer. An implicit solution technique with coupling between the two conservation equations was used to solve the equations. A FORTRAN 77 computer program, called MOIST, with a tridiagonal-matrix solution algorithm was prepared. At each time step, the calculation proceeds by first solving for the temperature distribution, followed by a set of moisture contents. By choosing a sufficiently small time step (i.e., 1 h), there is no need to iterate between the temperature and moisture solutions.

Model MOIST was used to analyze the moisture accumulation in roof construction sub-, jected to both a cold climate and a hot and humid climate. For the baseline roof construction (Fig. 1), two nodes were used for the gypsum board, sixteen in the roof sheathing, and two in the roofing paper and asphalt shingles. The kraft paper, glass fiber insulation, and the air space were combined into a single nonstorage layer. When the computer program was run on a Model 386 personal computer with a 33 MHz clock speed, equipped with a math coprocessor, about 60 min of computer time was required to simulate one year of real time.

Description of Baseline Roof Construction

The roof construction shown in Fig. 1 was analyzed. An insulation thermal resistance of $R - 2.5$ m² · K/W (14 h · ft² · °F/Btu) was used in the cold climate analysis and $R - 1.9$ m² · K/W (11 h · ft² · °F/Btu) in the hot and humid climate analysis. Two roof sheathing materials were considered: 12 mm (¹⁵⁄₃₂ in.) exterior-grade plywood and 11 mm (⁷⁄₁₆ in.) oriented strand board (OSB). In addition, a metal roof consisting of a single layer of 0.33 mm (0.013 in.) galvanized steel was analyzed.

Since the mathematical model used for the analysis was one dimensional, it could not include the effect of wood-framing members. The moisture content of the roof sheathing of actual construction would tend to be a little lower than the theoretical predictions of the present report, due to the additional storage of moisture provided by wood-framing members.

Parameters Used in Analysis

The following diffusion properties and boundary conditions were used as input for the model MOIST.

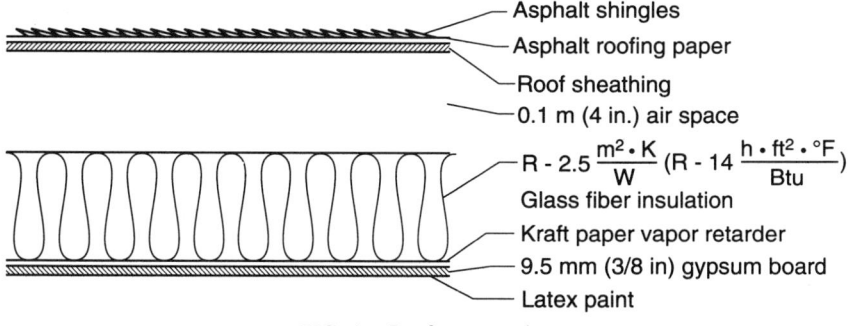

FIG. 1—*Roof construction.*

Parameters for Baseline Construction

Significant features of the baseline construction shown in Fig. 1 include a ceiling vapor retarder having a nominal permeance of 5.7×10^{-11} kg/Pa · s · m² (1 perm) and a roof solar absorptance of 0.7.

Ventilation Rate for Roof Cavity—For the baseline roof construction, it was assumed that the roof cavity had no ventilation openings. For this case, it was assumed that outdoor air infiltrated into the roof cavity through cracks in the construction at a rate of 0.25 air changes per hour (ach). Multiplying by the cavity volume and dividing by the ceiling area gives a volumetric infiltration rate per unit area of ceiling of 1.2×10^{-5} m³/s per m² (0.14 ft³/h per ft²).

The following four cavity air exchange rates were analyzed:

	Air Exchange Rate	
Description	m³/s per m²	(ft³/h per ft²)
Perfectly Sealed Cavity	0.0	(0.0)
Cavity without Ventilation Openings	1.2×10^{-5}	(0.14)
Cavity with Ventilation Openings	8.5×10^{-5}	(1.0)
Mechanically Ventilated Cavity	4.2×10^{-4}	(5.0)

A perfectly sealed attic is an idealization that is virtually impossible to achieve in practice. The air exchange rate for a cavity with ventilation openings and a mechanically ventilated cavity are based on assumed volumetric ventilation rates of 2 and 10 ach, respectively.

Indoor Air Exfiltration Rate into Roof Cavity—The ceiling of a manufactured house usually contains a light fixture in each room. During the winter, indoor air is warmer and therefore lighter than colder outdoor air. As a result, the stack effect on the indoor air causes indoor air to exfiltrate through air leakage sites around the light fixtures and other air leakage sites.

In selecting an exfiltration rate for the baseline roof construction, it was assumed that the total house infiltration rate was 0.25 ach. In addition, it was assumed that half was induced by temperature-difference (i.e., stack effect) driving force, while the remainder was induced by wind-speed driving force. This 50/50 percent breakdown is consistent with measurements reported by Goldschmidt and Wilhelm [6]. Furthermore, it was assumed that about one third of the stack effect portion exfiltrates into the roof cavity, while the other two thirds exfiltrate through the upper portion of the walls to the outdoor environment. This gives an exfiltration rate into the roof cavity of 0.042 ach. Multiplying by the house volume and dividing by its ceiling area gives a volumetric rate per unit ceiling area of 2.5×10^{-5} m³/s per m² (0.30 ft³/h per ft²).

The following four rates of exfiltration are analyzed in the paper:

	Exfiltration Rate	
Description	m³/s per m²	(ft³/h per ft²)
Perfectly Sealed Ceiling	0.0	(0.0)
Well Sealed Ceiling	1.3×10^{-5}	(0.15)
Typical Ceiling	2.5×10^{-5}	(0.30)
Leaky Ceiling	5.1×10^{-5}	(0.60)

A perfectly sealed ceiling is an idealization that is very difficult to achieve in practice. The typical ceiling corresponds to the baseline roof construction discussed above. It was assumed that the exfiltration rate for a well-sealed ceiling was one half that for the typical ceiling, while it was assumed that the value for a leaky ceiling was twice that of the typical ceiling.

Outdoor Boundary Conditions

The outdoor temperature, relative humidity, and solar radiation were based on weather year for energy calculations (WYEC) hourly weather data for each of the five locations [7].

Diffusion Properties

A plot of the equilibrium moisture content versus relative humidity (called a sorption isotherm) for the construction materials is given in Fig. 2a. The sorption isotherm data were measured at NIST by Richards et al. [8]. The water-vapor permeability of the materials were

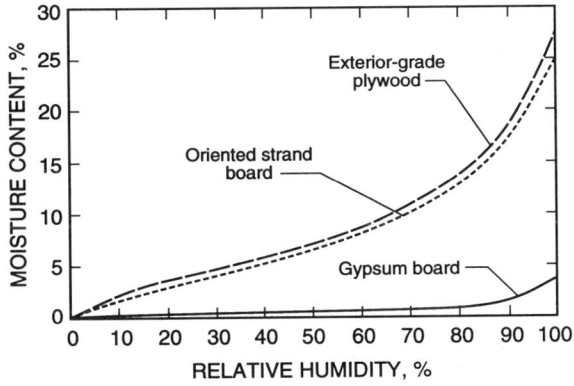

a. Sorption isotherms

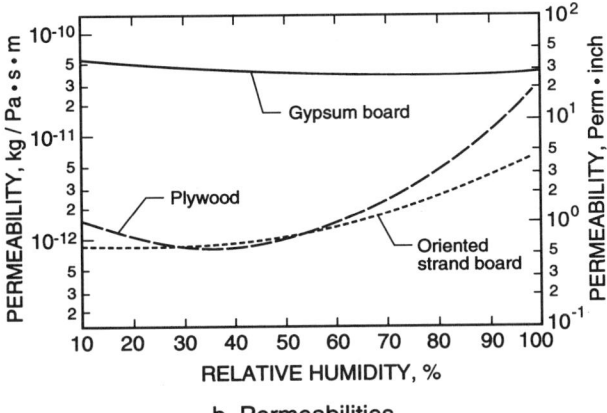

b. Permeabilities

FIG. 2—*Diffusion properties of materials used in the analysis.*

also measured at NIST by Burch et al. [9]. Permeance is equal to the permeability of the material divided by its thickness. A plot of the permeance of the materials versus relative humidity is given in Fig. 2b.

In the analysis, latex paint, roofing paper and shingles, glass-fiber insulation, kraft paper, and the air space were treated as nonstorage layers. The permeance values for these materials are given in Table 1. In actual practice, the permeability of the roofing paper and shingles would have a very small finite value. However, treating these materials as impermeable has very little effect on the predicted moisture content of the roof sheathing. This is because the outflow of moisture from the roof sheathing to the outdoor environment is very small compared to the inflow of moisture from the air within the roof cavity, due to the small temperature difference between the roof sheathing and the outdoor environment and the low permeance of roofing paper and shingles.

Cold Climate Analysis

In the analysis, the indoor temperature and relative humidity are maintained at 21°C (70°F) and 50%, respectively, unless otherwise indicated. The use of an indoor relative humidity of 50% is warranted because manufactured homes are tighter and have lower infiltration rates and therefore tend to have higher indoor relative humidities than site-built homes. In fact, Zieman and Waldman report that manufactured homes have indoor relative humidities above 50% [3].

The outdoor temperature, relative humidity, and solar radiation are derived from weather year for energy calculations (WYEC) hourly weather data [7] for a mild winter heating climate (Atlanta, GA); and intermediate winter climate (Boston, MA); a cold winter climate (Madison, WI); and a Pacific northwest climate (Portland, OR). The heating degree days for these cities are 4228°C · days (7610°F · days) for Madison; 3207°C · days (5773°F · days) for Boston; 2579°C · days (4642°F · days) for Portland; and 1706°C · days (3071°F · days) for Atlanta. Figure 3 shows weekly average outdoor temperatures for these cities.

For each simulation, the performance of the roof cavity is predicted for a 1.5 year period. Six months of weather data are used to initialize the reported one-year simulations so that the initial moisture content and temperature would have a small effect on the results.

Results for Baseline Roof Construction

The moisture content of the plywood roof sheathing is plotted versus time of year in Fig. 4 for the climate of Madison, WI. During cold winter periods, indoor moisture is transferred into the roof cavity by way of diffusion and air exfiltration through the ceiling construction.

TABLE 1—*Permeance for nonstorage layers.*

Material	10^{-10} kg/s · m² · Pa
Latex Paint	5.7
Roofing Paper and Shingles	Impermeable
110 mm Glass Fiber Insulation	15
Kraft Paper (Asphalt Impregnated)	0.86
100 mm Air Space	17

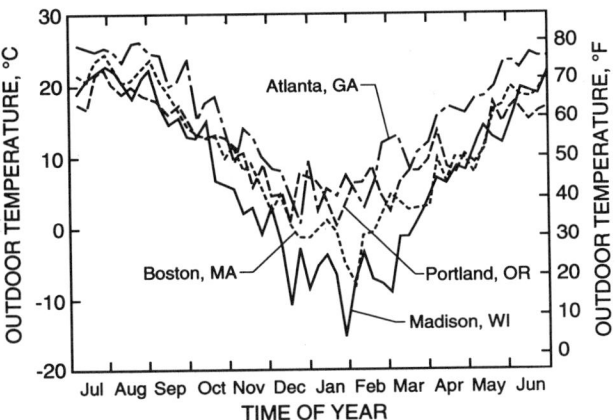

FIG. 3—*Weekly average outdoor temperature for the four winter climates.*

Moisture is adsorbed and accumulates at the plywood roof sheathing. During the spring, the elevated temperatures promote drying and the moisture content decreases.

The solid horizontal line depicts fiber saturation in the plywood roof sheathing. As previously mentioned, when the moisture content rises above fiber saturation, free liquid water exists within the pores of the material and a significant potential for material degradation exists. The moisture content of the plywood roof sheathing rises above fiber saturation for almost a three-month period. The plywood roof sheathing experiences repeated expansion and contraction cycles due to seasonal fluctuations in moisture content.

In subsequent sections, model MOIST is used to analyze the effect of various parameters on the performance of the baseline roof construction. Unless otherwise indicated, the roof sheathing is plywood and the climate is Madison, WI.

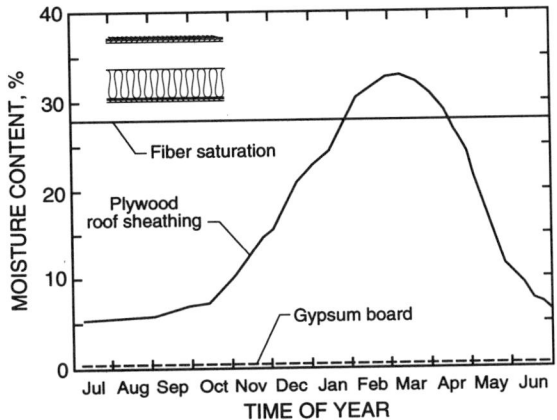

FIG. 4—*Moisture content of plywood roof sheathing plotted versus time of year for baseline roof construction located in Madison, WI.*

Effect of Significant Parameters

Outdoor Climate—The moisture content of the plywood roof sheathing is plotted versus the time of year in Fig. 5 for the four winter climates. Comparing the four curves, higher moisture contents occur in colder climates. In the two coldest climates (i.e., Madison and Boston), the moisture content rises above fiber saturation. These results indicate that climate is a very significant parameter affecting moisture accumulation in the roof sheathing of manufactured housing.

Comparing Figs. 3 and 5, the peak moisture content lags behind the minimum winter temperature by several months. This lag is due to moisture storage within the plywood roof sheathing.

It should be pointed out that the effect of climate may be less pronounced in actual manufactured houses because the indoor relative humidity does not remain constant and tends to decrease in colder climates as a result of increased moisture losses by window condensation and infiltration of drier outdoor air.

Indoor Air Exfiltration into Roof Cavity—Next, model MOIST is used to investigate the effect of indoor air exfiltration in the roof cavity. Results for four exfiltration rates are given in Fig. 6. The results show that air exfiltration into the roof cavity has a profound effect on the amount of moisture buildup. As the rate of air exfiltration increases, both the peak and breadth of the profiles increase. In addition, the peak moisture content for a well-sealed ceiling reaches fiber saturation for a brief period. These results illustrate the importance of sealing air leakage paths in the ceiling construction of manufactured housing.

Outdoor Ventilation of Roof Cavity—Next, model MOIST is used to investigate the effect of outdoor ventilation of a roof cavity with a ceiling vapor retarder. The moisture content of the plywood roof sheathing is plotted versus time of year for four outdoor ventilation rates in Fig. 7. These results reveal that providing natural ventilation significantly reduces both the peak and breadth of the profile. In fact, the peak moisture content is maintained slightly below fiber saturation. It will be shown later that natural ventilation, used in combination with sealing air leakage paths in the ceiling construction, maintains the peak moisture content considerably below fiber saturation. On the other hand, mechanical ventilation,

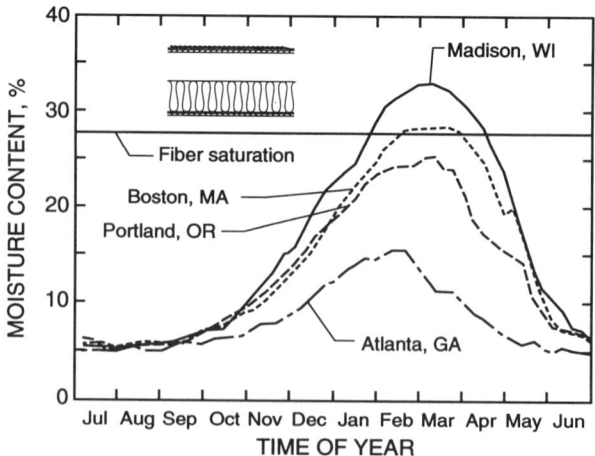

FIG. 5—*Moisture content of plywood roof sheathing plotted versus time of year for four winter climates.*

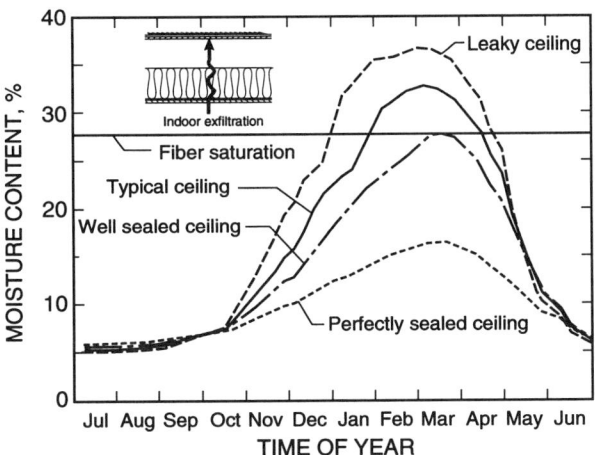

FIG. 6—*Moisture content of plywood roof sheathing plotted versus time of year for four exfiltration rates (Madison, WI).*

used in combination with a ceiling vapor retarder, reduces the peak moisture content well below fiber saturation.

Figure 8 shows similar simulation results for Portland. Here, it is seen that ventilation is still effective in reducing moisture accumulation in the roof sheathing. In this climate, the author was concerned that the outdoor air has higher relative humidity during the winter and therefore less drying capacity.

It should be pointed out that the above analysis assumes that providing outdoor ventilation for a roof cavity does not increase the exfiltration of indoor air into the roof cavity. TenWolde and Carll have recently shown that outdoor ventilation sometimes increase the exfiltration

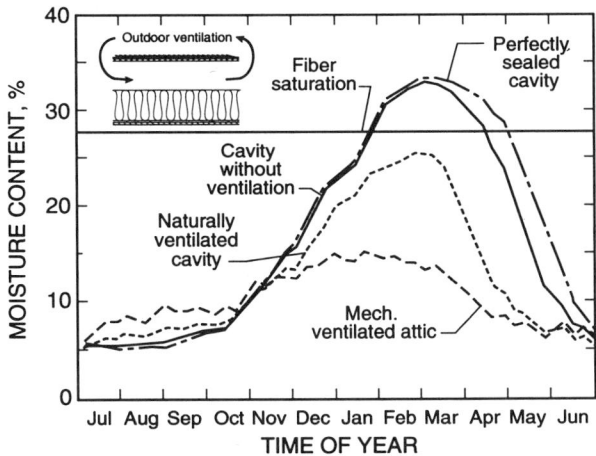

FIG. 7—*Moisture content of plywood roof sheathing plotted versus time of year for four roof cavity ventilation rates (Madison, WI).*

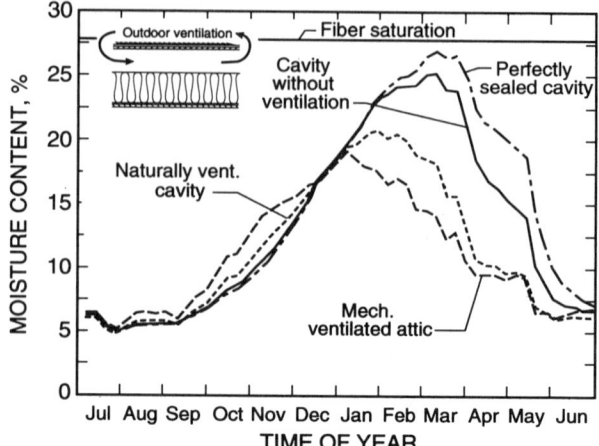

FIG. 8—*Moisture content of plywood roof sheathing plotted versus time of year for four roof cavity ventilation rates (Portland, OR).*

of indoor air into building cavities [10]. Since air exfiltration transports a considerable amount of moisture into building cavities, ventilation may in these instances be counterproductive, and a better strategy would be to improve the ceiling airtightness rather than to increase attic ventilation.

Indoor Relative Humidity—Separate computer runs were conducted for an indoor relative humidity of 35% and 50%. The results, given in Fig. 9, reveal that indoor relative humidity has an important effect on both the peak and breadth of the moisture content profile. For an indoor relative humidity of 35%, the peak moisture content is maintained below fiber saturation. On the other hand, for an indoor relative humidity of 50%, the moisture content is above fiber saturation for almost a three-month period.

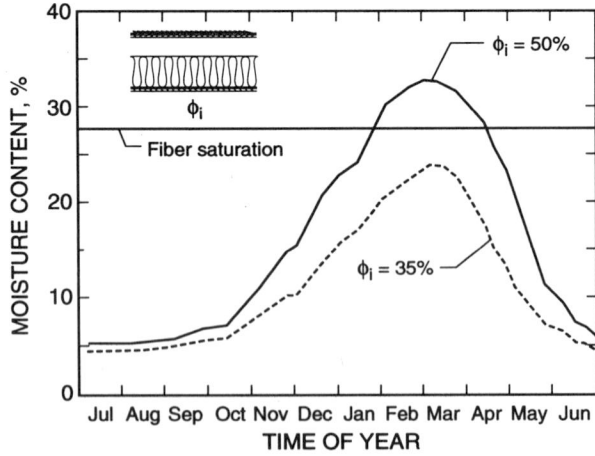

FIG. 9—*Moisture content of plywood roof sheathing plotted versus time of year for two indoor relative humidities (Madison, WI).*

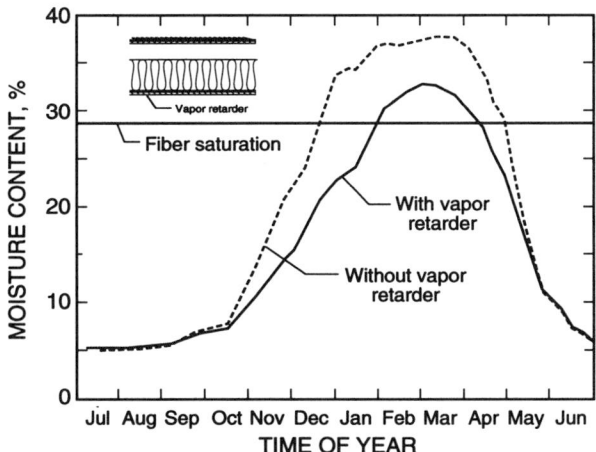

FIG. 10—*Moisture content of plywood roof sheathing plotted versus time of year for cases with and without a ceiling vapor retarder (Madison, WI).*

Ceiling Vapor Retarder—A computer simulation was conducted without a ceiling vapor retarder in the baseline roof construction. The results of this simulation are compared to the baseline construction with a vapor retarder in Fig. 10. As expected, the removal of the vapor retarder increases substantially both the peak and breadth of the profile. The HUD Standards currently require a ceiling vapor retarder. The results indicate this practice should continue.

Type of Roof Construction—Model MOIST was next used to analyze the moisture buildup for three roofing materials: exterior-grade plywood, oriented strand board (OSB), and galvanized steel roofing.

Figure 11a shows the results for plywood. The solid curve depicts the moisture content of a 2.4 mm (0.094 in.) thin surface layer, while the broken curve depicts the moisture content for the remaining bulk of sheathing. The moisture content of the thin surface layer is seen to follow closely that of the interior bulk layer, thereby indicating a small gradient in moisture content across the sheathing thickness.

Figure 11b shows similar results for OSB. This figure shows the thin surface layer has a considerably higher moisture content than the bulk layer during the winter. During this period, a significant gradient in moisture content exists across the thickness of the OSB sheathing, thereby providing a potential for buckling and warping.

The difference in behavior for the plywood and OSB sheathing was attributed to a difference in the permeability functions for the two materials. Figure 2b indicates that the permeability of plywood becomes large as the moisture content approaches fiber saturation. On the other hand, the permeability of OSB is considerably smaller. As a result, moisture at the surface of OSB is not readily transferred to its interior.

The moisture buildup at the lower surface of a galvanized steel roof is given in Fig. 11c. The horizontal line depicts an estimated amount of liquid water that the metal surface can retain without dripping off the surface.[3] Note that the peak moisture accumulation is about 2.1 kg/m² (0.43 lb/ft²) which corresponds to about 2.1 mm (0.08 in.) water. If this moisture

---

[3] The maximum amount of retainable water was determined by spraying water onto the bottom surfaces of several pieces of galvanized steel roofing.

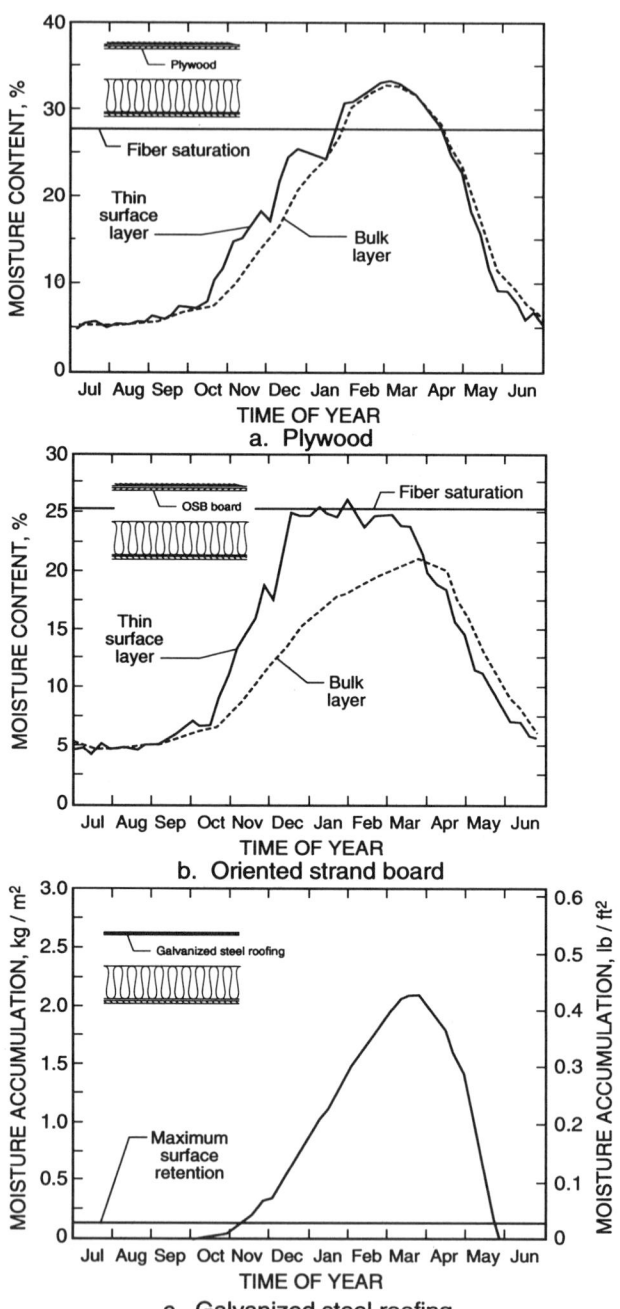

FIG. 11—*Moisture accumulation of roof plotted versus time of year for three roof constructions* (*Madison, WI*).

accumulates as frost and abruptly melts in the spring, it would drip through the fibrous insulation and puddle on the kraft paper. It is unlikely that the kraft paper could support this amount of puddled water. The gypsum board would then show stains where water leaked.

Effect of Other Parameters

Solar Absorptance of Roof—Three solar absorptances of the roof were analyzed: a light color ($\alpha = 0.3$), a typical medium-color ($\alpha = 0.7$), and a dark color ($\alpha = 0.9$). The results given in Fig. 12 indicate that lower moisture contents occur in darker roofs since they absorbed more solar radiation and dried more quickly.

Thermal Resistance of Ceiling Insulation—Two thermal resistance levels were analyzed: $R - 2.5$ m$^2 \cdot$ °C/W (14 h $\cdot$ ft$^2 \cdot$ °F/Btu) and $R - 5.3$ m$^2 \cdot$ °C/W (30 h $\cdot$ ft$^2 \cdot$ °F/Btu). The results given in Fig. 13 indicate that the addition of insulation has very little effect on roof moisture content.

Passive Moisture-Control Measures

The author followed the general approach of finding a combination of passive measures which would maintain the peak moisture content in hygroscopic roof sheathing and moisture accumulation at metal roofs below critical levels. The author gave preferential consideration to passive, as opposed to active, measures, because passive measures seem to be more likely to remain in effect during the life of the home. Passive measures also provide a lower first cost to the purchaser of the home.

For hygroscopic roof sheathing, a critical level was deemed to be fiber saturation. For metal roofs, we deemed a critical level to be a thickness of accumulated moisture (i.e., less than 0.8 mm or 1/32 in.) judged by the author to pose little or no condensation risk to the roof construction.

The following combination of passive measures was found to maintain the peak moisture content in the three roofs below critical levels:

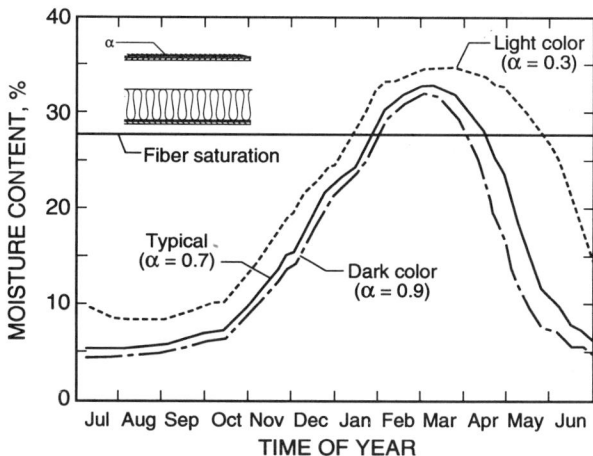

FIG. 12—*Moisture content of plywood roof sheathing plotted versus time of year for three solar absorptances (Madison, WI).*

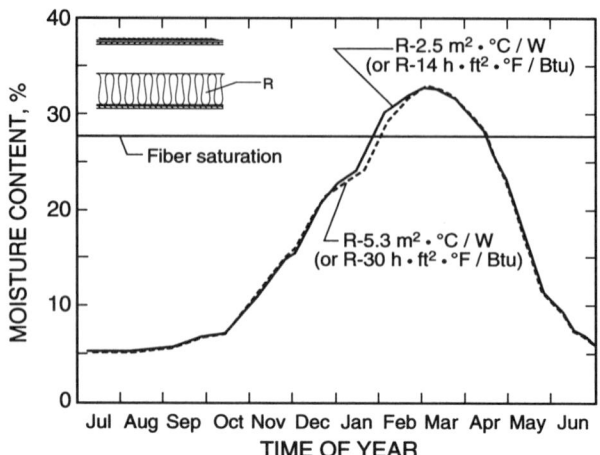

FIG. 13—*Moisture content of plywood roof sheathing plotted versus time of year for two insulation levels (Madison, WI).*

- a ceiling vapor retarder having a permeance less than 5.7×10^{-11} kg/Pa · s · m² (1 perm);
- sealing air leakage paths in the ceiling construction and attaining an exfiltration rate into the roof cavity less than 1.3×10^{-5} m³/s per m² (0.15 ft³/h per ft²);
- providing ventilation openings in the roof cavity and achieving a natural ventilation rate of 8.5×10^{-5} m³/s per m² (1.0 ft³/h per ft²).

The current HUD Standards require a ceiling vapor retarder, but do not require compliance with the other two recommended measures. The effectiveness of the recommended measures are analyzed below.

Plywood Roof Sheathing—Figure 14 shows the effectiveness of the three passive measures for plywood roof sheathing exposed to the four winter climates. In each plot, the upper curve is for a vapor retarder permeance of 5.7×10^{-11} kg/Pa · s · m² (1.0 perm), while the lower curve is for a vapor retarder permeance of 5.7×10^{-12} kg/Pa · s · m² (0.1 perm). A lower vapor retarder permeance was analyzed to investigate its merit.

In Fig. 14, the three passive measures are seen to maintain the peak moisture content considerably below fiber saturation. Since the three passive measures are very effective, it is unnecessary to obtain further reductions in the moisture content by decreasing the permeance of the vapor retarder.

Oriented Strand Board Roof Sheathing—Similar results for the OSB roof sheathing exposed to the climate of Madison are given in Fig. 15. Since moisture contents are highest in Madison, the results for only Madison are presented. These results indicate that the three passive measures maintain the peak moisture content below fiber saturation. Separate curves are given for a thin surface layer and a bulk interior layer, in order to illustrate that a gradient in moisture content still exists across the thickness of the OSB sheathing.

Galvanized Steel Roofing—Similar results are given in Fig. 16 for a galvanized steel roof exposed to the climate of Madison. The peak moisture accumulation is 0.41 kg/m² (0.082 lb/ft²) for a ceiling vapor retarder permeance of 5.7×10^{-11} kg/Pa · s · m² (1.0 perm). The fact that moisture accumulates above maximum surface retention for a three-month period means that the accumulated moisture will drip downwards onto the vapor retarder. However,

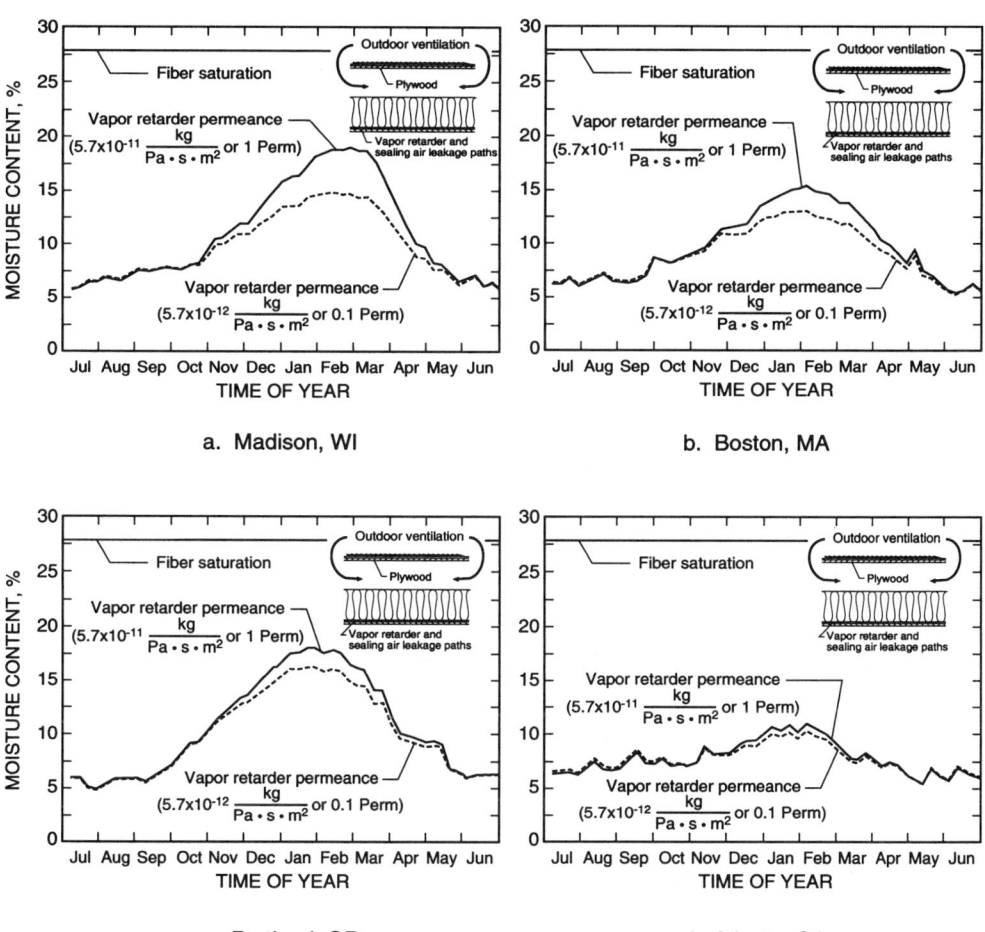

FIG. 14—*Effectiveness of recommended practices for plywood roof construction (i.e., sealing air leakage paths in the ceiling, naturally ventilating the roof cavity, and providing a ceiling vapor retarder).*

this amount of water corresponds to a thickness of 0.4 mm ($\frac{1}{64}$). It was believed that this amount of water poses little or no risk to the roof construction.

Hot and Humid Climate Analysis

Model MOIST was next used to predict the performance of the roof construction exposed to a hot and humid climate (i.e., Lake Charles, LA). In the analysis, the indoor temperature and relative humidity were 24°C (76°F) and 50%, respectively.

Unventilated Roof Construction

First, roof construction in compliance with the current HUD Standards was considered. The construction given in Fig. 1 was used. A kraft-paper vapor retarder is installed in the ceiling, but openings are not provided to naturally ventilate the roof cavity.

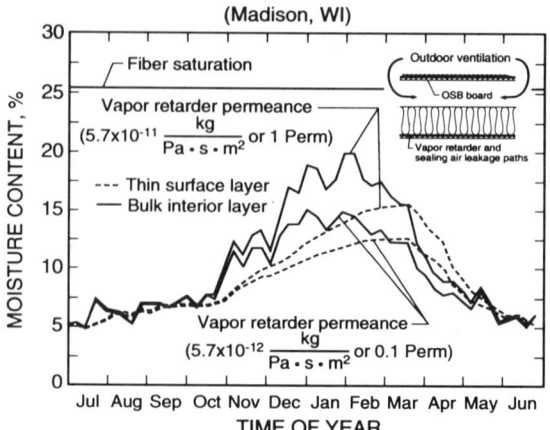

FIG. 15—*Effectiveness of recommended practices for OSB roof construction (i.e., sealing air leakage paths in the ceiling, naturally ventilating the roof cavity, and providing a ceiling vapor retarder).*

The weekly-average relative humidity at the upper and lower surfaces of the kraft-paper are plotted versus time of year in Fig. 17a. At the upper surface, the relative humidity rises and reaches a peak during the summer. On the other hand, the relative humidity at the lower surface departs very little from the indoor value of 50%.

The solid horizontal line in Fig. 17a depicts a critical 80% level believed to coincide with the "onset for mold and mildew growth." The International Energy Agency has recently published Guidelines and Practices (Volume 2) for preventing mold and mildew growth at building surfaces [11]. This consensus document indicates that a monthly-mean surface relative humidity above 80% is conducive to mold and mildew growth. Note that the peak relative humidity at the upper surface of the kraft paper is below the critical 80% level. Therefore, mold and mildew growth is unlikely to occur in an unventilated roof cavity.

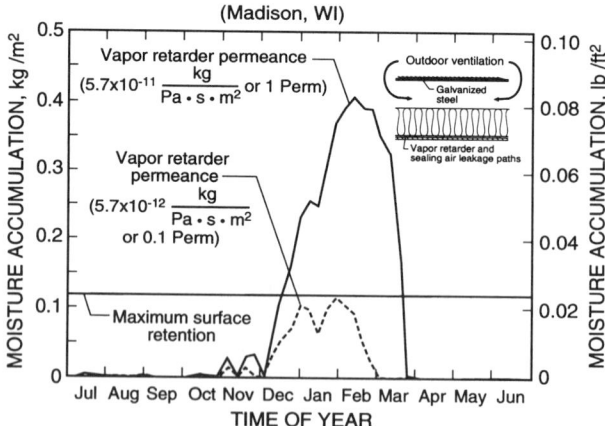

FIG. 16—*Effectiveness of recommended practices for galvanized steel roof construction (i.e., sealing air leakage paths in the ceiling, naturally ventilating the roof cavity, and providing a ceiling vapor retarder).*

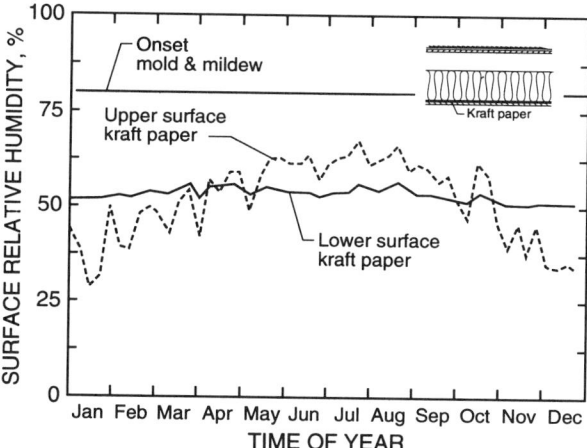

a. Weekly average surface relative humidity

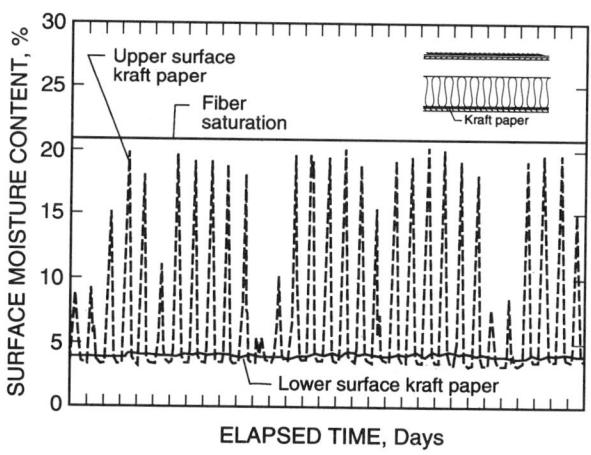

b. Hourly surface moisture content for July

FIG. 17—*Moisture content of kraft paper plotted versus time for baseline construction exposed to hot and humid climate (unvented roof cavity).*

The hourly moisture content at the upper and lower surfaces of the kraft paper is plotted versus time for the month of July in Fig. 17b. At the upper surface, the moisture content undergoes large diurnal fluctuations. During warm day periods, moisture is transferred downwards and accumulates at the upper surface of the kraft paper which is cooled by indoor air conditioning. However, the moisture content never reaches fiber saturation (21%), indicating that liquid water is never present. TenWolde and Mci experimentally observed similar diurnal humidity fluctuations in walls [12].

Ventilated Roof Cavity

Next, model MOIST was used to develop a similar pair of plots for a roof cavity ventilated at 2 ach. The results are given in Fig. 18.

The weekly-average relative humidity at the upper surface of the kraft paper rises above the critical 80% level for a two-month summer period (Fig. 18*a*). The moisture content at the upper surface sometime reaches fiber saturation (Fig. 18*b*). Such an environment is conducive to mold and mildew growth. It is possible that the operation of fans that ventilate the interior will cause air from the roof cavity to infiltrate and transport fungal spores to the indoors. This could result in an indoor air quality problem (i.e., musty odor).

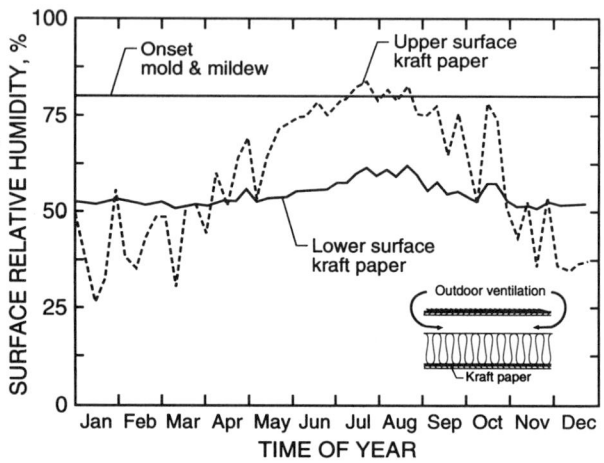

a. Weekly average surface relative humidity

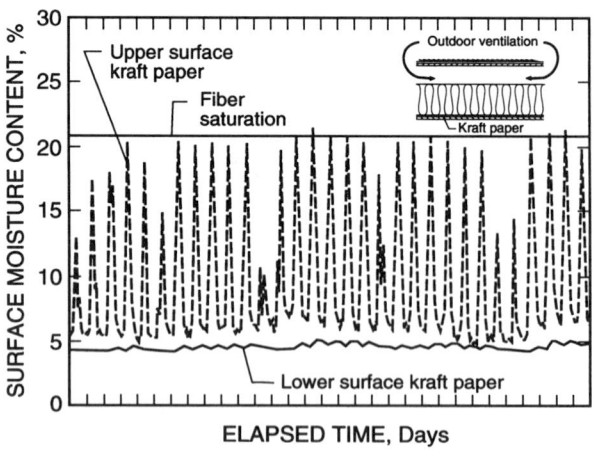

b. Hourly surface moisture content for July

FIG. 18—*Moisture content of kraft paper plotted versus time for baseline construction exposed to hot and humid climate (ventilated roof cavity).*

Based on the above results, the author recommends that roof cavities of manufactured housing not be ventilated in hot and humid climates.

Needs for Future Research

The author recommends that the findings of this theoretical study be corroborated by a comprehensive experimental study prior to implementation of rule changes for the HUD Standards for manufactured housing. Important aspects of this research would be to (1) measure the seasonal variation in moisture content of the roof sheathing in two side-by-side manufactured houses: one, a current practice house; the other, an identical house with the recommended moisture-control measures implemented, (2) measure outdoor air exchange rates for a roof cavity under a range of outdoor temperature and wind speeds and establish a relationship between the net free ventilation opening and the corresponding air exchange rate, and (3) measure the indoor air exfiltration rate into the roof cavity.

Summary and Conclusions

A detailed computer analysis was conducted of the combined transfer of heat and moisture in the roof construction of manufactured housing using hourly weather data for four winter climates: a cold winter climate (Madison, WI), an intermediate winter climate (Boston, MA), a mild winter climate (Atlanta, GA), and a Pacific northwest climate (Portland, OR).

The current HUD Standards for manufactured housing require that a ceiling vapor retarder be installed, but they do not require that the roof cavity be ventilated with outdoor air. In homes constructed to this standard, the computer predictions revealed that a detrimental amount of moisture accumulated at the roof sheathing of homes located in cold winter and intermediate winter climates. In plywood and oriented strand board roof sheathing, the peak moisture content during the winter rose above fiber saturation, indicating the presence of free liquid water in the pore structure of the materials. In this situation, degradation of the roof sheathing may occur. In metal roofs, a significant amount of moisture accumulated at the underside of the roof surface, which may drip downward wetting and staining the ceiling construction.

A sensitivity analysis was conducted to determine the effect of various parameters on roof moisture accumulation in cold climates. Parameters having a significant effect on the moisture accumulation in the roof sheathing included coldness of the climate, the airtightness and permeability of the ceiling construction, the outdoor ventilation rate of the roof cavity, indoor relative humidity, and roof type. Factors having a less important effect were the thermal resistance of the ceiling insulation and the solar absorptance of the roof.

The analysis revealed that the following combination of passive practices will prevent detrimental winter moisture accumulation at the roof sheathing of manufactured housing:

1. providing a ceiling vapor retarder having a permeance of 5.7×10^{-11} kg/Pa $\cdot$ s $\cdot$ m^2 (1 perm) or less,
2. sealing penetrations and openings in the ceiling construction, and
3. providing natural ventilation openings in the roof cavity.

Computer simulations were also carried out to investigate the performance of a roof cavity exposed to a hot and humid climate using weather data for Lake Charles, LA. In an unventilated attic, the accumulation of outdoor moisture at a ceiling vapor retarder, cooled by indoor air conditioning, was not a problem. However, when the roof cavity was naturally ventilated, intermittent wetting of the vapor retarder occurred during warm day periods. In this situation, the monthly mean relative humidity at the upper surface of the vapor retarder rose above 80% during the summer, thereby posing a risk for mold and mildew growth. For this reason, the author recommends that the roof cavities of manufactured homes not be ventilated in hot and humid climates.

It is recommended that a comprehensive experimental study be conducted to corroborate the theoretical findings of the present study.

Acknowledgments

The author would like to thank William Freeborne of the Division of Innovative Technology of the U.S. Department of Housing and Urban Development for funding this project and making many helpful suggestions during the review of the report. The author would also like to thank Felix Castillo of the National Conference of States on Building Codes and Standards, Inc. for providing information on the construction of manufactured housing.

References

[1] Harrje, D. T., Gibson, R. G., Jacobson, D. J., et al., "Field Measurement of Seasonal Wood Moisture Variations in Residential Attics," Princeton University Center for Energy and Environmental Studies, 1984.

[2] Department of Housing and Urban Development, "Manufactured Home Construction and Safety Standards," *Code of Federal Regulations,* Title 24, Part 3280, 1987.

[3] Zieman, M. L. and Waldman, J. D., "Moisture Problems in Mobile Homes," *Final Report for HUD Contract H-10992,* Department of Housing and Urban Development, 10 July 1984.

[4] Lee, T. G., "Condensation in Manufactured Housing," *Report 0-88654-173-5,* Alberta Municipal Affairs Housing Division, Edmonton, Alberta, Canada, February 1987.

[5] Burch, D. M. and Thomas, W. C., "An Analysis of Moisture Accumulation in a Wood Frame Wall Subjected to Winter Climate," *NISTIR 4674,* National Institute of Standards and Technology, October 1991.

[6] Goldschmidt, V. W. and Wilhelm, D. R., "Relation of Infiltration to Weather Parameters for a Mobile Home," *ASHRAE Transactions,* Vol. 87, Pt. 2, 1981.

[7] Crow, L. W., "Development of Hourly Data for Weather Year for Energy Calculations (WYEC)," *ASHRAE Journal,* Vol. 23, No. 10, October 1981, pp. 37–41.

[8] Richards, R. F., Burch, D. M., and Thomas, W. C., "Water Vapor Sorption Measurements of Common Building Materials," *ASHRAE Transactions,* Vol. 98, Pt. 2, 1992.

[9] Burch, D. M., Thomas, W. C., and Fanney, A. H., "Water Vapor Permeability Measurements of Common Building Materials," *ASHRAE Transactions,* Vol. 98, Pt. 2, 1992.

[10] TenWolde, A. and Carll, C., "The Effect of Cavity Ventilation on Moisture in Walls and Roofs," *Proceedings of the ASHRAE/DOE/BTECC Conference on Thermal Performance of Exterior Envelopes of Buildings V,* Clearwater Beach, FL, 7–10 December 1992.

[11] International Energy Agency, "Guidelines and Practice," *Annex XIV (Condensation and Energy),* Vol. 2, March 1990.

[12] TenWolde, A. and Mei, H. T., "Moisture Movement in Walls in a Warm Humid Climate," *Proceedings of the ASHRAE/DOE/BTECC Conference on Thermal Performance of the Exterior Envelopes of Buildings III,* ASHRAE SP 49, Clearwater Beach, FL, 2–5 December 1985.

Discussions

Stephen N. Flanders[1] (written discussion)—You advocate three measures for cold climates. One of them, roof ventilation, is counterproductive in warm climates. What happens when one provides a vapor retarder, seals leaks, but omits ventilation in cold conditions?

---

[1] USA CRREL, Hanover, NH 03755.

D. Burch (author's closure)—In response to this question, the author used MOIST to investigate the effectiveness of an alternate set of practices which included installing a ceiling vapor retarder, providing supplemental ventilation of the indoors to comply with the ASHRAE Ventilation Standard 62 (i.e., 0.35 ach), and sealing air leakage sites in the ceiling construction. Here the roof cavity was assumed not to be ventilated, and the roof sheathing was plywood. The climate of Madison, WI was used as a worst-case cold climate.

The simulation results revealed that the alternate set of practices maintained the peak moisture content of the plywood roof sheathing considerably below fiber saturation, thereby posing little or no risk of material degradation. In particular, ventilating the indoors to comply with the ASHRAE Ventilation Standard 62 reduced the relative humidity of the indoor space, making it unnecessary to ventilate the roof cavity.

Anton TenWolde[2] (written discussion)—What was the orientation of the roof in your analysis?

Author's closure—The roof was treated as a horizontal surface.

G. Proskiw[3] (written discussion)—Does your model assume uniform air exfiltration across the ceiling or does it assume point or linear air exfiltration?

Author's closure—Air exfiltration across the ceiling was modeled as a constant flow rate of indoor air to a finite-difference node at the lower surface of the roof sheathing.

---

[2] Forest Products Laboratory, Madison, WI, 53705-2398.
[3] Proskiw Engineering Ltd., Winnipeg, Manitoba, Ontario, Canada.

D. E. Claridge,[1] M. Liu,[1] and S. Bhattacharyya[2]

Impact of Air Infiltration in Frame Walls on Energy Loads: Taking Advantage of the Interaction between Infiltration, Solar Radiation, and Conduction

REFERENCE: Claridge, D. E., Liu, M., and Bhattacharyya, S., **"Impact of Air Infiltration in Frame Walls on Energy Loads: Taking Advantage of the Interaction between Infiltration, Solar Radiation, and Conduction,"** *Airflow Performance of Building Envelopes, Components, and Systems, ASTM STP 1255,* Mark P. Modera and Andrew K. Persily, Eds., American Society for Testing and Materials, Philadelphia, 1995, pp. 178–196.

ABSTRACT: The energy impact of air leakage has classically been treated independently of conduction gains and losses through building envelopes. However, an increasing body of evidence shows that heat exchange in envelope components resulting from interactions between air leakage and conduction and solar radiation systematically changes (generally reducing) the energy load of air infiltration. A systematic study performed by the authors shows that this heat exchange can substantially reduce the air infiltration energy requirements in frame structures or walls under laboratory conditions depending on airflow rate and leakage configurations. The interaction of airflow with solar radiation is an important consideration—perhaps as important as the interaction of airflow with conduction under outdoor conditions, and depressurizing a house during winter and pressurizing during summer using a fan may reduce energy consumption significantly. This paper reviews earlier theoretical and experimental work in this area and summarizes results of a series of experiments and theoretical work conducted by the authors.

KEYWORDS: energy calculation, heat recovery, airflow, conduction, radiation

Building components sustain air infiltration due to natural forces or mechanically induced pressure differences. This air infiltration affects component heat loss, temperature distribution, and moisture distribution. If cold air infiltrates through a porous material to a room, its temperature may be close to room temperature before it enters the room. Therefore, air infiltration energy consumption may be substantially smaller than the classical value, $MC_p(T_{room} - T_{out})$, although the conduction loss may be increased due to a steeper temperature gradient near the inner surface caused by airflow. If warm air exfiltrates from a room to outside, the temperature gradient will be smaller than that when there is no air exfiltration. Therefore, this airflow can reduce the conduction heat loss. Airflow through building components always results in heat loss reduction compared with the classical calculation method when solar radiation is not present.

[1] Professor and assistant research scientist, Energy Systems Laboratory, Texas A&M University, College Station, TX 77843-3123.

[2] Assistant professor, Department of Mechanical Engineering, Indian Institute of Technology, Kharagpur, India 721302.

The authors have systematically studied air infiltration energy consumption since 1986, carried out a number of theoretical and experimental studies, and introduced the concept of air Infiltration Heat Exchange Effectiveness (IHEE), which is defined as:

$$\text{IHEE} = \frac{UA_0 + MC_p - UA_{\text{actual}}}{MC_p} \tag{1}$$

where UA_0 is the heat loss factor (W/K) due to conduction only, M is airflow rate (Kg/s), C_p is the specific heat $(J/Kg\ K)$ of air, and UA_{actual} (W/K) is the actual total heat loss factor at airflow rate M. IHEE represents air infiltration heat recovery as a fraction of classical infiltration energy consumption. Consequently, it is suggested that air infiltration energy consumption be calculated as:

$$Q_{\text{inf}} = (1 - \text{IHEE})MC_p(T_{\text{room}} - T_{\text{out}}) \tag{2}$$

When solar radiation is present, however, infiltration reduces heat loss in winter, but may increase cooling energy consumption in summer; exfiltration reduces cooling energy, but may increase heating consumption because infiltration carries more solar energy from the exterior surface into the buildings while exfiltration rejects more solar energy outside. A better understanding of this combined effect can help: (1) heating, ventilation and air-conditioning (HVAC) load determination and system sizing; (2) infiltration retrofitting to save energy; (3) develop new techniques for building energy conservation; and (4) improve air quality and reduce building material damage associated with moisture condensation.

This paper gives a brief review of air infiltration energy consumption research, and presents the results of infiltration heat exchange measurements under controlled steady-state and dynamic conditions. It then presents results from outdoor measurements which show the influence of the solar interaction with infiltration, and concludes with modeling results which quantify the impact of the combined interaction of air leakage with conduction and absorbed solar radiation.

Overall Review

The energy impact of air infiltration on double-frame windows, porous insulation, and wood-frame walls was studied and reported by a number of researchers. Bursey and Green [1] measured the overall heat loss factor of the double frame window with a hot box, where both infiltration and temperature difference were present. It was found that the overall heat loss factor U was 10% to 30% lower than the classical value. Unfortunately, due to the lack of a logical way to express this air infiltration heat recovery, it was concluded that it would be difficult to incorporate it into an industry standard. Guo and Liu [2] performed both theoretical analysis and experimental testing on double frame windows. Their work established correction equations for U and air infiltration energy consumption, and good agreement was found between the test data and the model prediction. But neither of these studies accounted for solar radiation.

Anderlind [3] proposed concentrated-flow and diffuse-flow concepts, in which concentrated-flow (CF) was defined as a short pass flow, such as airflow through doors, windows, and large cracks or holes. Diffuse-flow (DF) was defined as long pass flow in which the air travels some distance (metres) before leaving the wall; alternatively, uniform flow where both air and solid wall material have the same temperature at any position also corresponds to diffuse flow. Anderlind demonstrated theoretically that air infiltration heat recovery was influenced by the type of the leakage configuration (CF or DF). The combined conduction/ air leakage heat transfer models for diffuse walls and roofs can be found in a number of sources [3–7] and are almost identical. The models assume uniform infiltration, identical air and solid temperature at any position, and one-dimensional steady-state conditions. The models are generally expressed by a simple explicit analytical formula which predicts significant heat recovery in walls and roofs.

Airflow windows, where either fresh or exhausted air is made to pass through the air space between panes of glass, originated in Scandinavia and have been successfully used in northern climates for almost three decades. These windows have proven useful in these cold regions by providing maximum space comfort and energy savings to building owners throughout northern and central Europe, and more recently in the northern states of the United States. Ripatti [8] investigated the application of the airflow window in a hot climate, and found that the air flow window could provide comfortable indoor thermal conditions in both arid and tropical climates, with lower cooling load and energy consumption.

A dynamic insulation system was proposed and studied [5,9–11] where the interaction of heat and mass transfer was used to reduce the house heating energy consumption. Prototype dynamic insulation houses were built in France and Canada [5,7]. Heating energy savings of up to 23% were measured, but no detailed theoretical analysis was presented, and the importance of each of the weather and insulation parameters was not clear.

The OPTIMA (Optimum) ceiling and floor, which evolved from the dynamic insulation concept, was tested by Andersson and Wadmark [12]. Fresh air was taken into the house through the roof structure and infiltrated through the ceiling insulation and a control film to a collecting space in the bottom of the ceiling; from there it was introduced in the living room and bedrooms by special air supply units. After use in these rooms the air was passed on to the bathroom and lavatory through slots under the doors. From there, it was drawn to the main ventilator and distributed under the entire house in the upper part of the floor structure. Then it leaked through a polymer fabric to the lower part, where it was drawn upwards by an exhaust fan. The test showed that the OPTIMA ceiling and roof could recover most of the ceiling heat loss in a single-family house.

Masoero and Aghemo [13] analyzed a ventilated curtain-wall, which consisted of an internal concrete load-bearing slab and an external concrete finish panel separated by a ventilated air cavity that could be put in communication either with the outdoor or the indoor environment through a set of thermostatically controlled dampers. During the heating season, in the presence of sufficient solar radiation, natural circulation of indoor air takes place in the cavity. Alternatively, the component may be coupled to a mechanical ventilation system, acting as a preheater of incoming outdoor air. In the summer the air cavity is open to the outside, and the natural air circulation through the cavity removes part of the solar heat gain. The analysis showed significant energy savings, and improved comfort conditions in the unconditioned space (no heating, no cooling, no mechanical ventilation).

Homma and Guy [14] investigated the effects of ventilation in the space between a main wall and exterior siding on the cooling load. The results showed that ventilation of this air space had the potential to reduce the radiation heat gain of both opaque walls and of triple glass windows.

Air infiltration heat recovery has been observed in numerous houses where it reduced loads. Beyea et al. and Harrje et al. [15,16] found that the temperature in attics was higher than that predicted by resistance models of attic insulation. Claridge et al. [17] found that attic temperature in nine of 25 houses had less than half the temperature drop expected across attic insulation, resulting from the air bypassing the insulation. Claridge et al. [18] also found that the overall house loss factor determined by engineering calculation was about

50% higher than that obtained by regressing the measured data; these observations stimulated the investigation reported in the present chapter.

Sonderegger, Condon, and Modera [19] measured the air infiltration load with the electric co-heating method; the air infiltration rate was stepped up or down and the load response was monitored. It was found that the heat load responded to the infiltration rate change very slowly, with a delay of about two hours, and, moreover, that both the peak load and total energy consumption during this dynamic period were less than the calculated values. This effect was explained as a regenerative heat exchanger. However, this result appears to be consistent with air infiltration heat recovery due to interaction between conduction and airflow possibly supplemented by additional solar energy gain due to the infiltrating air. Because of the relatively long time delay of the load response, it seems obvious that the air infiltration was diffuse-flow dominated, so significant heat recovery should be expected for this kind of natural infiltration.

The effects of combined heat transfer on the overall building have also been studied. Guo, Xu, and Li [20] corrected classical air infiltration energy consumption by a factor of 0.45 in a four-story dormitory energy balance measurement, where total heat supply and electricity were measured, and energy consumption from conduction, radiation and air infiltration rate was also measured. Measurement showed a $\pm 5\%$ agreement between consumption and heat supply in a heating season. Kohonen, Tech, and Virtanen [22] studied numerically and experimentally the overall energy recovery effect of air infiltration in a house. They suggested a correction factor of 0.8 for the opaque envelope conduction to account for the reduced energy consumption during the heating season caused by the interaction between conduction and air leakage.

Recently, the authors have performed a systematic study of air infiltration energy consumption which includes the following measurements: (1) indoor test cell experiments [23] in which the impact of airflow rate and leakage configuration on the energy consumption were investigated under controlled laboratory conditions; (2) frame wall tests [24] in which the impact of airflow rate and leakage configuration on energy load were investigated in a section of standard frame-wall under steady-state conditions; (3) indoor cell dynamic tests [25] in which heat recovery was investigated under dynamic temperature conditions; (4) outdoor test cell measurements [26] in which the impact of airflow rate and direction, leakage configuration and solar radiation on energy load were investigated under outdoor conditions; (5) house tests in which the impact of airflow direction and airflow rate and retrofit measures on energy consumption are investigated; (6) a theoretical model study [27] in which combined heat conduction/air leakage/solar radiation heat-transfer models were established for building systems, and these models were used to simulate houses under different climate conditions; (7) a measurement methodology study [28,29] in which short-term measurement methods were developed for air infiltration heat recovery measurement under dynamic weather conditions; and (8) an analysis methodology study [30] in which the systematic error of the steady-state methodology due to nonlinearity in the combined heat-transfer process was investigated. This systematic series of studies has emphasized a fundamental understanding of air infiltration energy consumption, energy performance under different weather conditions, and the impact of individual weather parameters in traditional frame wall construction. This paper describes the experimental and theoretical work (topics 1–4 and 6) and presents the major findings of this work.

Indoor Cell Test

A set of experiments has been conducted to measure the energy impact of controlled amounts of infiltration air in a small test cell under steady-state conditions. The test cell, 1.4

m wide by 1.2 m high by 2.4 m long (4.6 by 4 by 8 ft), was constructed using standard frame construction for the six wall, ceiling and floor surfaces, and consisted of:

- 9.5 mm (0.37 in.) plywood sheathing
- 3.8 × 8.9 (1.5 × 3.5 in.) cm studs
- R-11 fiberglass bat insulation between studs
- 9.5 mm (0.37 in.) plywood sheathing

One of the 1.4 m by 1.2 m (4.6 by 4 ft) end-walls contained a removable 0.6 m square window glazed with 9.5 mm Plexiglas.® This aperture served as the door to the test cell between experiments. All joints between the walls and all visible cracks in the wood were tightly caulked to minimize air leakage. The test cell had an air change rate of 0.25 per hour under this leakage configuration when the pressure difference was about 4 Pa (0.000 039 2 atm).

Air inlet, outlet, and inside cell temperatures are measured by nine thermocouples, the airflow rate is measured by a rotameter before it enters the test cell, and the power input for the test cell is determined by measuring the voltage, current, and power factor.

The various holes provided in the test cell for air inlet and outlet are illustrated in Fig. 1. Holes P [1.3 cm (0.5 in.) in diameter] and E [3.8 cm (1.5 in.) in diameter] penetrate the wall, while hole B [1.3 cm (0.5 in.) in diameter] penetrates exterior and interior plywood only at well-separated locations. When no exit hole is indicated, air exfiltrates through nat-

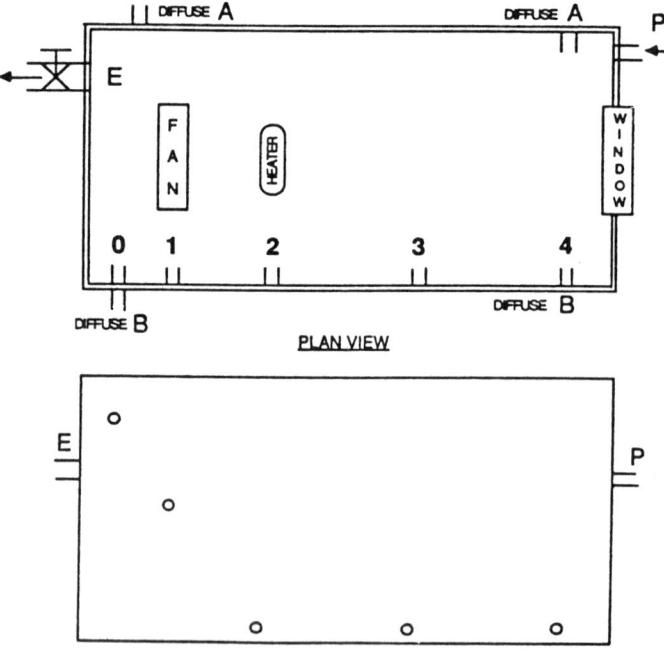

FIG. 1—*Schematic of indoor test cell.*

urally occurring leaks in the test cell. The air infiltration energy consumption was investigated for the airflow configurations given in Table 1.

The heat loss factors (*UA*) were measured for these leakage configurations with different airflow rates under steady-state conditions. The heat loss factor (UA_0) at zero airflow rate was also measured. Then the heat recovery due to air infiltration was determined according to Eq 1.

Figure 2*a* shows IHEE for case 1 as a function of nondimensional airflow rate, which was defined as the ratio of MC_p to UA_0. This leakage configuration simulates a tight house where air infiltrates through part of the envelope and exfiltrates through the other part diffusely. This figure shows that IHEE decreases from 0.8 to 0.6 when the nondimensional airflow rate varies from 0.05 to 0.25. Cases 2 through 6 gave similar results.

Figure 2*b* shows IHEE for case 10 as a function of nondimensional airflow rate. This leakage configuration has concentrated infiltration and diffuse exfiltration, which may simulate normal houses where both concentrated and diffuse airflows exist. The figure shows that IHEE changes from 0.5 to 0.3 when the nondimensional airflow rate varies from 0.05 to 0.25. Cases 7, 8 and 9 have similar IHEE results. Cases 7 and 8 have diffuse infiltration but concentrated exfiltration, while case 9 again has concentrated infiltration and diffuse exfiltration.

Figure 2*c* shows IHEE for case 12 as a function of nondimensional airflow rate. This leakage configuration has largely concentrated infiltration and exfiltration, which simulates a very leaky house. IHEE is close to 0.2 for nondimensional airflow from 0.05 to 0.2. The results for case 11 are similar to this case since concentrated airflow dominated both infiltration and exfiltration.

The measurements show that airflow through frame construction can substantially reduce the energy requirements due to infiltration, and IHEE depends on airflow rate and leakage configurations.

Frame Wall Test

A modified calibrated hot box with the associated instrumentation was designed and constructed for testing a single stud-cavity. The standard ASTM Test Method for Thermal Performance of Building Assemblies by Means of a Calibrated Hot Box (ASTM C 976-82) was modified to incorporate airflow through the test specimen.

A single 2.4-m-high (8 ft) stud wall section filled with glass fiber blanket insulation (Fig. 3 served as the test specimen. The design of the enclosure (Fig. 3) around the stud-cavity incorporated extruded polystyrene and poly-isocyanurate board insulation covered with plywood casing. A chilled brine cooling system and a radiant heating panel provided the conditioning needed for the hot box. Air was circulated using small fans to maintain temperature constant within 0.5°C over the area of the test specimen on either side. A proportional control loop was employed to control the cold space temperature within ±0.5°C of the set point. The air supply to the stud-cavity wall specimen was metered and controlled by a mass flow controller.

TABLE 1—*Airflow configuration tested in indoor test cell measurement.*

Case	1	2	3	4	5	6	7	8	9	10	11	12
Entry	B	B0	B1	B2	B3	B4	B	B	P	P	P	P
Exit	A	···	···	···	···	···	E	E&A	···	A&B	E	A,B,E

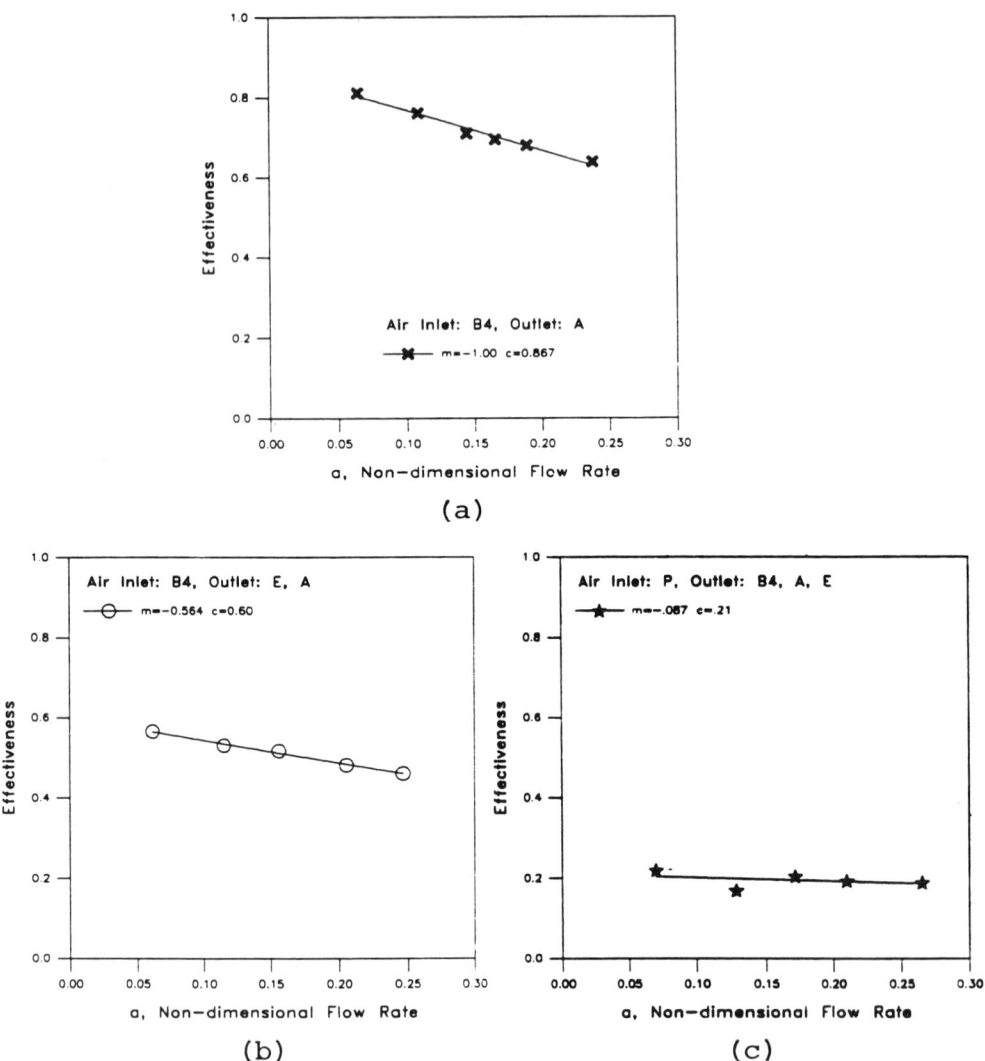

FIG. 2—*IHEE versus nondimensional airflow rate (indoor test cell results).*

IHEEs were measured for five different leakage configurations (Table 2) under steady-state conditions, after temperature distributions within both the specimen and hot box cavity had reached quasi-steady state.

Leakage configurations A–D were expected to provide diffuse flow since the air had to flow more than 1 m (3.28 ft) to pass through the specimen. For configuration E, air may cross the specimen following a path as short as 0.1 m (0.33 ft) (thickness of the wall).

IHEE is shown as a function of the airflow rate for configurations B–E in Fig. 4. Figure 4a shows IHEE results for configuration B, which uses the whole height of the test specimen for heat exchange. IHEE varies from 0.65 to 0.37 when the airflow rate ranges from 4 to 14 SLPM. Configuration A had a similar flow path/area and similar IHEE results.

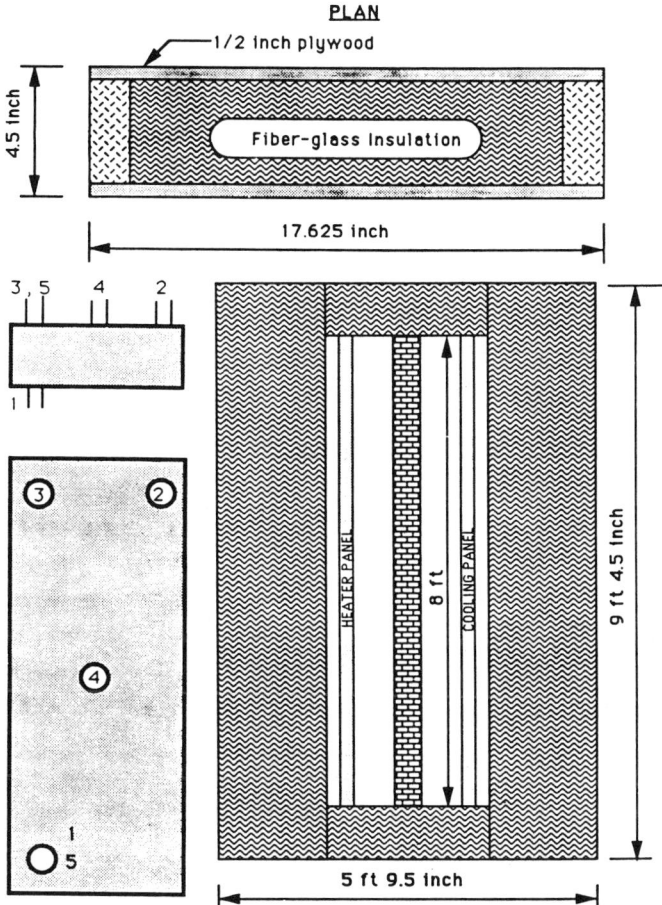

FIG. 3—*Schematic of hot box and specimen.*

Configuration C provides about half of the heat exchange path/area of A and B and hence lower IHEEs are shown in Fig. 4d. Configuration D has three exit holes as shown and shows heat exchange between that of configurations B and C. Configuration E has a very short pass, where the outlet is located directly across the wall thickness from the inlet. However, IHEE values are not as low as might be expected (Fig. 4c). This demonstrates that heat exchange not only occurs within the hole area but also results from diffuse circulation through a significant volume of the insulation. This suggests that heat exchange may be important for similar holes in buildings.

TABLE 2—*Airflow configurations in frame wall test.*

Case	A	B	C	D	E
Entry (hole)	1	1	1	1	1
Exit (hole)	2	3	4	2,3,4	5

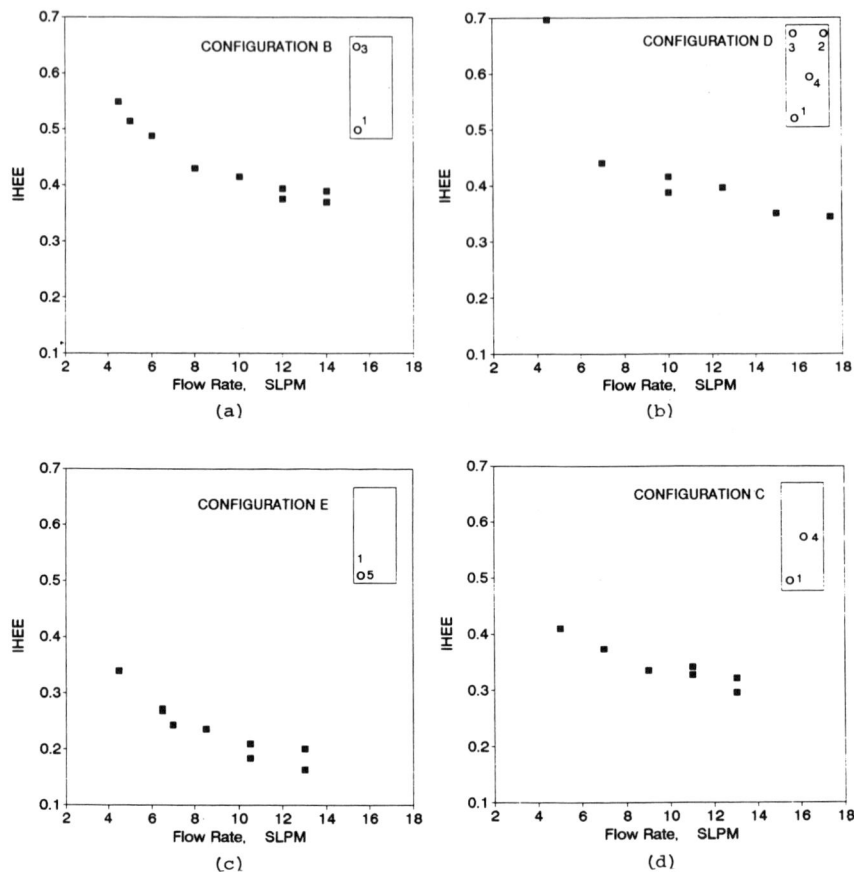

FIG. 4—*IHEE versus airflow rate (frame wall results).*

Measurement shows that airflow through a frame wall can exhibit significant heat exchange; IHEE increases as the airflow rate decreases and as the affected area or "path length" increases.

Dynamic Indoor Cell Tests

To validate the steady-state measurements under dynamic temperature conditions, the indoor test cell was modified, a test methodology was developed, and IHEE was measured for four different air leakage configurations and different airflow rates under dynamic conditions.

The rotameter used for flow measurement in the earlier tests was replaced by a calibrated orifice to permit continuous measurement of airflow rate, and the internal heating system was modified to provide a more uniform temperature distribution. The room temperature was measured by ten thermocouples surrounding the test cell to provide a better record of mean room temperatures.

IHEEs were measured by using a standard test profile for four different leakage configurations (Table 3). The basic dynamic test profile imposed on the indoor cell is shown in Fig. 5. The profile consists of three periods: (1) a quasi-steady period (SS-1), where 250 W heat

TABLE 3—*Air leakage configuration in indoor cell dynamic test.*

Case	1	2	3	4
Entry (hole)	P	P	P	A&B
Exit (hole)	E	...	A&B	...

input is supplied for 24 h; (2) a dynamic period, where heat input is switched off and on at 3-h intervals for a total of 15 h; and (3) a second quasi-steady period (SS-2), where 250 W heat input is supplied for 36 h.

Two steady-state periods, where the test cell temperature was stable, were chosen from SS-1 and SS-2. Then IHEE was determined from each of these periods using a procedure described in Ref 25. These IHEE values are called steady-state values because they are measured from quasi-steady-state periods. The dynamic IHEE value is determined from a period which started at the end of the steady-state period SS-1 and ended at the start of the steady-state period SS-2.

The measured IHEEs for case 1 from both steady-state and dynamic periods are shown as a function of nondimensional airflow rate in Fig. 6. Figure 7 shows results for configurations 1–4. Note that configuration 4 provides nearly twice the heat recovery since both the infiltration and exfiltration recover heat. These results show that IHEEs from the dynamic period are similar to the steady-state results within ± 12%; it is believed that these differences are due to measurement error, possibly influenced by time-dependent moisture content of the test cell. It is concluded that steady-state measurement results are valid under the dynamic temperature conditions when constant airflow is present.

Outdoor Cell Test

The outdoor test cell was built in 1990 especially for the measurement of air infiltration heat recovery. The cell is a 2.4 m (8 ft) cube topped with a small attic. The walls are standard

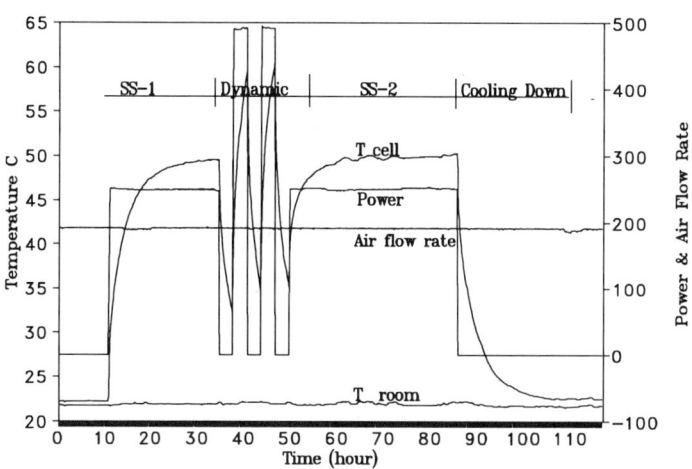

FIG. 5—*Schematic of dynamic test of indoor test cell.*

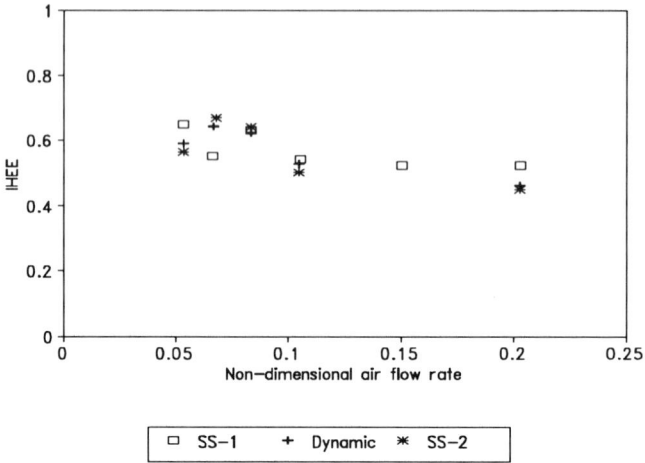

FIG. 6—*IHEE comparison for diffuse leakage configuration SS-1 = first steady-state period; Dynamic = dynamic period; SS-2 = second steady-state period.*

5 × 10 cm (2 × 4 in.) frame construction with exterior plywood, R-11 insulation, and plywood sheathing. The ceiling and roof were built with plywood, R-19 insulation, an attic air space, plywood sheathing, and asphalt roofing. A schematic of the test cell is shown in Fig. 8.

There are two concentrated flow holes in the north wall (Fig. 9): hole A, which is 10 cm (4 in.) in diameter, serves as the fan outlet or inlet; hole B, which is 5 cm (2 in.) in diameter, serves as the concentrated flow path. The north wall also contains a 61 cm wide by 205 cm high (24 by 80 in.) door to simulate crack flow. The construction of the "door" is similar to that of the wall. Both the interior and exterior plywood panels of each wall have eight "diffuse" holes [four located 13 cm (5.1 in.) from the ceiling, four located 13 cm (5.1 in.) from the floor] 1.3 cm (0.5 in.) in diameter and equally spaced across the wall.

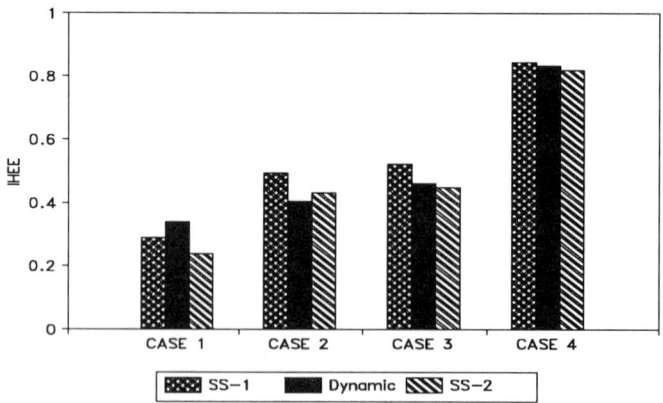

FIG. 7—*IHEE comparison for different leakage configuration case 1 = concentrated flow; case 2 = diffuse flow; case 3 = quasi-diffuse flow; case 4 = double flow.*

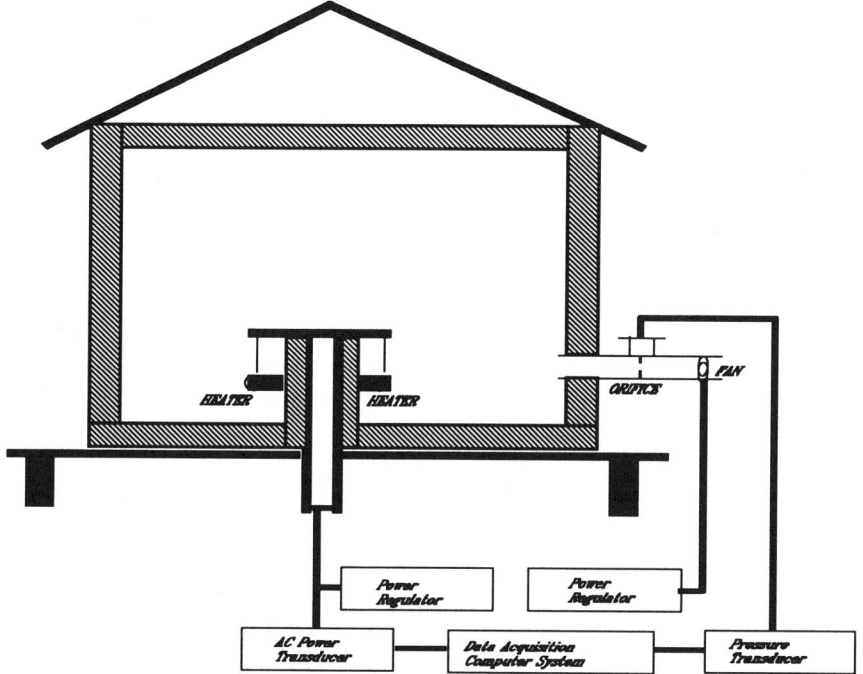

FIG. 8—*Schematic of test cell.*

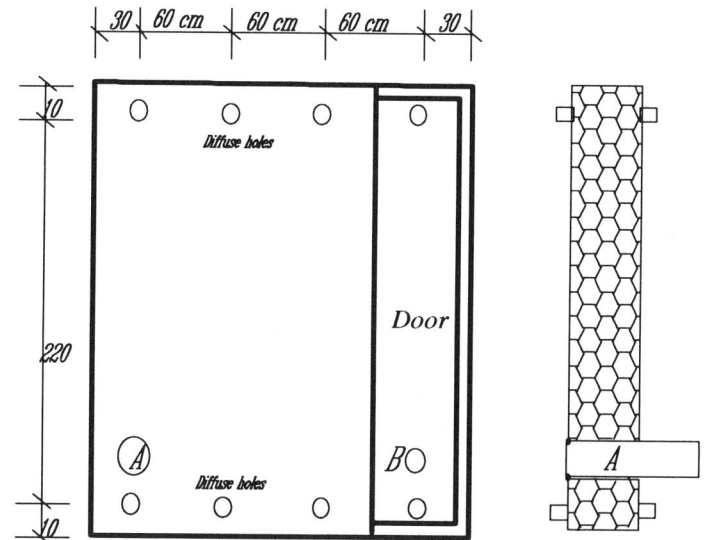

FIG. 9—*Schematic of north wall.*

Four leakage configurations were defined: (1) mixed concentrated/diffuse leakage (air introduced or exhausted through hole A, hole B open, "diffuse" holes blocked, and the "cracks" sealed); (2) diffuse leakage (air introduced or exhausted through hole A, hole B closed, diffuse holes blocked, and cracks sealed); (3) quasi-diffuse leakage (air introduced or exhausted through hole A, hole B closed, diffuse holes open, and cracks sealed); and (4) mixed crack/diffuse leakage (air introduced or exhausted through hole A, hole B closed, diffuse holes blocked, and cracks unsealed).

The test cell mass temperatures are monitored by 220 thermocouples, the test cell air temperature by 12 thermocouples, and the outside temperature by two well-shielded thermocouples located north of the cell. The horizontal solar radiation is measured by an on-site pyranometer and the airflow rate is controlled by a fan with the flow rate measured by an orifice. The power is supplied by a heater and is measured by a power transducer. The pressures across the walls, due to both natural forces and mechanically induced airflow, is measured separately by five pressure transducers (four for natural force, one for mechanical force).

A data acquisition system is used to couple the sensors to a computer, where the signals are sampled, precalculated, and finally recorded on a hard disk. All the signals are sampled, converted to necessary engineering units every 10 s, and 10-min time average values are recorded.

Tests were carried out for the four different leakage configurations with different air flow rates under both infiltration and exfiltration flow. Each test at a constant airflow rate took three to five days where a constant power input was used. The short term average measurement (STAM) method [28] was used to determine IHEE values from these data.

Both the measured and the diffuse-wall model IHEEs (three leakage configurations only) are shown in Fig. 10 as a function of the nondimensional airflow rate.

These results show much larger scatter than the indoor test cell results due to numerous uncontrollable parameters. However, closer examination shows that IHEE is considerably higher for infiltration than for exfiltration flows. This appears to show that the airflow is interacting with the solar radiation absorbed on the exterior surface of the test cell as well as with the conduction heat flow. It can be interpreted as follows: for exfiltration flow, the airflow pushes solar radiation out, resulting in less heat gain than given by the classical conduction/solar radiation procedure and in a lower IHEE; for infiltration flow, the process is reversed and the airflow carries some of the solar heat along with it, increasing the solar gain. Since these effects do not occur in the absence of airflow, it is appropriate to include them in IHEE. This now permits IHEE to have values larger than unity or less than zero. In fact, we note that several values less than zero were measured.

The data points in Fig. 10 generally have a number shown next to the symbol. This number shows the time-averaged equivalent solar radiation incident on the cell during the test when the IHEE value was measured. The time-averaged equivalent solar radiation is defined as the ratio of the time-averaged solar radiation (incident on all surfaces of the test cell) to the test cell surface area. The total solar radiation incident on the cell was calculated from measured horizontal solar radiation data using the ASHRAE model [31]. Several of the data points shown were determined from slightly different periods of data taken during the same test and hence show the same solar radiation values. The variation in IHEE shown for these points provides one measure of experimental repeatability.

Examination of the IHEE values together with the solar radiation data shows that larger values of solar radiation generally correspond to lower values of IHEE for the exfiltration configuration. The data for infiltration are less consistent.

We conclude from Fig. 10 that the interaction of airflow with solar radiation is an important consideration. The following conclusions may also be inferred: (1) IHEE values outside the

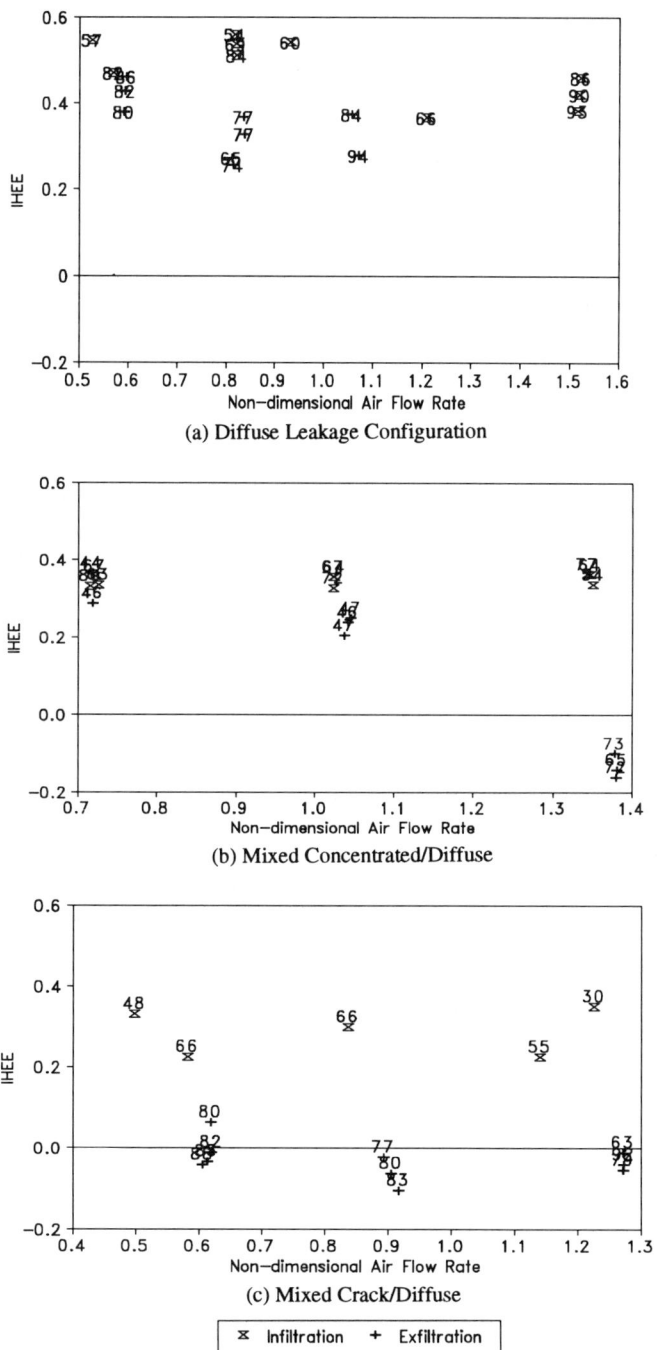

FIG. 10—*IHEE versus nondimensional airflow rate for different air leakage configurations (time-averaged equivalent solar radiation values are marked beside IHEE values).*

range of 0 to +1 are possible when solar interaction occurs; (2) IHEE is dependent on leakage configuration; the diffuse leakage configuration has the highest IHEE value; (3) infiltration has a higher IHEE than does exfiltration when the test cell is heated above outside temperature; (4) IHEE decreases when solar radiation increases and exfiltration airflow is present under heating conditions.

Theoretical Model Study with Air Leakage/Conduction/Solar Interaction

The outdoor cell test showed that both solar radiation and airflow direction can strongly impact IHEE values. However, no available model includes these factors. Hence, IHEE models for an ideal diffuse wall and an attic with solar radiation absorbed on the exterior surfaces were developed [27]. These models were used to simulate three different model houses under differing weather conditions and airflow rates:

(a) An "idealized" house, which is depressurized during heating conditions and pressurized under cooling conditions. This house is expected to have lower energy use than a standard house with the same rate of air leakage.

(b) A "normal" house, which is assumed to provide infiltration heat recovery; half of the wall area has infiltration and the other half has exfiltration while the ceiling has infiltration in summer and exfiltration in winter.

(c) A "classical" house, where energy use is modeled in the classical way with no interactions considered between air leakage and solar radiation or conduction.

Additional characteristics of the basic model and of the houses are described in Table 4, followed by the modeling results. The ideal diffuse wall and attic models assume that leakage air and solid materials both have the same temperature at any position, and one-dimensional steady-state heat transfer. Weather conditions are represented by a weather indicator, which is defined as the ratio of the average solar radiation absorbed by the house surfaces to the average inside-outside temperature difference during a modeling period. The weather indicator has positive values for heating and negative values for cooling. The airflow rate is nondimensionalized as the ratio of MC_p to the conduction heat loss factor.

The simulation output is presented as the ratio $[Q_{cl} - Q_i]/Q_{cl}$ for the comparison of the idealized house with the classical house. This ratio is called the "energy recovery ratio" since it represents the fraction of the total heating or cooling load of the classical baseline house which is recovered by the solar and conduction interaction with the air leakage in the idealized house. Similar energy recovery ratios are defined for the other comparisons matching the idealized versus the normal house and the normal house versus the classical house.

The simplified simulation assumed the attic has no cross-ventilation, and the average solar radiation on all walls is two thirds of the solar radiation on the roof. The simulation explored

TABLE 4—*Summary of thermal and associated parameters for house model.*

External dimensions of house	12 × 15 × 2.4 m (39 × 49 × 8 ft)
Area of ceiling	180 m² (1938 ft²)
Area of opaque walls	104 m² (1116 ft²)
Area of windows	26 m² (279 ft²)
R_w for walls	2 m² K/W (11 ft²h°F/Btu)
R_w for ceiling	3.5 m² K/W (20 ft²h°F/Btu)
R_l for ceiling	0.15 m² K/W (0.85 ft²h°F/Btu)
R_{b0}	0.05 m² K/W (0.28 ft²h°F/Btu)

nondimensional airflow rates from 0 to 1 with a step of 0.05 and the weather indicator ranged from 0 to 40 for winter and −100 to 0 for summer with a step of 2.

A comparison of the idealized and normal house models is given in Fig. 11a. The results show that the energy recovery ratio of the idealized house varies from −0.15 to greater than 2 during heating and from −0.15 to 0.52 during cooling depending on the weather indicator values and nondimensional airflow rates. When the absolute value of the weather indicator is greater than 10, the energy recovery ratio is positive. If the weather indicator is 20 for winter and −40 for summer, then the idealized house model saves from 21% to 35% of the heating energy and from 7% to 31% of the cooling energy as the nondimensional airflow varies from 0.2 to 1. These numbers demonstrate the potential energy savings that could result from "organizing" the airflow through the house envelope to take advantage of the solar/airflow interaction.

A comparison of the idealized and classical house models is shown in Fig. 11b. The results show that the energy saving ratio varies from 0 to greater than 2 during heating and from 0 to 0.33 during cooling depending on the weather indicator values and airflow rate. If the weather indicator is 20 in winter and −40 in summer, the heating energy recovery ratio varies from 0.38 to 0.52 and the cooling energy recovery ratio varies from 0.09 to 0.27 as the nondimensional airflow varies from 0.2 to 1. These results demonstrate that the classical method can significantly overestimate "ideal" house energy consumption.

A comparison of the normal and classical house models is given in Fig. 11c. The results show that energy recovery ratio varies from 0 to 0.91 during heating and from −0.40 to 0.28 during cooling depending on weather indicator and airflow rate. If the weather indicator value is 20 in winter and −40 in summer, the energy recovery ratio varies from 0.21 to 0.26 and from −0.05 to −0.16 for heating and cooling, respectively, when the nondimensional airflow varies from 0.2 to 1. These results demonstrate that the classical calculation can significantly overestimate "normal" house heating energy consumption but may underestimate "normal" house cooling energy consumption.

Note that extremely high heat recovery values are observed within a small area where the weather indicator ranges from 25 to 40 and the nondimensional airflow rate ranges from 0 to 0.5. These abnormally high values are due to the very small heating load for the classical house model under these conditions.

This model study suggests that the ideal house, defined as overpressurized in winter and underpressurized in summer, can save significant amounts of energy without sacrificing fresh air; a normal house consumes less heating energy compared with a classical house when both of them have identical airflow rates; and the classical load calculation method can result in significant load estimation error when air infiltration heat recovery is present. While the model results are consistent with the trends of the outdoor cell measurements, insufficient data were available to validate the model.

Conclusions

Air infiltration heat recovery has been systematically investigated by the authors using a number of experimental facilities and theoretical modeling. The indoor steady-state experimental results show that airflow through frame construction or a frame wall can substantially reduce the energy requirements due to infiltration, and IHEE depends on airflow rate, and leakage configurations. The outdoor measurement results suggest that the interaction of airflow with solar radiation is an important consideration. The model study suggests that depressurizing a house during winter and pressurizing it using a fan during summer may reduce energy consumption significantly. Moreover, this study also showed that buildings can be

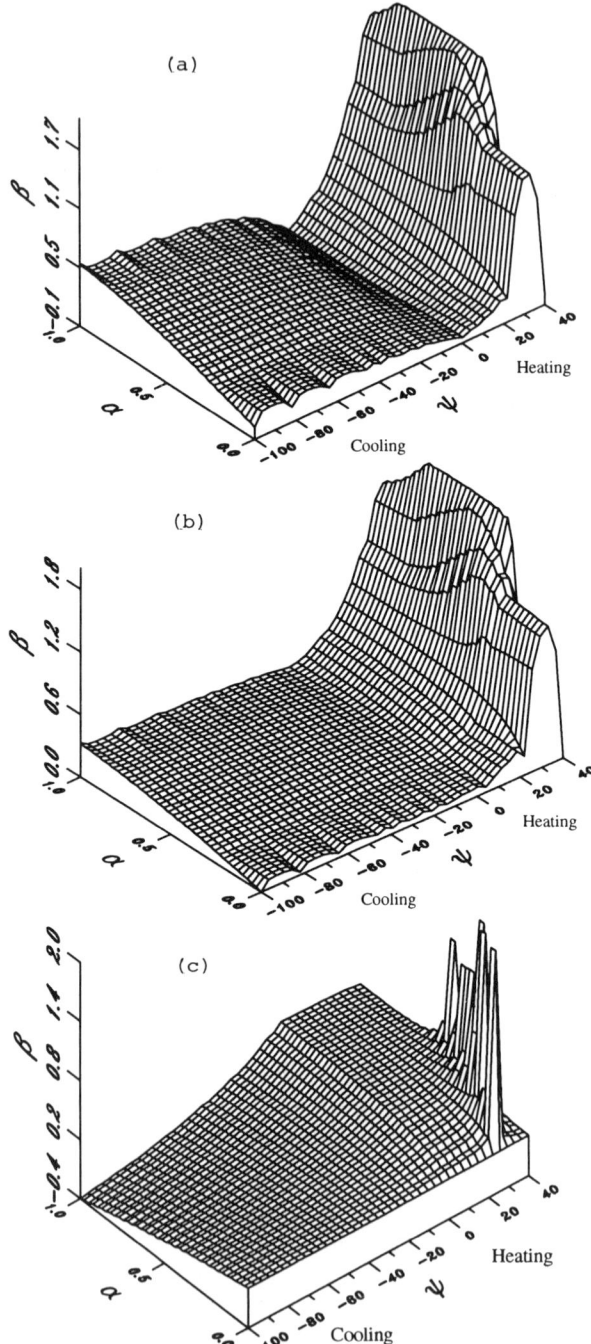

FIG. 11(a)—*Energy recovery ratio of idealized to normal house models:* β = *energy recovery ratio;* ψ = *weather indicator* (W/m²°C); α = *nondimensional airflow rate;* (b) *energy recovery ratio of idealized to classical house models:* β = *energy recovery ratio;* ψ = *weather indicator* (W/m²°C); α = *nondimensional airflow rate;* (c) *energy recovery ratio of normal to classical house models:* β = *energy recovery ratio;* ψ = *weather indicator* (W/m²°C); α = *nondimensional airflow rate.*

treated as a linear system for load calculation when airflow is constant, and the classical calculation method can result in significant load estimation error when air infiltration heat recovery is present.

The heat recovery effects of air infiltration require further study under the following conditions: direct flow through/around single windows and short path flow around doors, etc.

Acknowledgment

The experimental work and theoretical studies reported here were sponsored by the Texas Higher Education Coordinating Board under Energy Research and Application Program (ERAP) Project #227, the Center for Energy and Mineral Resources, Texas A&M University, and the U.S. Department of Energy.

References

[1] Bursey, T. and Green, G. H., "Combined Thermal and Air Leakage Performance of Double Windows," *ASHRAE Transactions,* Vol. 73, 1970, pp. 215–226.

[2] Guo, J. and Liu, M., "The Effective & Ineffective Heat Loss By Infiltration—Field Measurement in a Dormitory," *Proceedings of the 3rd International Congress on Building Energy Management ICBEM,* Vol. 2, Lausanne, Switzerland, 1987.

[3] Anderlind, G., "Energy Consumption Due to Air Infiltration," *Proceedings of the 3rd ASHRAE/DOE/BTECC Conference on Thermal Performance of the Exterior Envelopes of Buildings,* Clearwater Beach, FL, 1985, pp. 201–208.

[4] Liu, M., "Thermal Techniques in Buildings," *Harbin Architectural and Civil Engineering Institute Press,* Harbin, China, 1987, pp. 3.85–3.90.

[5] Bailly, N. R., "Dynamic Insulation Systems and Energy Conservation in Buildings," *ASHRAE Transactions,* Vol. 93, Part 1, 1987, pp. 447–466.

[6] Arquis, E. and Langlais, C., "What Scope for Dynamic Insulation?," *Batiment International Building Research and Practice,* Vol. 19, 1986, pp. 84–93.

[7] Golovkin, N. A., "New Cold Store Insulation System," *Bulletin de l'IIF,* Annexe 1960-3 Commission V, Conference de Marseilles, France, 1960.

[8] Ripatti, H., "Air Flow Windows in Hot Climates," *Proceedings of the CLIMA 2000 World Congress in Heating, Ventilating and Air-Conditioning,* Copenhagen, Vol. 2, 1985, pp. 45–50.

[9] Timusk, J., Seskus, A. L., and Lio, M., "Performance Evaluation of the Dynamic Wall House," Project report prepared by Center for Building Science, University of Toronto, Toronto, ON, Canada, 1987.

[11] Dubois, P., "Energy Efficient Building Walls," *Proceedings of the First E. C. Conference on Solar Collectors in Architecture,* Venice, 1983, pp. 144–158.

[12] Andersson, K. A. and Wadmark, T., "Use of Ventilation and Insulation in Single Family Houses for Thermal Comfort, Energy Conservation and Protection Against Moisture, Dust and Radon, the OPTIMA Concept," *Proceedings of the 3rd International Congress on Building Energy Management ICBEM,* Vol. 3, Lausanne, Switzerland, 1987.

[13] Masoero, M. and Aghemo, C., "Performance of an Innovative Ventilated Curtain-Wall Component," *Proceedings of the 3rd ASHRAE/DOE/BTECC Conference on Thermal Performance of the Exterior Envelopes of Buildings,* Clearwater Beach, FL, 1985, pp. 1004–1018.

[14] Homma, H. and Guy, R. W., "Ventilation of Back Space of Building Enclosure Siding for Solar Heat Gain Reduction," *Proceedings of the ASHRAE/DOE-ORNL Conference on the Thermal Performance of the Exterior Envelopes of Buildings,* Kissimmee, FL, 1979.

[15] Beyea, J., Dutt, G., and Wotecki, T., "Critical Significance of Attics and Basements in the Energy Balance of Twin Rivers Townhouses," *Energy and Buildings,* Vol. 1, 1977, pp. 261–269.

[16] Harrje, J., Dutt, G., and Beyea, J., "Locating and Eliminating Obscure but Major Energy Losses in Residential Housing," *ASHRAE Transactions,* Vol. 85, Part 2, 1979, pp. 521–559.

[17] Claridge, D. E., Jeon, H., and Bida, M., "Performance Analysis of the Colorado 50/50 Retrofit Program," Vol. 1, submitted to the Solar Energy Research Institute by the University of Colorado, Department of Civil and Architectural Engineering, Boulder, CO, 1984.

[18] Claridge, D. E., Jeon, H., and Bida, M., "A Comparison of Traditional Degree-Day and Variable-Base Degree-Day Prediction with Measured Consumption of 20 Houses in the Denver Area," *ASHRAE Transactions*, Vol. 91, Part 2, 1985, pp. 865–874.

[19] Sonderegger, R. C., Condon, P. E., and Modera, M. P., "In-situ Measurements of Residential Energy Performance Using Electric Co-heating," *ASHRAE Transactions*, Vol. 86, Part 1, 1980, pp. 394–407.

[20] Guo, J., Xu, J.-X., and Li, X.-D., "Heat Balance Tests and Energy Conservation Studies for Buildings in Harbin," *Proceedings of the CLIMA 2000 World Congress in Heating, Ventilating and Air-Conditioning*, Copenhagen, Vol. 2, 1985.

[21] Guo, J. and Liu, M., "The Energy Saving Effect of Double Frame Windows," *Proceedings of the CLIMA 2000 World Congress in Heating, Ventilating and Air-Conditioning*, Copenhagen, Vol. 2, 1985.

[22] Kohonen, R., Tech, D., and Virtanen, M., "Thermal Coupling of Leakage Flows and Heating Load of Buildings," *ASHRAE Transactions*, Vol. 93, Part 2, 1987, pp. 2303–2318.

[23] Claridge, D. E. and Bhattacharyya, S., "The Measured Energy Impact of 'Infiltration' in a Test Cell," *Journal of Solar Energy Engineering*, Vol. 112, 1990, pp. 132–139.

[24] Bhattacharyya, S. and Claridge, D. E., "The Energy Impact of Air Leakage Through an Insulated Wall," *Proceedings of ASME Solar Energy Division Conference*, Miami, 1992.

[25] Liu, M. and Claridge, D. E., "The Measured Energy Impact of Infiltration under Dynamic Conditions," *Proceedings of the Eighth Symposium on Improving Building Systems in Hot and Humid Climates*, Texas A&M University, College Station, TX, 1992.

[26] Liu, M. and Claridge, D. E., "The Measured Energy Impact of Infiltration in an Outdoor Test Cell," *Proceedings of the Eighth Symposium on Improving Building Systems in Hot and Humid Climates*, Texas A&M University, College Station, TX, 1992.

[27] Liu, M. and Claridge, D. E., "The Energy Impact of Combined Solar-Radiation/Infiltration/Conduction Effects in Walls and Attics," *Proceedings of 5th ASHRAE/DOE/BTECC Conference on Thermal Performance of the Exterior Envelopes of Buildings*, Clearwater Beach, FL, 1992.

[28] Liu, M. and Claridge, D. E., "A Calorimetric Method for Heat Transfer Coefficient Identification of Thermal Enclosures," *Solar Engineering 1993: Proceedings of the ASME Solar Energy Conference*, Washington, DC, 1993.

[29] Liu, M. and Claridge, D. E., "A Non-calorimetric Method for Heat Transfer Coefficient Identification of Thermal Enclosures," *Solar Engineering 1993: Proceedings of the ASME Solar Energy Conference*, Washington, DC, 1993.

[30] Liu, M., "Study of Air Infiltration Energy Consumption," Ph.D. Dissertation, Texas A&M University, College Station, TX, 1992.

[31] *ASHRAE Handbook: 1981 Fundamentals*, American Society of Heating, Refrigerating and Air-Conditioning Engineers, Atlanta, GA, 1981.

David C. Jones,[1] *David G. Ober,*[2] *and John T. Goodrow*[2]

Thermal Performance Characterization of Residential Wall Systems Using a Calibrated Hot Box with Airflow Induced by Differential Pressures

REFERENCE: Jones, D. C., Ober, D. G., and Goodrow, J. T., **"Thermal Performance Characterization of Residential Wall Systems Using a Calibrated Hot Box with Airflow Induced by Differential Pressures,"** *Airflow Performance of Building Envelopes, Components, and Systems, ASTM STP 1255,* Mark P. Modera and Andrew K. Persily, Eds., American Society for Testing and Materials, Philadelphia, 1995, pp. 197–228.

ABSTRACT: ASTM E 283 and ASTM E 1424 in conjunction with ASTM C 976 were used to study the effect of airflow on thermal performance of the wall. A typical residential 2 × 4 stud wall was constructed and placed on top of a subfloor, making a 2.44 × 2.74 m (8 by 9 ft) test specimen. This base wall assembly was then covered with two types of XPS sheathing, various housewraps, a 15# felt, and a polyethylene vapor retarder film in 40 different configurations and tested individually per ASTM E 283 and per ASTM C 976. For 24 of the 40 C 976 tests, a differential pressure was induced across the test wall as per and ASTM E 1424. Airflows ranged from undetectable airflow at $0 \cdot$ Pa ΔP to 1.63 L/s $\cdot$ m² for the base wall assembly alone. Difference in airflow resistance performance between the ASTM E 283 and ASTM E 1424 test methods were noted. Thermal testing results incorporating both ASTM C 976 and ASTM E 1424 for tests 1–28 produced apparent thermal conductances (C-values) in the range of 0.40 W/m² $\cdot$ K for a nondetectable airflow level to 1.81 W/m² $\cdot$ K for an airflow of 1.53 L/s $\cdot$ m² for the base wall assembly alone with a 20-Pa ΔP. The calculated C-value for this base wall assembly was 0.40 W/m² $\cdot$ K. Test results reveal that airflow rates as low as 0.2 L/s $\cdot$ m² could produce a 46% increase in apparent C-value. Similar thermal performance differences were revealed when thicker shiplap XPS sheathing was used. Tests were also conducted using an Air-Tight Drywall configuration showing the effect of "wind washing" on thermal performance. By sealing the gypsum drywall on the base wall assembly tested, the apparent C-value, when exposed to a 12.5 Pa wind pressure, was found to be equivalent to a base wall assembly configuration which allows 0.15 L/s $\cdot$ m² airflow to penetrate completely through.

KEYWORDS: airflow, walls, thermal performance, blower door, housewraps, air retarder, air barrier, hot box, air leakage, infrared

The need for controlling airflow through opaque walls is recognized by the building science community as a way to limit air-transported moisture and increase the thermal efficiency of the wall and subsequently the home. The installation of an air retarder within the opaque wall is becoming a normal construction practice in many regions throughout the United States and Canada. Quirouette [*1*], Lux and Brown [*2*] and Perreault [*3*] discuss the value

[1] Research Associate, Tyvek Market & Product Development, Dupont
[2] Director of Testing Services Division and Project Engineer for Hot Box Testing, respectively, Holometrix, Inc.

of air leakage control and various methods of constructing these systems. There are various methods available to the builder to limit airflow through an opaque wall such as caulking seams, housewraps, interlocking sheathing, taping seams, sealing interior drywall or polyethylene vapor retarder, and foaming the stud cavities. In a wall system, the air retarder is regarded as the product or system which has the greatest resistance to airflow under pressure.

Typically this air resistance is measured using a test protocol outlined in ASTM Standard Test Method for Rate of Air Leakage Through Exterior Windows, Curtain Walls and Doors (E 283). Using E 283 requires a wall section to be built, typically 2.44 × 2.44 m (8 × 8 ft) sealed at the perimeter, which closely simulates actual field construction. Airflow measurements are taken at various pressure differential levels to define the air leakage rate of the test specimen. ASTM E 283 does not account for temperature differentials across the wall which can effect the airtightness of the structure due to differences in the thermal expansion and contraction of the various building materials that make up the wall. The ASTM Standard Test Method for Determining the Rate of Air Leakage Through Exterior Windows, Curtain Walls, and Doors Under Specified Pressure and Temperature Difference Across the Specimen (E 1424) may be a more appropriate test method since it provides for the same type of measurements made in E 283 but incorporates a temperature differential across the test wall as well. ASTM E 1424 has been used to characterize window and door air leakage at closer to "real world" conditions [4,5]. Another airflow measurement technique developed in Canada uses an apparatus which just measures the air retarder and does not include the other wall components [6]. The approximate sample size is about one square meter. This method utilizes the same pressure differentials as ASTM E 283 but is more suited to sheet type air retarders or housewraps.

The thermal performance of small specimens of wall cavity insulation such as fiber glass can be measured by using the ASTM Standard Test Method for Steady-State Heat Flux Measurements and Thermal Transmission Properties by Means of the Heat Flow Meter Apparatus (ASTM C 518). Larger wall sections can be tested by utilizing the ASTM Standard Test Method for Thermal Performance of Building Assemblies by Means of a Calibrated Hot Box (ASTM C 976). The rated R-value assigned to a certain thickness of mass insulation is determined by testing the specimen in a static environment or in a condition where no airflow is present. Since air is a very poor conductor of heat and the majority of a fibrous insulation such as fiber glass is made up of air pockets between the fibers, the majority of the thermal resistance of these types of products comes from maintaining still air. The object for any fibrous-type insulation is to trap as much still air as possible and make these air pockets as small as possible to avoid convection. This balance is achieved in excellent fashion by a fiber glass or mineral wool-type insulation; however, these materials by themselves do not resist air movement when forced by a pressure differential across their bulk.

The aim of this study is to further the understanding of relationships between airflow through a wall and its effect on the thermal performance of wall systems. The continued effort to better understand this relationship will further the efforts made to design more energy efficient wall systems. Desired outcomes from this study are to

- develop a test protocol which integrates ASTM E 1424 and ASTM C 976,
- define the relationship between thermal performance and airtightness of a wall system for a test wall assembly in a laboratory,
- characterize the airflow leakage paths for a test wall assembly,
- study the effect of a temperature differential across a wall on its airtightness, and
- demonstrate the effect of wind washing, or exchanges of air in and out of a wall along the same exterior plane, for an airtight drywall air retarder arrangement.

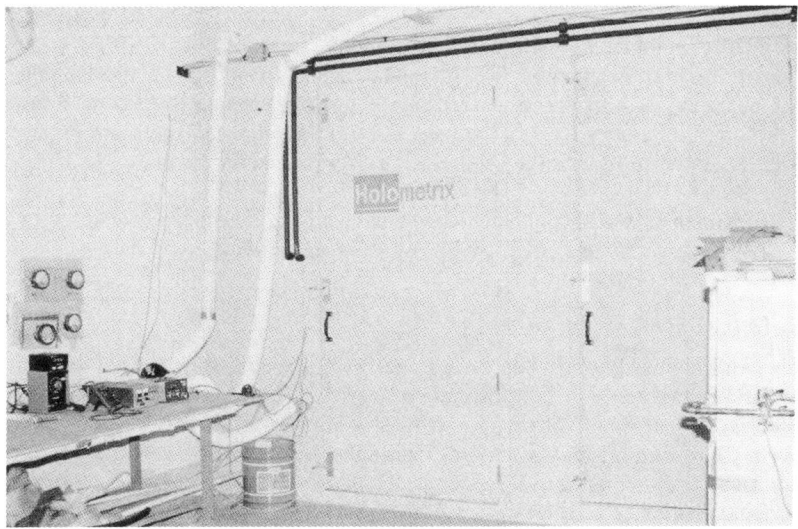

FIG. 1—*Exterior of hot box used at Holometrix, Inc.*

Test Apparatus Description

Standard R-value testing done by hot box methods in a laboratory does not include the effects of airflow through the specimen, and specimens are normally sealed to prevent air leakage from influencing the results. Since it was desired to incorporate airflow into this type of test, a calibrated hot box conforming to ASTM C 976 was modified to introduce a pressure differential across the specimen and a means of measuring air leakage through the test specimen (Fig. 1). A schematic of the test apparatus is shown in Fig. 2. The modifications also included reducing the 24 km/h (15 mph) perpendicular airflow in the climate chamber to the point of just providing enough to achieve proper mixing of the air to give a uniform temperature across the surface of the test wall. Had this 24 km/h (15 mph) not been reduced, there would most likely have been further reductions in thermal performance due to air

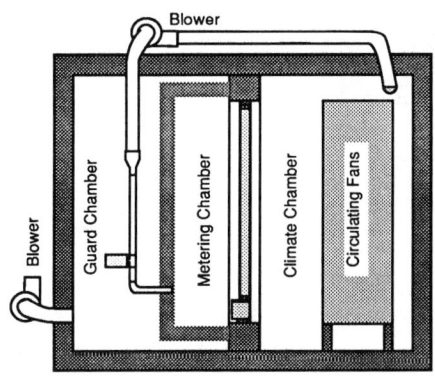

FIG. 2—*Modified calibrated hot box cross section.*

exchanges in and out of the wall strictly on the cold side due to wind washing. For each test the climate chamber was maintained at a temperature of $-18°C$ (0°F) at atmospheric pressure and the metering chamber was maintained at 21°C (70°F). For 24 of the 40 tests, the climate chamber was depressurized to induce airflow through the test wall assembly. In addition, the guard chamber was maintained at a temperature and pressure equal to that of the metering chamber to reduce the potential for thermal exchange via air leaks between the two chambers.

The test specimen was constructed using a base wall placed on top of a subfloor assembly having overall dimensions of 2.44 m width by 2.74 m height (8 × 9 ft).

Base Wall Description

The base wall assembly consisted of nominal 2 × 4 wood studs placed 40.6 cm (16 in.) on center with a single 2 × 4 bottom plate and two 2 × 4 top plates. In the stud cavities, inset stapled R-13 kraft-faced fiber glass insulation was installed and the warm side covered with 1.27 cm (0.5 in.) gypsum wall board. Inside the base wall, a standard 14 gage electrical service wire was installed from a lightswitch located in the far right stud cavity to a duplex receptacle located in the far left stud cavity. A 1.27 cm (0.5 in.) hole was drilled through each internal stud to facilitate running the wire through the base wall interior. Extra care was taken to fit the wire through the fiber glass insulation by splitting the insulation and wrapping it around both sides of the wire (Fig. 3). This avoided any reduction in insulation thickness. A cross section of the complete wall assembly is shown in Fig. 4.

Subfloor Assembly Description

The subfloor assembly consisted of nominal 2 × 10 floor joists installed 40.6 cm (16 in.) on center over a nominal 2 × 6 sill plate and topped by a piece of 1.9-cm-thick (0.75 in.) plywood. The ends of the 2 × 10 joists were secured to a nominal 2 × 10 belly band which flushed with the plywood and the sill plate. The subfloor assembly extended 5 cm (2 in.) into the interior warm side of the wall assembly when flushed with the exterior surface of

FIG. 3—*Wire through insulation arrangement.*

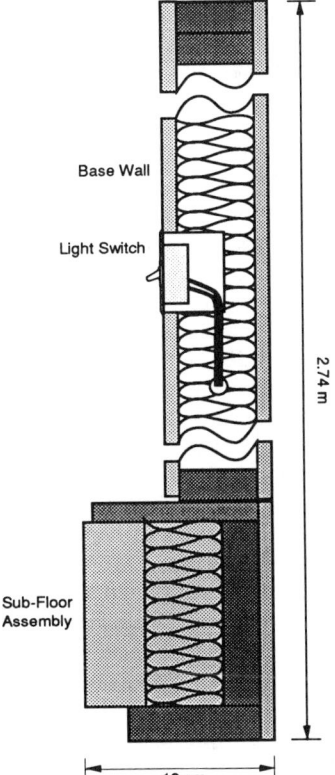

FIG. 4—*Base wall assembly cross section.*

the studs of the base wall. The cavities between the floor joists were fitted with R-13 kraft-faced fiber glass batts which were pushed into light contact with the band joist or bellyband.

Final Base Wall Assembly

The base wall was assembled on top of the subfloor assembly and fitted into the partition wall between the warm and cold side of the calibrated hot box with the nominal 2 × 4 stud portion of the wall centered in the partition wall. The perimeter of the base wall assembly was stuffed with fiber glass to fill in any small gaps that existed and then taped to prevent air leakage from occurring. The gypsum wallboard on the interior side of the wall was taped at the top interior seams and sides and left open at the interface between the bottom and the plywood on the subfloor as is typical in the field. The subfloor assembly was also sealed at the perimeter. The two electrical boxes were fitted and wired with a lightswitch and a duplex receptacle. The interface of the electrical box was sealed to the gypsum wallboard as might be typical when wallboard compound is applied. Cover plates were installed over the light-switch and duplex receptacle. No caulk or sealants were used in this assembly. While perhaps not the best construction practice, this is typical practice for many areas of the United States. Model energy codes such as the CABO Model Energy Code, which would promote measures to address air infiltration, have not been widely adopted by the building community; therefore it was desired with this wall assembly to capture typical construction practice and not the best case for the base wall. With this arrangement, leakage paths for air could occur along the bottom plate and plywood interface and through the knockout hole in the back of the electrical boxes and out through the cover plates. This configuration was the core for three distinct base wall assemblies.

The base wall assembly used in tests 1–28 had the exterior or cold side covered with 1.22 m × 2.44 m × 1.27 cm (4 ft × 8 ft × 0.5 in.) extruded polystyrene (XPS) foam sheathing following the manufacturer's suggested fastening schedule of 0.40 m on center within the field of the board and 0.15 m on center around the perimeter using screws fitted with 25-mm-diameter (1 in.) washers. The XPS sheathing was installed to create two vertical seams. A 1.22 × 2.44 m (4 × 8 ft) sheet was placed in the center of the wall with a 0.81 m × 2.44 m (32 in. × 8 ft) sheet on one side and a 0.41 m × 2.44 m (16 in. × 8 ft) sheet on the other. Two 1.22 m × 0.30 m × 1.27 cm (4 ft × 1 ft × 0.5 in.) final pieces of XPS foam were attached to the bellyband exterior and flushed up the XPS foam installed on the base wall as would be necessary where 1.22 m × 2.44 m sheets were installed using the tilt wall construction method. The exterior perimeter of the XPS foam sheathing was also sealed with tape to prevent any air leakage at the perimeter. With this arrangement, leakage paths for air could only occur through the seams in the XPS sheathing. The base wall assembly in tests 29–31 used the same exterior XPS sheathing arrangement but had all interior, warm side, potential leakage sites sealed to simulate the "airtight drywall" approach. These areas involved the additional sealing of the electrical switch and receptacle and the interface between the subfloor and the bottom of the gypsum drywall. For tests 32–40, the interior configuration of tests 1–28 was used but the exterior XSP sheathing was replaced with an interlocking shiplap sheathing. This XPS sheathing was approximately 2.79 cm (1.1 in.) thick and came in sizes of 0.61 m × 2.44 m (2 × 8 ft). It was installed horizontally per manufacturer's instructions with staggered vertical butt joints and fastened with screws fitted with 25 mm diameter (1 in.) washers on a schedule of every 0.15 m (0.49 ft) along the vertical joints and every 0.30 m (0.98 ft) within the field of the board. The perimeter was sealed with tape as in tests 1–31.

Thermocouples used to measure temperatures during the tests were 24 gage Type T with special limits of error. The interior and exterior surfaces on the base wall assembly were fitted with 16 thermocouples on each side covering approximately 0.42 m² (4.5 ft²) of test wall surface area each. These 32 thermocouples were placed to measure surface temperatures over both studs and insulation cavities. The outputs were area-weighted in the final data analysis to provide an average sample surface temperature for both the hot and cold sides of the test wall, which defined the temperature gradient across the wall used in calculating the effective thermal conductance (C-value) for a given test. An additional 47 thermocouples were located at various internal positions within the wall and subfloor. The relationship between XPS butt joint seams and the internal thermocouple locations is shown in Figs. 5 and 6.

No exterior siding was installed on the final base wall assembly due to the difficulty in keeping this variable constant from test to test and also due to the wide variety of sidings to choose from. Exterior siding should not become a variable in determining thermal performance versus airtightness unless the exterior siding becomes the wall component which allows the least amount of air to penetrate through and thus becomes the air retarder. Exterior sidings are not typically regarded as air retarders.

Instrumentation

Eight heat flow transducers were also used, located at expected leakage sites and other areas within the wall where leakage was not expected to occur (Fig. 7). The transducers

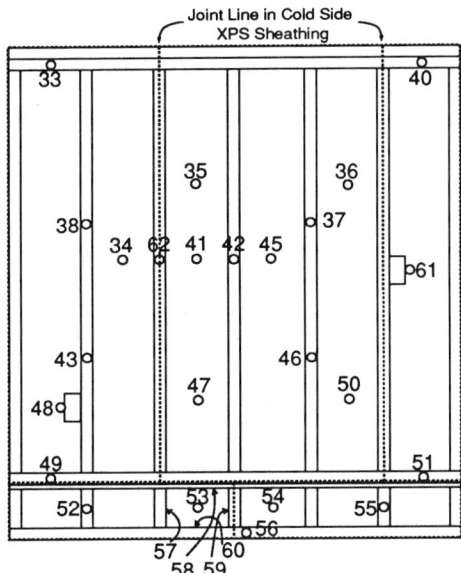

FIG. 5—*Internal thermocouple positions on warm side behind the 1.27 cm gypsum drywall and in the floor joist area.*

were taped to the stud cavity side of the gypsum drywall and their voltage output read at the steady-state condition of each test. The heat flow transducers had not been calibrated at the time of this writing so only relative comparisons can be made from the voltage readings.

Thermocouples and heat flow transducer outputs were fed to a Hewlett Packard Model #3852 data acquisition unit equipped with a 5½-digit voltmeter model #44701 and an ex-

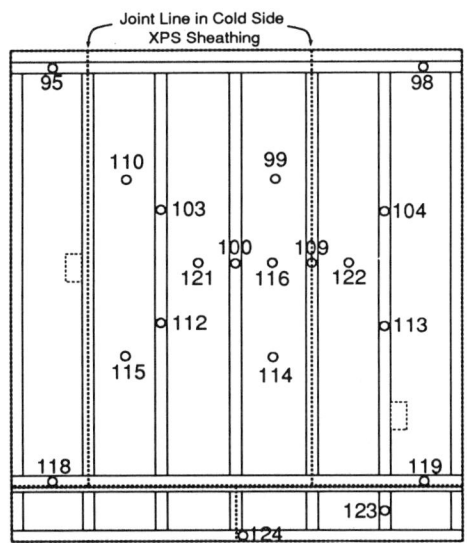

FIG. 6—*Thermocouple positions cold side located behind the 1.27 cm XPS sheathing.*

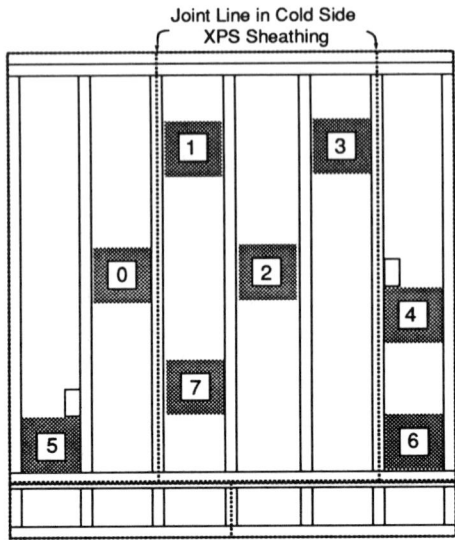

FIG. 7—*Heat flow transducer placement viewed from warm side.*

tension chassis model #3852A with sufficient input cards to handle 220 channels. The voltmeter was calibrated using a NIST traceable voltage source and has a manufacturer's stated accuracy of ±0.02% of reading + 6 μV for temperature measurements and ±0.016% of reading + 360 μV for power measurements. Actual calibration at the facility using a NIST traceable voltage source showed that the accuracy was ±2 μV total for temperature measurements and ±100 μV total for power measurements. The data acquisition unit was linked to a Macintosh IIcx Computer running LABview software which monitored the test conditions (Fig. 8). The LABview software was set up to take data at five-minute intervals throughout the test and output to a text file which could later by analyzed and reduced using

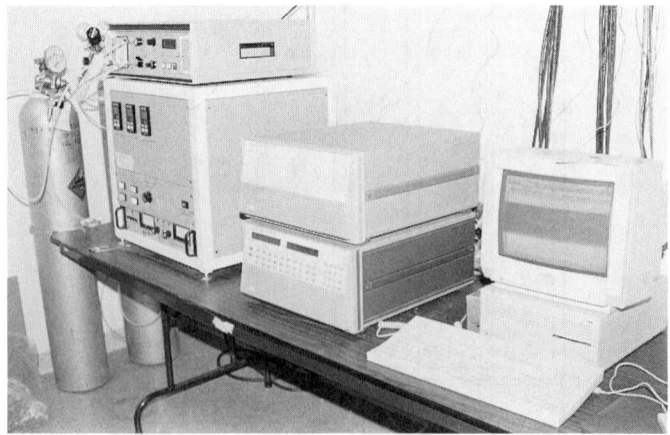

FIG. 8—*Data acquisition equipment and hot box control.*

Microsoft Excel software for final output. Temperature control for each of the three chambers was accomplished by using Eurotherm 808 controllers located within a hot box control console.

The heat flow into the metering chamber was computed as the sum of the power input to the heaters and the air circulating fans. The heat (power) inputs were calculated using

$$P = V \cdot I \tag{1}$$

where

P = power,
V = volts, and
I = amperage.

Voltage was measured by the Hewlett Packard voltmeter and current by measuring the voltage drop across a 0.01 Ω precision resistor that had been calibrated using a NIST traceable resistor network. The heat input measurement provided the total heat flow, which when corrected for other losses in the test apparatus gave the heat flow through the specimen.

Airflow was measured using one of three Kurz series 565 mass flowmeters (Fig. 9) providing the capability to measure airflows in three ranges 0 to 0.5 L/s, 0 to 2.4 L/s, and 0 to 14.2 L/s (0–1, 0–5, and 0–30 scfm). Before each test, the appropriate flowmeter was selected so that leakage would fall as close as possible to the midrange of the measurement capability of the device. Each flowmeter had been calibrated with a NIST traceable mass flowmeter and was accurate to ±3% of reading + 0.5% of full scale.

Pressure was measured in the metering chamber using a Dwyer series 605 Magnehelic indicating transmitter capable of sending a signal to the data acquisition unit or being read visually by a dial indicator. Pressures in the climate chamber and guard chamber were measured by a Dwyer series 2000 Magnehelic pressure gage that were read visually. All of the pressure gages were accurate to ±2% of full scale, which was 0 to 125 Pa (0–0.5 in. of water) and were readable in divisions of 2.5 Pa or 0.01 in. of water.

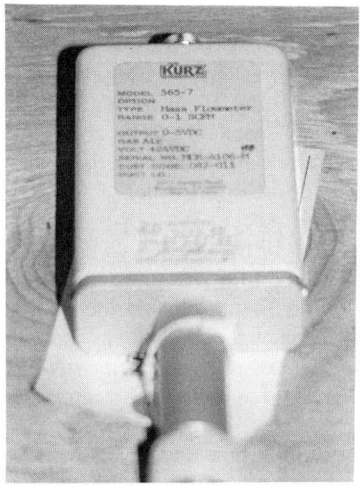

FIG. 9—*Kurz series 565 flowmeter.*

Hot Box Calibration

Prior to initiation of the test program, a homogeneous wall of known R-value was run in the test facility. This wall was constructed of a 101-mm-thick (4 in.) expanded polystyrene foam faced with 0.4-cm-thick (0.16 in.) Plexiglas® on both sides. Conductance measurements done per ASTM C 518 in a NIST traceable heat flowmeter were compared with the conductance measurements from the hot box at the beginning and end of the test program. The before and after calibration runs in the hot box differed by 0.6% and the average of the hot box runs differed from the ASTM C 518 heat flowmeter runs by 3.5%.

Also, as part of the hot box calibration, random individual thermocouples were calibrated by comparing their outputs to an ice reference. The differences observed were well within the accuracy stated for the thermocouples in the instrumentation section ($\pm 1°C$ or 1.8°F). Actual differences were all less than $\pm 0.55°C$ or 1.0°F.

Test Protocol

Calibration of the hot box was conducted both before and after testing the various wall configurations using the calibration standard described previously. Describing the thermal performance of the test wall at various airflow rates required finding a simple way of varying the airtightness of the structure at a given set of pressure differentials. It was decided to use a variety of housewraps to attempt to vary the airflow since various products and technologies are employed in the manufacturing of these types of products which yield different levels of airtightness. Interlocking XPS sheathing was also tested as well as sealing all the known interior leakage sites as would be done if the Air-Tight Drywall approach were used. The configurations tested are listed in Table 1.

Tests 1–28 were performed to provide data to help characterize the thermal performance of the wall with various air leakage rates. A relative feel for the performance of these products would also be achieved in terms of preventing air leakage into a wall and the resultant effect on thermal performance. Tests 29–31 were conducted to determine if sealing all known interior leakage sites would significantly enhance the thermal performance of the wall versus an externally applied air retarder. Tests 32–40 were carried out to determine the performance of an alternate air retarder approach.

In Tests 1–28 there were eight different subsets where either a different housewrap or the base wall assembly alone was being tested. Housewraps were installed using 3.8 cm (1.5 in.) screws with a 3.2 cm (1.25 in.) washer. All housewrap installations were taped around the perimeter to force air leakage to only occur through the product.

For each of eight subsets in Tests 1–28, the first test configuration always tested a condition where little to no airflow was permitted in or through the wall (static condition) by installing a 6-mil polyethylene film sheet over the entire exterior surface. This was done to ensure that no conditions had changed in the test specimen or operating equipment from subset to subset since it would be expected that each of these tests would yield similar thermal performance. Beyond this first test, the polyethylene film was removed and an air pressure differential was induced across the wall by pulling a vacuum inside the meter chamber. The suction side of a blower fan was connected to the meter chamber via flexible duct with a Kurz flowmeter in line to measure airflow. The exhaust side of the blower fan was routed into the climate chamber (cold side of the hot box) to minimize the introduction of moist outside air into the system. A second blower fan was used to equalize the pressure between the guard chamber and the meter chamber to minimize any air exchanges that might occur at the meter chamber seal to the dividing wall. Pressure measurements were visually monitored by Magnehelic gages and consistently maintained throughout each test.

TABLE 1—*Test specimen configurations.*

Test		Description	Δ P (Pa)
1	Subset 1	Base Wall Assembly sealed on exterior with polyethylene film	0
2		Base Wall Assembly no film	0
3		Base Wall Assembly no film	4
4		Base Wall Assembly no film	10
5		Base Wall Assembly no film	20
6	Subset 2	Test 1 with Flash Spunbonded Polyethylene under polyethylene film	0
7		Test 2 with Flash Spunbonded Polyethylene	0
8		Test 2 with Flash Spunbonded Polyethylene	10
9		Test 2 with Flash Spunbonded Polyethylene	25
10		Test 2 with Flash Spunbonded Polyethylene	50
11	Subset 3	Test 1 with Pin Perforated Extrusion Coated Spunbonded Polypropylene under polyethylene film	0
12		Test 2 with Pin Perforated Extrusion Coated Spunbonded Polypropylene	10
13		Test 2 with Pin Perforated Extrusion Coated Spunbonded Polypropylene	25
14	Subset 4	Test 1 with Pin Perforated Extrusion Coated Woven Polypropylene Slit film under polyethylene film	0
15		Test 2 with Pin Perforated Extrusion Coated Woven Polypropylene Slit film	10
16		Test 2 with Pin Perforated Extrusion Coated Woven Polypropylene Slit film	25
17	Subset 5	Test 1 with Pin Perforated Extrusion Coated Woven Polyethylene slit film[1] under polyethylene film	0
18		Test 2 with Pin Perforated Extrusion Coated Woven Polyethylene Slit film[1]	10
19		Test 2 with Pin Perforated Extrusion Coated Woven Polyethylene Slit film[1]	25
20	Subset 6	Test 1 with Pin Perforated Extrusion Coated Woven Polyethylene Slit film[2] under polyethylene film	0
21		Test 2 with Pin Perforated Extrusion Coated Woven Polyethylene Slit film[2]	10
22		Test 2 with Pin Perforated Extrusion Coated Woven Polyethylene Slit film[2]	25
23	Subset 7	Test 1 with Pin Perforated Extrusion Coated Woven Polyethylene Slit film[3] under polyethylene film	0
24		Test 2 with Pin Perforated Extrusion Coated Woven Polyethylene Slit film[3]	10
25		Test 2 with Pin Perforated Extrusion Coated Woven Polyethylene Slit film[3]	25
26	Subset 8	Test 1 with Pin Perforated Cross Laminated Polyethylene Film under polyethylene film	0
27		Test 2 with Pin Perforated Cross Laminated Polyethylene Film	10
28		Test 2 with Pin Perforated Cross Laminated Polyethylene Film	25
29	Subset 9	Test 1 with all known interior leakage points sealed "Air Tight Drywall"	0
30		Test 2 with all known interior leakage points sealed "Air Tight Drywall"	12.5
31		Test 30 Covered with Flash Spunbonded Polyethylene	12.5
32	Subset 10	Test 1 with 1/2" XPS foam sheathing replaced with R-5 interlocking shiplap XPS foam sheathing under the polyethylene film	0
33		Test 32 with no polyethylene film	10
34		Test 32 with no polyethylene film	20
35	Subset 11	Test 32 with Flash Spunbonded Polyethylene under the polyethylene film	0
36		Test 33 covered with Flash Spunbonded Polyethylene	10
37		Test 34 covered with Flash Spunbonded Polyethylene	25
38	Subset 12	Test 32 with 15# Building Felt under the polyethylene film	0
39		Test 33 covered with 15# Building Felt	10
40		Test 34 covered with 15# Building Felt	25

[1],[2],[3] Although the descriptions are identical, these three housewraps are manufactured and distributed by different entities and subtle differences exist in the perforations and finish making them unique.

After the last thermal test for each subset a pressure versus air leakage test was performed, conforming to ASTM E 1424, while there was still a temperature gradient across the wall. The wall was then allowed to reach thermal equilibrium where temperatures throughout the wall were the same, and the air leakage test was repeated as per ASTM E 283.

Tests 29–31 were designed to characterize the effect that wind washing might have on an Air-Tight Drywall installation. For these tests, a wind pressure was induced from the climate chamber air circulation fans. The perpendicular airflow against the surface of the XPS foam sheathing was sufficient to produce areas of higher and lower pressure across the cold surface of the base wall assembly. A wind pressure equivalent to 16.4 km/h (10.2 mph) or approximately 12.5 Pa (0.05 in. of water) dynamic pressure was measured at the surface of the XPS foam sheathing during the test by running a 0.63 cm (0.25 in.) plastic tube from that point to a Magnehelic gage on the exterior of the chamber. Again, Test 29 was set up with the exterior polyethylene film to ensure static conditions and to link to all the other test subsets. The exterior polyethylene film was removed for Test 30 and the wall was exposed to the potential for air exchange in and out from the same side. Airflow was monitored from the meter chamber throughout the test to ensure that no air penetration through the wall was occurring. Test 31 was designed to show the additive effect in thermal performance a housewrap might have to an Air-Tight Drywall installation.

Tests 32–40 employed the use of a shiplap interlocking XPS foam sheathing. The same protocol that was used for each of the eight subsets in Tests 1–28 was used on this arrangement. Tests 32–34 tested just the shiplap sheathing in lieu of a housewrap as the air retarder. In Tests 35–37, the wall was again covered with the flash spunbonded polyethylene housewrap used in Tests 6–10 to determine the additive effect in thermal performance. Finally, in Tests 38–40, the housewrap was removed and a widely used product, 15# felt, was tested to gain an understanding of its relative contribution in this area.

Visual Leakage Testing

For several wall configurations, the metering chamber was opened up and the climate chamber was pressurized to 25 Pa (0.1 in. of water). With a smoke pencil, leakage sites were identified and compared from configuration to configuration. On the warm side of the wall the leakage sites were identified to be at the electrical boxes and along the bottom plate and plywood subfloor. These findings were in agreement with field observations made by Nelson [5] and were also verified by using infrared thermography to capture the temperature profile of the wall (Fig. 10).

On the cold side of the wall, the leakage sites were identified along the joints between the XPS foam sheathing. Where the sheathing was fastened the leakage was reduced, but between the fasteners the leakage was very pronounced. Visual observation revealed a slightly bowed appearance between the fasteners which would account for the higher leakage in these areas. When housewraps were installed, although some leakage was occurring, the leakage was dispersed across the entire surface area and was not detectable with the smoke pencil.

Leakage Testing Results

For each of the 40 tests conducted, an airflow versus pressure profile was generated when the wall was at a 0°C ΔT per ASTM E 283, at the beginning and end of each subset, to ensure that the wall did not develop any additional leakage as a result of the test and while the test arrangement was at a 39°C ΔT (70°F ΔT) during the test. Since there is no protocol for testing an opaque wall in this fashion, pressures between 0 and 50 Pa were used. The 39°C ΔT (ΔT of 70°F) was used because of the requirement for this ΔT in ASTM C 976, and this temperature differential was in agreement with the cold temperature mode specified in ASTM E 1424.

The results comparing airflows at these two temperature differentials are shown in Table 2. Figures 11 and 12 show data from Tests 1–28 which had the same base wall assembly arrangement covered with various housewraps. A good spread of airflow resistance performance was available from the different housewraps, which aided in achieving a good spread of data points for the thermal performance testing. For most of the tests, there was no appreciable difference in airflow for a given pressure between tests run with the 39°C ΔT

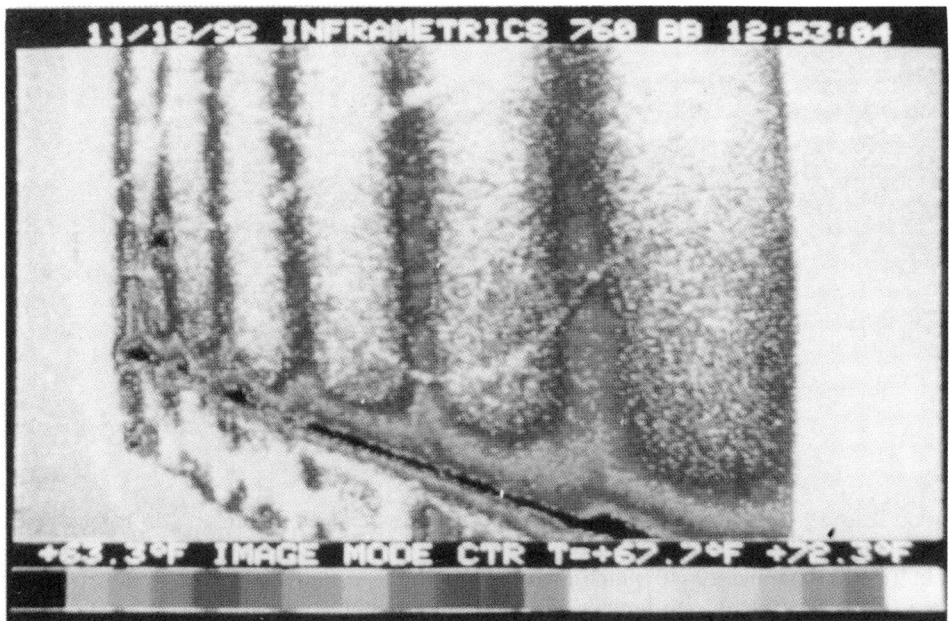

FIG. 10—*Infrared thermal photo of cold-air leakage* (dark) *through to the warm side of the test wall.*

(70°F ΔT) and the 0°C ΔT condition. In two of the four test subsets, however, which tested a perforated polyethylene or polypropylene extrusion coated woven slit film, a noticeable difference did occur (see bold italicized Tests 15, 16, 18, 19 in Table 2 and Figs. 13 and 14). No visual difference was noted in these products at the two ΔT's tested.

TABLE 2—*ASTM E 1424 and E 283 data.*

Subset	Test #	ΔP Pa	Air Flow 39°CΔT cfm/ft²	L/s.m²	Air Flow 0°CΔT cfm/ft²	L/s.m²	Subset	Test #	ΔP Pa	Air Flow 39°CΔT cfm/ft²	L/s.m²	Air Flow 0°CΔT cfm/ft²	L/s.m²
1	1	0	0.00	0.00	0.00	0.00	6	21	10	0.02	0.10	0.02	0.10
1	2	0	0.00	0.00	0.00	0.00	6	22	25	0.04	0.20	0.03	0.15
1	3	4	0.06	0.31	0.06	0.31	7	23	0	0.00	0.00	0.00	0.00
1	4	10	0.15	0.76	0.15	0.76	7	24	10	0.05	0.25	0.04	0.20
1	5	20	0.30	1.53	0.29	1.48	7	25	25	0.08	0.41	0.06	0.31
2	6	0	0.00	0.00	0.00	0.00	8	26	0	0.00	0.00	0.00	0.00
2	7	0	0.00	0.00	0.00	0.00	8	27	10	0.03	0.15	0.02	0.10
2	8	10	0.01	0.05	0.01	0.05	8	28	25	0.04	0.20	0.03	0.15
2	9	25	0.02	0.10	0.02	0.10	9	29	0	0.00	0.00	0.00	0.00
2	10	50	0.03	0.15	0.03	0.15	9	30	12.5	0.00	0.00	0.00	0.00
3	11	0	0.00	0.00	0.00	0.00	9	31	12.5	0.00	0.00	0.00	0.00
3	12	10	0.05	0.25	0.05	0.25	10	32	0	0.00	0.00	0.00	0.00
3	13	25	0.10	0.51	0.09	0.46	*10*	*33*	*10*	*0.13*	*0.66*	*0.09*	*0.46*
4	14	0	0.00	0.00	0.00	0.00	*10*	*34*	*20*	*0.23*	*1.17*	*0.13*	*0.68*
4	*15*	*10*	*0.09*	*0.46*	*0.06*	*0.31*	11	35	0	0.00	0.00	0.00	0.00
4	*16*	*25*	*0.16*	*0.81*	*0.12*	*0.61*	11	36	10	0.01	0.05	0.01	0.05
5	17	0	0.00	0.00	0.00	0.00	11	37	25	0.02	0.10	0.02	0.10
5	*18*	*10*	*0.07*	*0.36*	*0.04*	*0.20*	12	38	0	0.00	0.00	0.00	0.00
5	*19*	*25*	*0.11*	*0.56*	*0.06*	*0.31*	*12*	*39*	*10*	*0.12*	*0.61*	*0.07*	*0.36*
6	20	0	0.00	0.00	0.00	0.00	*12*	*40*	*25*	*0.23*	*1.17*	*0.14*	*0.71*

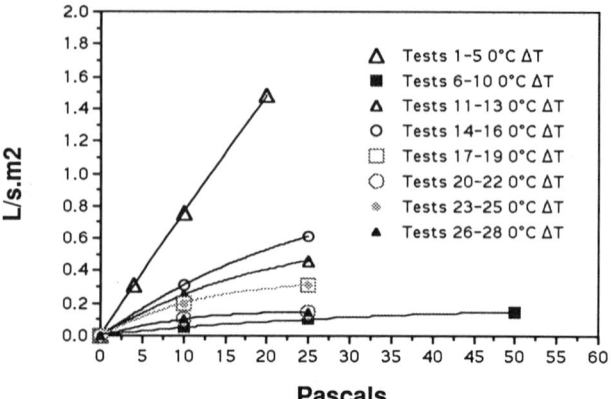

FIG. 11—*Airflow for Tests 1–28 at a 0°C ΔT.*

For Tests 29–31, there was no measurable airflow as expected with an Air-Tight Drywall arrangement.

In Tests 33–34 and 39–40 which involved the shiplap interlocking XPS foam sheathing, there was also a notable difference in air flow depending on the temperature differential across the wall (Figs. 15 and 16). Because the shiplap XPS sheathing in tests 32–40 was installed horizontally using 0.61 × 2.44 m (2 × 8 ft) sheets, approximately 60% more linear butt joint seam was present versus the XPS sheathing used in Tests 1–28. The shiplap sheathing installation arrangement in relationship to internal thermocouple locations is shown in Figs. 17 and 18. The shiplap joints incorporate an overlapping seal which is not present in the squared-off butt joints used in Tests 1–28. Figure 19 shows an arrangement of the butt joints and a cross section of these two joint types. Airflow rates for the shiplap XPS sheathing were very close to the larger, non-shiplap sheet used in Tests 1–28 (see Table 2 Tests 4 versus 33). It appears that the 60% increase in linear butt joint seam, which would intuitively produce higher leakage rates, was compensated for in the tighter butt joint seams created by the interlocking overlap. The 15# felt building paper, used in Tests 38–40, did

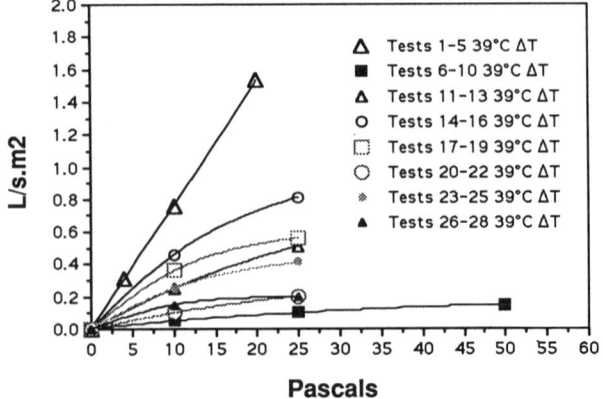

FIG. 12—*Airflow for Tests 1–28 at a 39°C ΔT.*

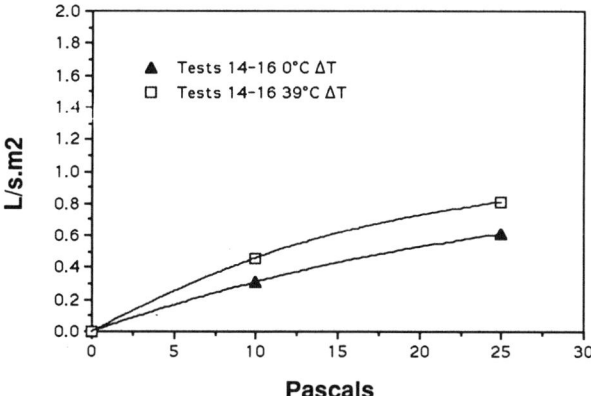

FIG. 13—*Airflow for Tests 14–16 at a 39°C and 0°C ΔT.*

not offer sufficient airflow resistance to significantly upgrade the performance attributed to the shiplap XPS foam sheathing. The flash spunbonded polyethylene housewrap used in Tests 35–37, however, was sufficiently airtight to significantly upgrade the overall airflow resistance of the base wall assembly at both 10 and 25 Pa ΔP. At either of these two pressures, the flash spunbonded polyethylene housewrap reduced the airflow by at least 93% versus the basewall performance and was unaffected by the 39°C temperature differential (Figs. 20 and 21). Airflow test results for Tests 36 and 37 were nearly identical to Tests 8 and 9 as would be expected due to the same housewrap being employed.

Thermal Testing

In conjunction with the airflow testing, thermal tests as per ASTM C 976 were performed. Similar testing was performed in the past by Henning [7] on a small one-by-one-metre specimen. The thrust of this work concerning wall tests simulated "wind washing" by measuring wind velocity versus thermal performance but did not measure actual full penetration

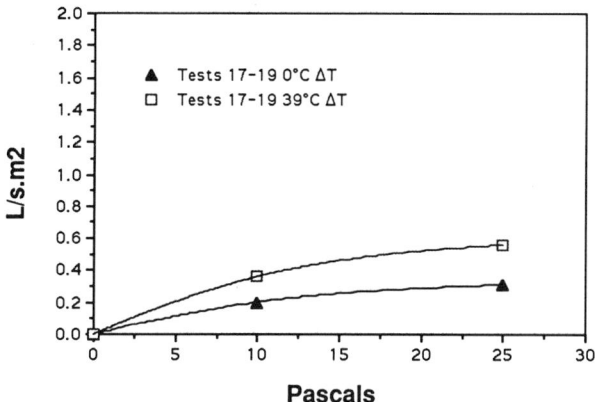

FIG. 14—*Airflow for Tests 17–19 at a 39°C and 0°C ΔT.*

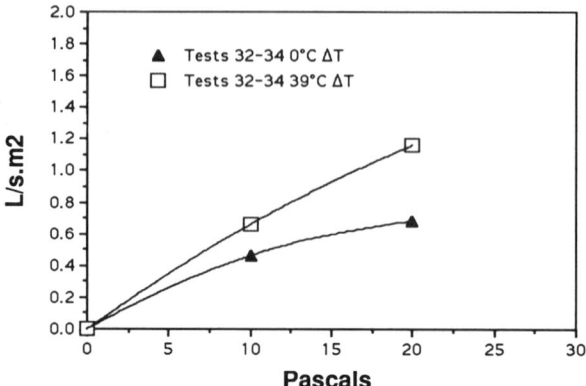

FIG. 15—*Airflow for Tests 32–34 at a 39°C and 0°C* ΔT. (*Note increased airflow as* ΔT *was increased*).

of air through a wall. Thermal performance loss via convection has been well documented by Schulyer and Solvason [8], Lecompt [9], Wolf et al. [10], Berlad [11,12], and in a literature survey done by Powell et al. [13].

The main purpose of this study was to measure the effect of air infiltration on the thermal performance of a typical residential wall. In this type of test, normal thermal losses measured by the C 976 method would be enhanced by direct transfer of cold air into the metering chamber through seams and cracks in the base wall assembly. While the static condition where no airflow is present, normally employed with C 976, is certainly a way to test various forms of thermal insulation under identical conditions for comparative purposes, this condition does not typically occur in the field with wall assemblies. Therefore, thermal performance in the field will most likely not be the same as measured under static conditions in a laboratory.

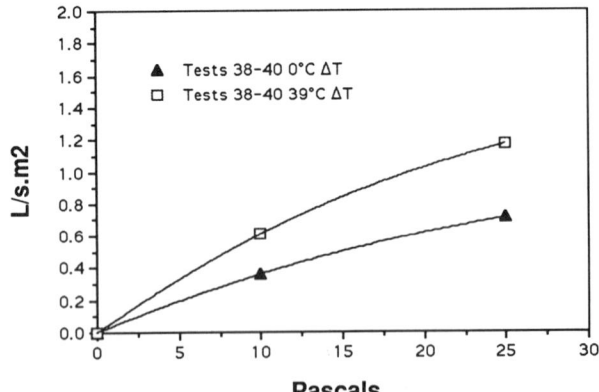

FIG. 16—*Airflow for Tests 38–40 at a 39°C and 0°C* ΔT. (*Note increased airflow as* ΔT *was increased*).

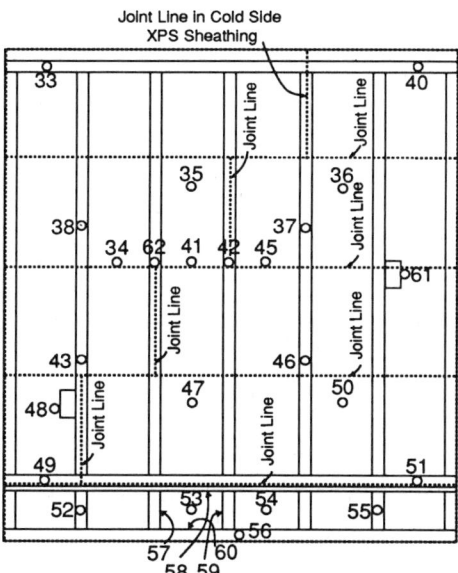

FIG. 17—*Shiplap XSP sheathing butt joint relationship to warm-side internal thermocouple placement.*

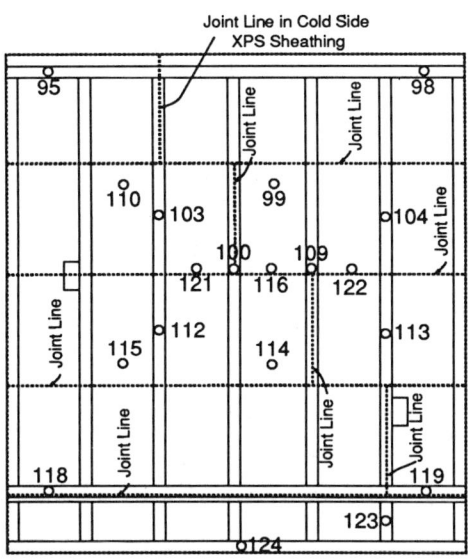

FIG. 18—*Shiplap XPS sheathing butt joint relationship to cold-side internal thermocouple placement.*

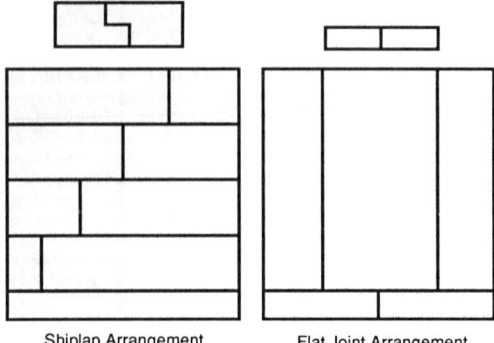

Shiplap Arrangement   Flat Joint Arrangement

FIG. 19—*XPS sheathing arrangement and butt joint cross-section.*

Calculated R-Value of Base Wall Assembly

A parallel heat flow calculation was done on the base wall assembly components described in Table 3 using equations formulated from the 1989 *ASHRAE Handbook Fundamentals,* Chapter 20.8.

Applying the values listed in Table 3 to these equations yields the following calculated R-value for the basewall used for Tests 1–31:

$$0.40 \text{ W/m}^2 \cdot {}^\circ\text{K}$$
$$14.1 \text{ }^\circ\text{F} \cdot \text{ft}^2 \cdot \text{h/Btu}$$
$$0.07 \text{ Btu/h} \cdot \text{ft}^2 \cdot {}^\circ\text{F}$$

Using these equations for the shiplap XPS base wall assembly used in Tests 32–40 the calculated R-value was calculated to be:

$$0.34 \text{ W/m}^2 \cdot {}^\circ\text{K}$$
$$16.9 \text{ }^\circ\text{F} \cdot \text{ft}^2 \cdot \text{h/Btu}$$
$$0.06 \text{ Btu/h} \cdot \text{ft}^2 \cdot {}^\circ\text{F}$$

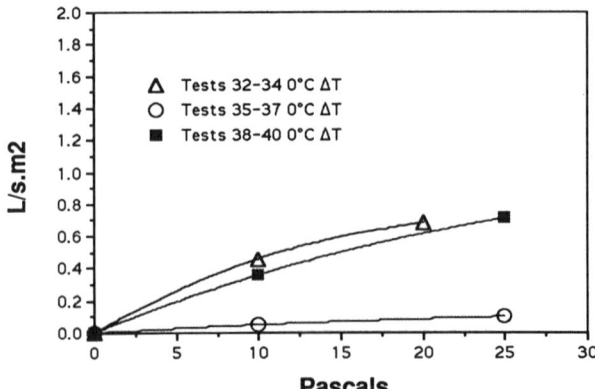

FIG. 20—*Airflow for Tests 32–40 at a 0°C ΔT.*

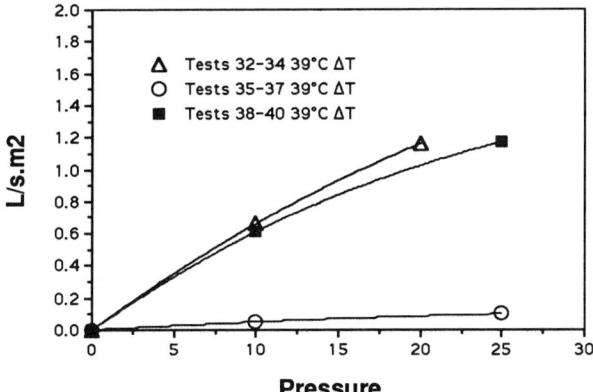

FIG. 21—*Airflow for Tests 32–40 at a 39°C* ΔT.

Thermal Testing Results

As described in the "Test Protocol" section, the ability to vary the degree of airflow through the base wall assembly was achieved by utilizing a variety of housewrap products with different air resistance characteristics. C-value performance was also measured for each of the conditions described in Table 1. Table 4 summarizes the results for each of the 40 tests. In Tests 1–28, the base wall assembly and seven different housewraps were tested at combinations of pressure differentials from 0 to 50 Pa. A relationship between airflow through the wall and apparent C-value can be examined by plotting all the airflow rates and associated heat flow rates from each of the 28 tests (Fig. 22). The curve fit for Fig. 22 is linear and is described by

TABLE 3—*Thermal properties and percentage contribution of base wall assembly components.*

Base Wall (area = 64.2 ft² or 6.0 m²) (89.1% of total area)

Component Element	R-Value per ASTM C518 °F•ft²•h/Btu	°K•m²/W	Area ft²	m²	Percent of Total Area
XPS Sheathing (1.27 cm)	2.8	0.5	64.2	6.0	100%
XPS Shiplap Sheathing (2.794 cm)	5.5	1.0	64.2	6.0	100%
Fiberglass	12.9	2.3	54.5	5.1	84.9%
Vertical Wood Studs	4.6	0.8	6.7	0.6	10.5%
Horizontal Wood Stud Plates	4.6	0.8	4.0	0.4	4.7%
Gypsum Board	0.4	0.1	64.0	6.0	99.7%
Gap at Bottom of Gypsum Board	0.0	0.0	0.2	0.02	0.3%

Sub-Floor (area = 7.8 ft² or 0.7 m²) (10.9% of total area)

Component Element	R-Value per ASTM C518 °F•ft²•h/Btu	°K•m²/W	Area ft²	m²	Percent of Total Area
XPS Sheathing (1.27 cm)	2.8	0.5	7.8	0.7	100%
XPS Shiplap Sheathing (2.794 cm)	5.5	1.0	7.8	0.7	100%
Band Joist	2.0	0.4	6.3	0.6	80.8%
Fiber Glass	12.9	2.3	5.7	0.5	72.3%
Floor Joists	7.2	1.3	0.7	0.6	8.6%
Plywood Floor	9.2	1.6	0.5	0.1	6.4%
Sill Plate	7.2	1.3	1.0	0.1	17.8%

Total Sample Area 72 ft² (6.7 m²)

TABLE 4—*ASTM C 976 data with airflow induced by a ΔP across the wall.*

Subset	Test #	ΔP Pa	Apparent C-Value W/m²·°K	Apparent C-Value Btu/h·ft²°F	Heat Load Btu/h·ft²	Subset	Test #	ΔP Pa	Apparent C-Value W/m²·°K	Apparent C-Value Btu/h·ft²°F	Heat Load Btu/h·ft²
1	1	0	0.42	0.07	4.60	6	21	10	0.47	0.08	5.30
1	2	0	0.44	0.08	4.88	6	22	25	0.55	0.10	6.19
1	3	4	0.71	0.12	7.78	7	23	0	0.41	0.07	4.43
1	4	10	1.15	0.20	12.40	7	24	10	0.61	0.11	6.74
1	5	20	1.81	0.32	19.11	7	25	25	0.69	0.12	7.64
2	6	0	0.41	0.07	4.37	8	26	0	0.40	0.07	4.37
2	7	0	0.41	0.07	4.57	8	27	10	0.50	0.09	5.63
2	8	10	0.43	0.08	4.85	8	28	25	0.59	0.10	6.59
2	9	25	0.46	0.08	5.18	9	29	0	0.40	0.07	4.51
2	10	50	0.50	0.09	5.57	9	30	12.5	0.50	0.09	5.59
3	11	0	0.41	0.07	4.39	9	31	12.5	0.41	0.07	4.59
3	12	10	0.59	0.10	6.55	10	32	0	0.34	0.06	3.84
3	13	25	0.81	0.14	8.86	10	33	10	0.97	0.17	10.52
4	14	0	0.41	0.07	4.41	10	34	20	1.46	0.26	15.48
4	15	10	0.72	0.13	7.91	11	35	0	0.35	0.06	3.83
4	16	25	1.13	0.20	12.24	11	36	10	0.38	0.07	4.28
5	17	0	0.41	0.07	4.44	11	37	25	0.42	0.07	4.72
5	18	10	0.73	0.13	8.08	12	38	0	0.35	0.06	3.74
5	19	25	0.88	0.15	9.67	12	39	10	0.87	0.15	9.38
6	20	0	0.40	0.07	4.37	12	40	25	1.44	0.25	15.06

$$y = 0.391\ 24 \cdot 0.910\ 07x \qquad (R^2 = 0.987) \qquad (2)$$

where

y = C-value, W/m² · K, and
x = airflow, L/s · m².

The apparent thermal conductance of the base wall assembly with the various housewraps applied can also be plotted against pressure differential (Fig. 23). For a given pressure differential, air leakage performance varied between the base wall assembly and the different housewrap products resulting in varying degrees of thermal performance. For a 10 Pa pressure differential, the base wall assembly apparent C-value was decreased from 37% to 63% depending on which housewrap was used. Pressure differentials for the base wall above 20 Pa were not possible due to the degree of leakage and inability of the test equipment to

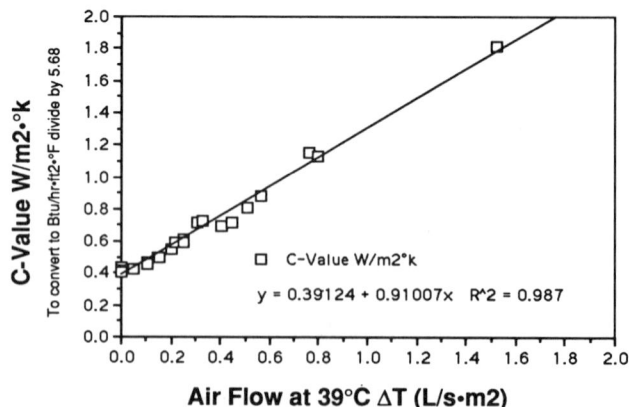

FIG. 22—*C-value versus airflow rate through the test wall for Tests 1–28.*

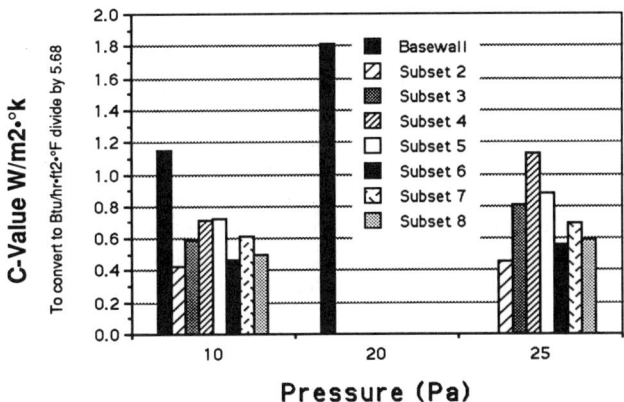

FIG. 23—*C-value versus ΔPressure for various housewraps used in subsets 2–8 and basewall.*

produce a sufficient volume of supply air. At the 0 Pa pressure differential in Tests 1, 6, 11, 14, 17, 20, 23, and 26, a polyethylene film covered the cold side of the base wall assembly and housewrap. There was excellent repeatability in thermal performance for these tests. Figure 24 shows a plot of these tests opposite the calculated C-value for the base wall assembly. The level of repeatability achieved is a good indication that no significant changes in the base wall assembly occurred from test to test. This allows for the ability to make direct comparisons between the first eight test subsets.

Airtight Drywall Simulation Testing Results

Test subset 9, which contained Tests 29–31, was designed to simulate an Airtight drywall installation. Since no air was allowed to completely penetrate the wall, only the effect of wind washing or convective looping or both became a factor in the degradation of the base wall assembly's thermal performance. As with all test subsets, a sheet of polyethylene film

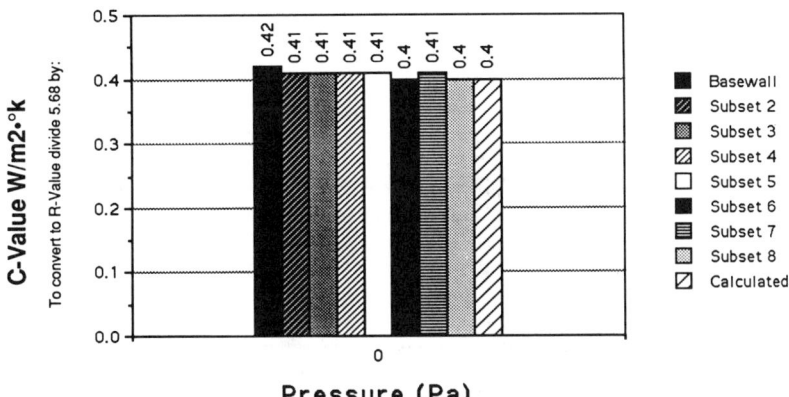

FIG. 24—*C-value versus 0 Pa ΔPressure (covered with polyethylene film) for subsets 1–8 and calculated.*

was placed over the base wall assembly in Test 29 for the 0 Pa ΔP test to ensure that no significant change had occurred to the base wall assembly. In Test 30, the polyethylene film was removed and a 12.5 Pa or 16.4 km/h (10.2 mph) wind pressure was exerted against the exterior XPS foam sheathing. Between Tests 29 and 30, a 25% increase in apparent thermal conductance was measured. In Test 31, a housewrap made from flash spunbonded polyethylene was applied over the base wall assembly configuration used in Test 30. At the 12.5 Pa deadheaded pressure, the apparent thermal conductance increased only 2.5% in relation to Test 29. For this test wall arrangement, using a flash spunbonded polyethylene housewrap in conjunction with an airtight drywall installation decreased the apparent thermal conductance by 18% at a 12.5 Pa deadheaded pressure versus no housewrap. The results of these three tests are shown in Fig. 25.

Shiplap XPS Sheathing Test Results

Test subsets 10–12 containing Tests 32–40 were also tested for thermal performance and, as in Tests 1–28, air was allowed to penetrate the wall. Thermal testing results coincided with the airflow rates as expected from conducting Tests 1–28. A comparison of the relationship between airflow and associated C-value for Tests 1–28 and 32–40 are shown in Fig. 26. The curve fit for Tests 32–40 is described by

$$y = 0.335\ 72 + 0.948\ 68x \qquad (R^2 = 0.998) \tag{3}$$

Since the shiplap XPS sheathed base wall assembly had a higher calculated R-value due to thicker XPS sheathing and since XPS sheathing, being closed cell, tends to be unaffected by airflow, it would be expected that the airflow versus C-value relationship would parallel the curve generated using the thinner XPS sheathing but be skewed downward. Figure 27 shows the relationship between using shiplap XPS sheathing alone versus using a housewrap or 15# felt building paper. There was only a slight thermal performance difference when the 15# felt building paper was applied over the shiplap XPS sheathing as expected since the airflow rates, listed in Table 2, were just slightly less comparing the two arrangements. When a flash spunbonded polyethylene housewrap was applied over the shiplap XPS sheathing, the apparent thermal conductance (C-value) was decreased from C-0.97 to C-0.38 W/m² ·

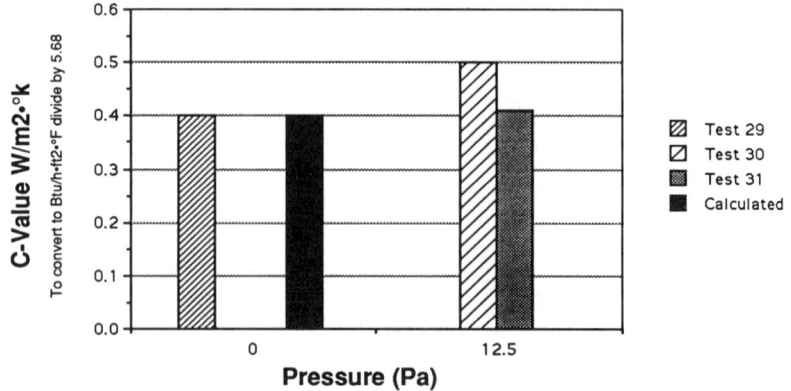

FIG. 25—*Airtight drywall thermal test results.*

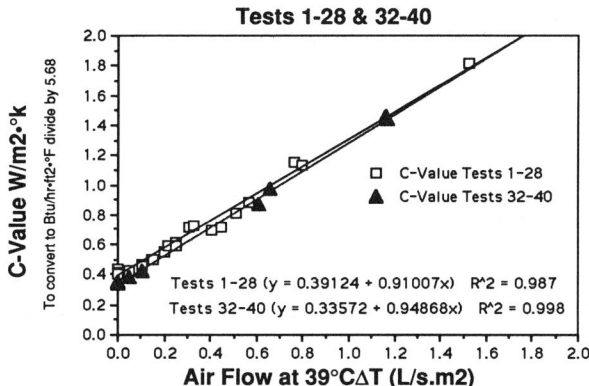

FIG. 26—*C-value versus flow rate relationships between Tests 1–28 and Tests 32–40.*

K for a 10 Pa ΔP and C-1.46 to C-0.42 W/m² · K for a 25 Pa ΔP. This represents a C-value decrease of 61% and 71% respectively.

Airflow Leakage Analysis

The 40 tests that were conducted produced a variety of air leakage rates that adversely affected the thermal performance of the test wall assembly. The 47 internal thermocouples were installed to aid in identifying air leakage paths by measuring temperature in specific locations. The thermocouple locations selected, shown in Figs. 5–6 and 11–12, were chosen based on suspected leakage areas inherent in the wall system design and locations where there would be little chance of airflow interfering with the performance of the insulation. For Tests 1–28 and 32–40, suspected leakage areas would be along butt joint seams in the XPS sheathing, at the electrical boxes, and along the bottom of the gypsum drywall. For

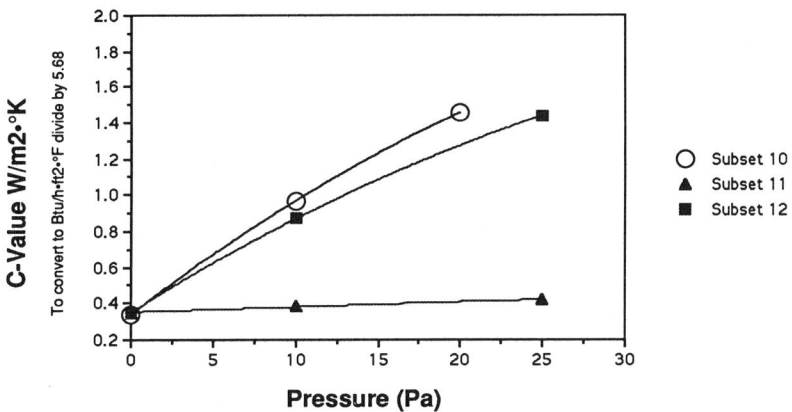

FIG. 27—*Thermal performance comparison between basewall assembly covered with shiplap XPS sheathing, and addition of a 15# felt building paper (subset 12) or flash spunbonded polyethylene housewrap (subset 11).*

each of the three distinct basewall assembly arrangements, an average of all the temperature readings was compiled for each of the 28 warm-side and 19 cold-side thermocouples. Figure 28 shows the average for each of the 28 warm-side internal thermocouples for Tests 1–28. Referring to Fig. 5 for thermocouple locations and examining Fig. 28 for averages with large standard deviations helps define the locations most affected by the airflow through the wall. Temperature averages with large standard deviations are indicative of locations where airflow was most prominent since these areas would be subjected to a larger temperature drop as airflow is increased than areas that were not within the mainstream airflow paths. As expected, thermocouples 61, 48, 51, and 49 had large standard deviations and were located at suspected leakage sites around the electrical boxes and at the bottom of the gypsum drywall. There also appears to be some leakage occurring between the XPS butt joint around thermocouple 62 and possibly around No. 40. Thermocouples located in a mid-stud cavity such as 34, 35, and 45 stayed at a fairly constant temperature regardless of airflow rate.

The 19 thermocouples on the opposite side of the wall (see Fig. 6) reveal similar findings. Figure 29 shows a different but equally explainable profile. Again, thermocouples located at butt joint seams such as 109, 118, and 119 measured large differences over the 28 tests. Thermocouples 95 and 98 also measured large differences but, surprisingly, the associated thermocouples on the opposite (warm) side (Nos. 33 and 40) did not measure extraordinarily large swings in temperature. Perhaps this might indicate that some convection was taking place in the interface between the fiber glass and the inside surface of the XPS sheathing. As with the warm side, the thermocouples located away from the butt joint seams had much less temperature variability from test to test.

Similar results were found for Tests 32–40 depicted in Figs. 30 and 31. Since there was approximately 60% more linear butt joint seam with these smaller shiplap XPS sheathing boards (see Figs. 17 and 18), a higher number of thermocouples on both sides of the wall had large standard deviations. On the warm side the thermocouples closest to electrical boxes or opposite butt joint seams had the higher variability in temperatures. Areas away from seams or known leakage paths had the lower degrees of temperature variability as expected. On the cold side, almost all thermocouples were affected by the airflow having three standard deviation ranges above 3°C.

For Tests 29–31, which simulated the airtight drywall approach, much fewer locations were found where the temperature varied. Figures 32 and 33 reveal only four locations where

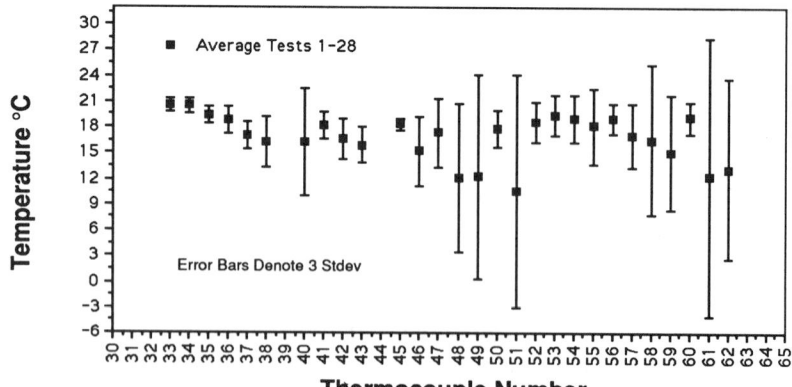

FIG. 28—*Warm-side internal thermocouple averages with 3 standard deviation error bars for Tests 1–28.*

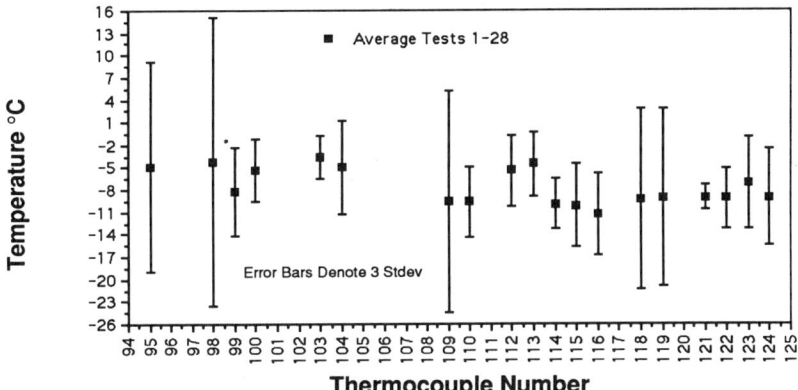

FIG. 29—*Cold-side internal thermocouple averages with 3 standard deviation error bars for Tests 1–28.*

there was high variability in temperature. Refer again to Figs. 5 and 6 for XPS sheathing placement. On the warm-side thermocouple 61, located at the electrical switch box, appears to have been supplied by a flow of air through a butt joint seam directly opposite on the cold side. Thermocouple 62 also appears to be affected by the same type of path. On the cold side the locations with the greatest temperature variability, 109 and 116, are also located at or near a seam.

Heat Flow Transducer Analysis

Table 5 gives the voltage outputs for each transducer for each test. Values over 450 μV have been highlighted in bold and italicized to help identify patterns. Since the transducers were not calibrated prior to the testing, it is not possible to equate the voltage outputs to heat flow. There is, however, patterns which occur that relate to the locations described in Fig. 7. Transducer 6, located in the bottom right corner facing the inside of the wall, was

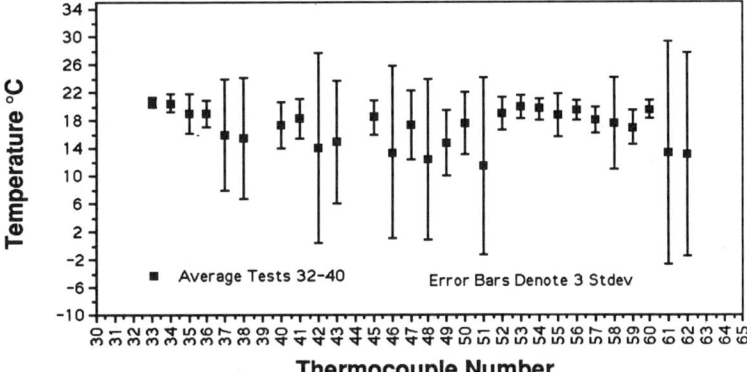

FIG. 30—*Warm-side internal thermocouple averages with 3 standard deviation error bars for Tests 32–40.*

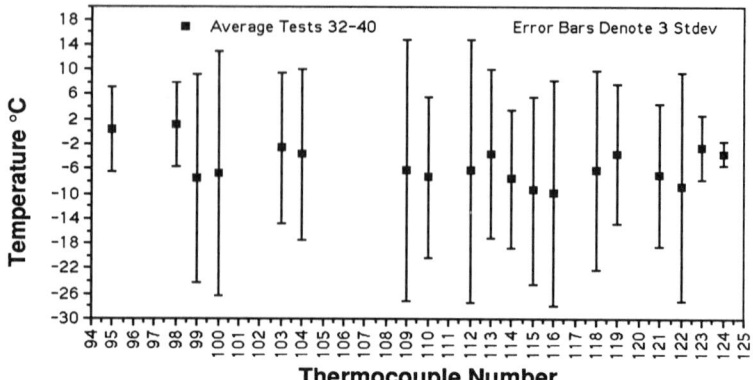

FIG. 31—*Cold-side internal thermocouple averages with 3 standard deviation error bars for Tests 32–40.*

in a location where an air leakage path existed between the sole plate and the plywood subfloor. This transducer averaged much higher overall voltages and hence heat flow throughout the 40 tests. This was also true to a lesser extent for transducer 5.

In Tests 4, 5, 33, 34, 39, and 40 where there was just the base wall or the base wall and a 15# felt, the majority of transducer locations registered high voltages (in excess of 450 μV), indicating that air leakage was occurring at various locations.

Conclusions

A careful integration of ASTM E 1424 and ASTM C 976 has demonstrated that an adverse effect in thermal performance of a wall system can occur when varying degrees of airflow, driven by a pressure differential, are allowed to fully penetrate. Test results for the wall assemblies reveal that airflow rates as low as 0.2 L/s · m^2 can produce a 46% increase in

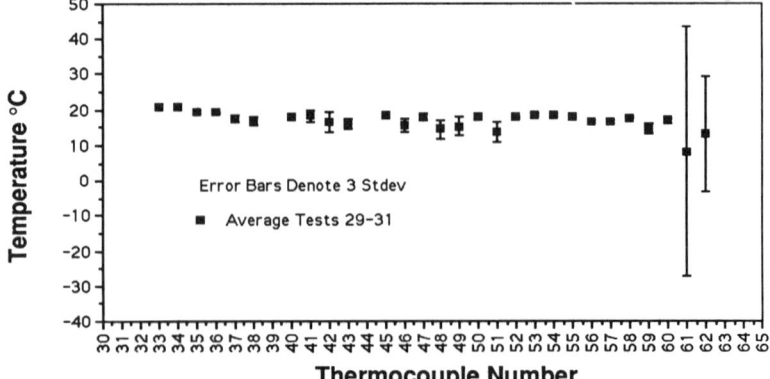

FIG. 32—*Warm-side internal thermocouple averages with 3 standard deviation error bars for Tests 29–31.*

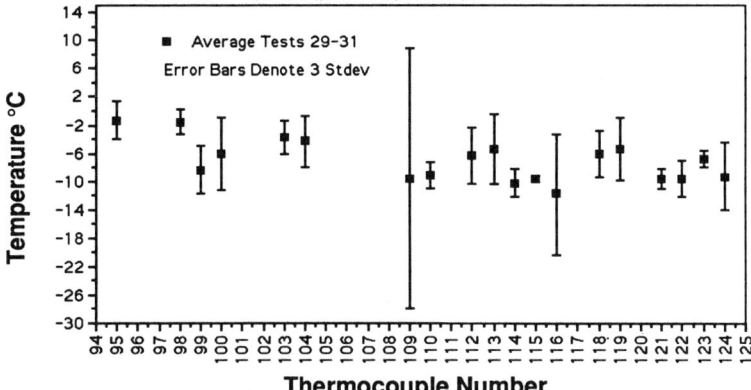

Thermocouple Number

FIG. 33—*Cold-side internal thermocouple averages with 3 standard deviation error bars for Tests 29–31.*

apparent thermal conductance (C-value). It has been demonstrated that sheet-type air retarders or housewraps can decrease apparent thermal conductance (C-value) by 37% to 63% when installed over XPS sheathing by slowing the flow of air into and through the wall for a pressure differential of 10 Pa. It was also demonstrated that these housewraps preserve thermal performance at different levels depending on the air penetration resistance capability of the product.

For some housewraps installed in conjunction with the base wall assembly, there appears to be a difference in the values obtained for airflow resistance using the E 1424 test protocol versus the E 283 protocol for the same ΔPressure. This may be caused by temperature-driven dimensional changes which could open voids in the sheet and diminish airflow resistance.

TABLE 5—*Voltage readings from heat flow transducers.*

Test	Heat Flow Transducer Number Units in μVolts									Test	Heat Flow Transducer Number Units in μVolts								
	0	1	2	3	4	5	6	7	Avg.		0	1	2	3	4	5	6	7	Avg.
1	300	319	339	326	346	377	392	317	340	21	336	359	336	379	377	386	430	348	369
2	368	358	320	362	395	373	414	351	368	22	343	364	343	392	399	413	446	359	382
3	364	345	337	364	*454*	*577*	*518*	392	419	23	315	330	322	350	345	371	401	338	347
4	*531*	423	345	*451*	*579*	*742*	*597*	*457*	*516*	24	350	361	346	400	419	451	*457*	365	394
5	*669*	496	354	*537*	*693*	*967*	*698*	*554*	*621*	25	365	370	356	428	447	*508*	*474*	377	416
6	298	317	326	333	338	339	376	305	329	26	315	330	319	349	336	358	395	331	342
7	326	343	314	357	332	367	400	330	346	27	342	356	336	378	389	338	434	348	365
8	330	349	313	373	344	376	421	346	357	28	349	365	342	391	408	417	*453*	354	385
9	341	360	315	375	364	383	426	349	364	29	324	344	331	356	363	379	421	338	357
10	326	345	337	365	380	382	415	331	360	30	432	387	359	370	717	284	379	374	413
11	321	339	305	355	329	366	409	335	345	31	335	353	332	367	367	372	414	343	360
12	359	371	345	374	349	431	*463*	356	381	32	260	288	278	290	287	314	354	299	296
13	435	439	324	419	*525*	*569*	*522*	395	*454*	33	*741*	*543*	*515*	*576*	*581*	346	*544*	*689*	*567*
14	312	329	317	344	335	375	399	324	342	34	*906*	*690*	*626*	*827*	*705*	372	*611*	*828*	*696*
15	355	377	333	414	449	*459*	481	372	405	35	269	292	278	295	294	313	346	295	289
16	382	419	340	*491*	*554*	*579*	*531*	408	*463*	36	294	314	296	323	332	317	369	310	319
17	311	333	314	345	335	363	398	326	341	37	309	328	309	361	365	326	394	333	341
18	363	381	347	413	445	*612*	*532*	390	435	38	268	266	274	263	297	309	367	303	293
19	381	394	352	430	*478*	*711*	*564*	414	*466*	39	*735*	*579*	*428*	*709*	*567*	319	*584*	*613*	*567*
20	315	333	322	351	345	369	396	330	345	40	*982*	*747*	*542*	*913*	*748*	342	*682*	*798*	*719*

On tests done without housewraps, temperature-driven dimensional changes in the wall components were found to adversely affect the seal created by the interlocking shiplap XPS sheathing joints.

Creating an air retarder on the interior by sealing the gypsum drywall on the base wall assembly appears to perform when the wall system is exposed to a 12.5 Pa wind pressure at a C-value equivalent to a base wall assembly configuration which allows 0.15 L/s · m^2 airflow to penetrate completely through. It was demonstrated for the wall assembly tested that an 18% decrease in apparent thermal conductance at a 12.5 Pa deadheaded pressure can be achieved by utilizing a flash spunbonded polyethylene housewrap on the exterior to reduce the effect of "wind washing" in conjunction with sealing the gypsum drywall on the interior.

Based on the smoke pencil testing, infrared thermography, and internal thermocouple readings, two major leakage pathways can be identified through the test wall assemblies. One of the major air leakage pathways can be identified as butt joint seams in the exterior XPS sheathing which channeled air to various pathways within the wall, exiting at the electrical boxes and out through the fixtures. The other major air leakage pathway identified was along the interface between the sole plate at the bottom of the wall and the plywood on top of the subfloor.

References

[1] Quirouette, R. L., "The Air Barrier Defined," *An Air Barrier for the Building Envelope, Proceedings of Building Science Insite '86,* National Research Council Canada, 1989, pp. 1–5.

[2] Lux, M. E. and Brown, W. C., "Air Leakage Control," *An Air Barrier for the Building Envelope, Proceedings of Building Science Insite '86,* National Research Council Canada, 1989, pp. 13–19.

[3] Perreault, J. C., "Air Barrier Systems: Construction Applications," *An Air Barrier for the Building Envelope: Proceedings of Building Science Insite '86,* National Research Council Canada, 1989, pp. 20–24.

[4] Kehrli, D. W., "Window and Door Air Leakage Tests Predict Real World Performance," *Fenestration,* Nov.-Dec. 1989, pp. 18–21.

[5] Kehrli, D. W., "Window Air Leakage Performance as a Function of Differential Temperatures and Accelerated Environmental Aging," *Thermal Performance of the Exterior Envelopes of Buildings III,* 2–5 Dec. 1985, pp. 872–890.

[6] "Air Permeance of Building Materials," Air-Ins, Inc., Prepared for Canada Mortgage and Housing Corp., 1988.

[7] Henning, G. N., "Energy Conservation with Air Infiltration Barriers," *Thermal Insulation, Materials, and Systems for Energy Conservation in the '80s, ASTM STP 789,* F. A. Govan, D. M. Greason, and J. D. McAllister, Eds., American Society for Testing and Materials, Philadelphia, 1983, pp. 551–558.

[8] Schuyler, G. D. and Solvason, K. R., "Effectiveness of Wall Insulation," *Thermal Insulation, Materials, and Systems for Energy Conservation in the '80s, ASTM STP 789,* F. A. Govan, D. M. Greason, and J. D. McAllister, Eds., American Society for Testing and Materials, Philadelphia, 1983, pp. 542–550.

[9] Lecompte, J. G. N., "The Influence of Natural Convection in an Insulation Cavity on Thermal Performance of a Wall," *Insulation Materials Testing and Applications,* American Society for Testing and Materials, Philadelphia, 1987.

[10] Wolf, S., Solvason, K. R., and Wilson, A. G., "Convective Air Flow Effects with Mineral Wool Circulation in Wood-Framed Walls," *ASHRAE Transactions,* Vol. 72, Part II, 1966, pp. III 3.1–III 3.8.

[11] Berlad, A. L., Tutu, N., Jaung, R., and Yeh, Y-J., "Energy Transport in Porous Insulator Systems," ORNL/Sub-7551/1, Oak Ridge National Laboratory, TN, 1979.

[12] Berlad, A. L., Jaung, R., Joshi, N., and Westerinen, J., "Energy Transport in Porous Insulator Systems," ORNL/Sub-7551/2, Oak Ridge National Laboratory, TN, 1982.

[13] Powell, F., Krarti, M., and Tuluca, A., "Air Movement Influence on the Effective Thermal Resistance of Porous Insulations: A Literature Survey," *Journal of Thermal Insulation,* Vol. 12, Jan. 1989, pp. 239–251.

Discussion

Phil Hendrickson[1] *(written discussion)*—A portion of the discussion highlighted the difference in air leakage performance of wall assemblies covered on the exterior with (1) two different widths of extruded polystyrene XPS sheathings and (2) the same assemblies covered with "housewrap" fabrics. However, it is clear that the initial wall construction was not put together in accordance with common construction practice as outlined in building energy codes or with the same level of workmanship as that displayed when the housewrap materials were attached. For example, there was no attempt to seal the plate of the frame wall to the plywood subfloor and this was emphasized as being a major problem area for the leakage of air through a wall assembly. It is a well-known fact and leaving this area unsealed is generally considered poor workmanship. In contrast to this, it is quite clear if you examine a building energy code such as the Council of American Building Officials Model Energy Code that this joint should be sealed. Common practice will result in this area being sealed with gasketing, caulking, etc. Because a joint in the sheathing was also located at this joint, leaving this joint unsealed had a significant impact on the assembly with the XPS sheathing. In comparison, the housewrap fabrics were installed with good construction practice because they were sealed around the entire perimeter as commonly suggested by the manufacturers. The level of workmanship between the two tested assemblies is inconsistent and as a result is not an apples-to-apples comparison. What is the explanation for changing the level of workmanship from poor to good?

D. C. Jones, D. G. Ober, and J. T. Goodrow (authors' closure)—The basewall assemblies used in this testing were built to common practice for the materials used. Whenever possible, manufacturers' installation instructions were used regarding the installation of their products as was the case with the installation of the XPS sheathing. Tilt wall is the predominant method of constructing vertical walls and exterior sheathing is normally applied prior to tilting the wall up to its vertical position. Exterior sheathing is typically sized in 4 ft by 8 ft (1.3 m by 2.4 m) sheets and aligned with the bottom plate of the wall. This causes a horizontal seam in the exterior sheathing that aligns with the joint between the plywood subfloor and the bottom plate of the vertical wall. While a prudent builder might seal this area, it is not necessarily common practice and not typically enforced by the building officials regardless of which energy related code or program might specify this practice. Air leakage through joints between sheathing is also addressed by several prominent building energy codes and it is highly improbable that a builder would address these areas by any means other than using a housewrap. If a housewrap was used, the amount of joint area to address is significantly reduced. Nevertheless, Mr. Hendrickson does raise an interesting question. Would there be a significant decrease in air flow through a wall if the joint between the bottom plate and the plywood subfloor were sealed? To answer this question, additional testing at Holometrix, Inc. was conducted in February 1994 to measure the impact that sealing this area has on air leakage rate. The same test protocols were used as described in this paper except that the basewall was 2 × 6 construction and 4 ft by 8 ft by ½ in. (1.3 by 2.4 by 12.7 mm) foil faced polyisocyanurate exterior sheathing was used. Figure 34 shows a cross-section of the basewall assembly and the locations where caulk was used to seal seams and joints. Ten tests were conducted with the arrangements described in Table 6.

---

[1] The Dow Chemical Company, Danville, Ohio.

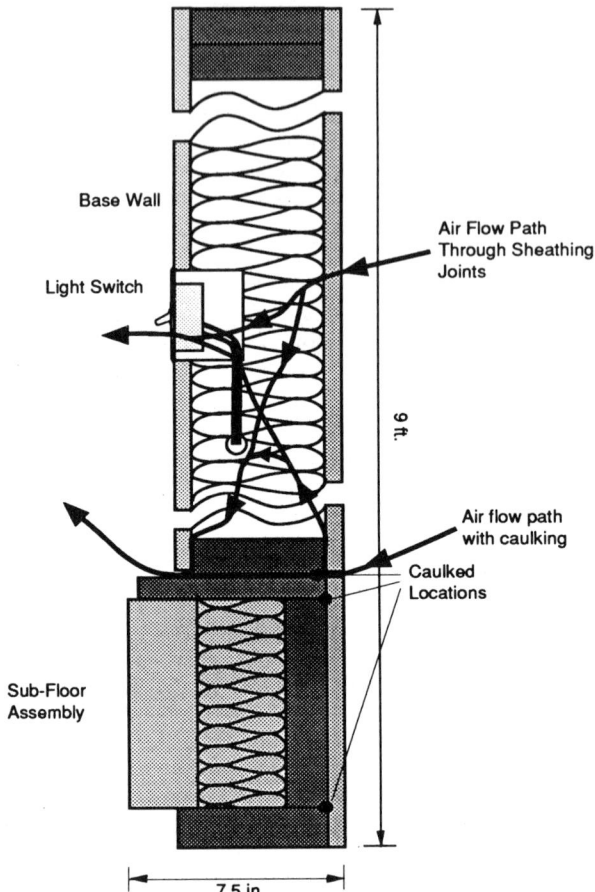

FIG. 34—*Cross-section of basewall showing airflow leakage paths.*

ASTM E 1424 testing on the ten test configurations is reported in Table 7. The tests results clearly show that the addition of caulking in the areas shown in Fig. 34 did not significantly impact the overall air leakage rate of the wall. Comparing the results of Test 2 with Test 7, and Test 3 with Test 8, show only minor air leakage reductions with the caulked joints. The addition of the caulked joints served only to alter the path of least resistance for air flowing through the wall assembly. The arrows drawn on Fig. 34 illustrate the potential air leakage pathways. If the caulking were not present, undoubtedly, air would leak through the joint between the plywood subfloor and the bottom plate. A comparison between XPS sheathing and polyisocyanurate sheathing with regard to joint movement due to thermal expansion reveal that the coefficient of linear expansion of polyisocyanurate is about half of the XPS sheathing tested in this paper. It is therefore likely that less joint movement occurred with the polyisocyanurate sheathing, yielding less potential for air leakage at the butt joints with the most recent ten tests than would have occurred if an XPS sheathing had been used. If these butt joints were sealed with caulk or tape, the coefficient of thermal expansion for either polyisocyanurate or XPS sheathing could create difficulty in maintaining a seal due to continuous thermal cycling and the relatively low elasticity of typical caulk and tape

TABLE 6—*Test configurations for evaluating the impact of caulking joints in a wall assembly.*

Test	Description	Δ Pressure (Pa)
1	Basewall assembly (see Fig. 34) without caulk but sealed with plastic film on the exterior.	0
2	Same basewall without the plastic film on the exterior.	10
3	Same basewall as in test 2.	25
4	Same basewall as test 2 with a flash spunbonded polyethylene housewrap on the exterior.	10
5	Same basewall as in test 4.	25
6	Same basewall as in test 1 but with seams and interfaces caulked as shown in Fig. 34.	0
7	Same basewall as in test 6 without the plastic film on the exterior.	10
8	Same basewall as in test 7.	25
9	Same basewall as test 7 with a flash spunbonded polyethylene housewrap on the exterior.	10
10	Same basewall as in test 9.	25

products. The addition of the flash spunbonded polyethylene housewrap in Tests 4, 5, 9, and 10, however, did cause a significant and consistent reduction in air leakage through the wall assembly either with or without the caulked areas illustrated in Fig. 34. Typically, housewrap products are sufficiently elastic to change with the dimensions of the wall from thermal cycling which reduces the stress on the taped seams.

In conclusion, it appears that selecting areas in the framing joints of a wall assembly to caulk or seal does little to reduce air infiltration into and through a wall without addressing the joints and interfaces between the framing and the sheathing and the sheathing to sheathing butt joints. The use of a flash spunbonded housewrap with taped seams can provide sealing continuity over all these framing and sheathing joints and interfaces effectively reducing the overall joint/seam area to a manageable level.

Anton TenWolde[2] (written discussion)—All the tests were done with infiltrative air pressures. What would be the effect of reverse (exfiltrative) air pressures on the results?

TABLE 7—*Test results from ASTM E 1424 testing on configurations described in Table 1.*

	Air Leakage		Δ Pressure,
Test	L/s · m²	cfm/ft²	(Pa)
1	0	0	0
2	0.52	0.10	10
3	1.05	0.21	25
4	0.05	0.01	10
5	0.09	0.02	25
6	0	0	0
7	0.46	0.09	10
8	0.76	0.15	25
9	0.05	0.01	10
10	0.11	0.02	25

D. C. Jones, D. G. Ober, and J. T. Goodrow (authors' closure)—Without having performed these tests with air flow in the reverse direction (exfiltration), we can only speculate as to the impact on the basewall. With air infiltration, there is a tendency to push some building components together such as the exterior extruded polystyrene (XPS) sheathing to the stud framing or a housewrap to the sheathing. This should serve to decrease air flow at a given pressure. When pressure is reversed, there could be an impact on the tightness of the XPS/framing and the housewrap/sheathing interfaces. Since the XPS sheathing is discontinuous, we should speculate that there might be an increase in airflow due to the opening up of joints between the sheathing. The airflow would also tend to push the housewrap away from the sheathing and expose more surface area at pressure which can lead to higher leakage rates with some housewraps. The air leakage rate of the flash spunbonded polyethylene housewrap tested in this paper measured alone without the basewall at 25 Pa was 0.06 L/s-m^2 (0.01 cfm/ft^2) using a test protocol developed at Air-Ins. Inc. in Quebec Canada. The test results in this paper for the flash spunbonded polyethylene housewrap, using ASTM E 283 for air infiltration with a basewall, leaked at a rate of 0.10 L/s-m^2 (0.02 cfm/ft^2) which is comparable to the nonbasewall test results. However similar performance in the infiltration versus exfiltration air leakage mode was not consistent between housewraps. For example, the air leakage rate of the pin perforated extrusion coated woven polyethylene slit film housewrap (subset 5) measured alone without the basewall at 25 Pa was 0.48 L/s-m^2 (0.10 cfm/ft^2). The test results in this paper for the pin perforated extrusion coated woven polyethylene slit film housewrap, using ASTM E 283 for air infiltration with a basewall, leaked at a rate of 0.31 L/s-m^2 (0.06 cfm/ft^2) so it would appear that this pin perforated housewrap benefited by being in intimate contact with the sheathing in the infiltration test condition. This same benefit was also achieved by another pin perforated housewrap. The pin perforated extrusion coated spunbonded polypropylene housewrap (subset 3) had an alone versus basewall air leakage performance of 1.33 L/s-m^2 (0.26 cfm/ft^2) versus 0.46 L/s-m^2 (0.09 cfm/ft^2).

The presence of pin perforations in a housewrap appears to create a performance difference between infiltration versus exfiltration. This performance difference was not observed in the nonperforated flash spunbonded polyethylene housewrap.

---

[2] Forest Products Laboratory, Madison, WI.

Envelope and
Distribution System Leakage

.

Per A. Levin,[1] *David J. Wilson,*[2] *and Mark Y. Ackerman*[2]

Air Leakage in the Perspective of International Standards

REFERENCE: Levin, P. A., Wilson, D. J., and Ackerman, M. Y., **"Air Leakage in the Perspective of International Standards,"** *Airflow Performance of Building Envelopes, Components, and Systems, ASTM STP 1255,* Mark P. Modera and Andrew K. Persily, Eds., American Society for Testing and Materials, Philadelphia, 1995, pp. 231–247.

ABSTRACT: There exist a number of national standards and a draft international standard for the fan pressurization method for measuring air leakage. Although the standardized methods in principle are the same, the way of interpreting and presenting the results is different.

In previous studies, houses that have a relatively large leakage area at a low pressure difference (4 to 10 Pa) still can seem comparatively airtight at a high pressure difference (50 Pa). This fact is a consequence of differences in the flow exponent in the power-law equation, which is the normal equation used to fit to the data points, and can be a source of error when trying to compare the relative airtightness of houses. Extrapolating results from high pressure differences to low pressures, which are out of the measured range, can thus result in substantial errors. Air leakage testing of windows normally starts at 50 Pa, which should be accounted for when trying to use these results as inputs in network air infiltration models. Measurement results on low pressure air leakage are discussed in the paper and compared with high pressure air leakage.

Pressurization test data from 105 tests in one house at the Alberta Home Heating Research Facility are used for the study. The tests were made automatically over a seven-month period in low wind conditions. A wide range of pressure differences were tested and the results cover the test specifications for most standards. In addition to comparing standards, these tests were used to measure seasonal effects on air leakage in a wood-frame house with a plastic film air/vapor barrier.

The results show some significant differences between the standards, and also a variation with month of test, indicating a seasonal variation in air leakage.

KEYWORDS: air leakage measurements, fan pressurization method, standards, wood-frame house, seasonal variation, pressure differences

The most common technique of measuring the overall air leakage of a building is the fan pressurization method. The measurement principle is shown in Fig. 1. A fan capable of creating more than 50 Pa pressure difference over the building envelope is mounted in a doorway or in a window opening. The airflow through the fan is measured for incremental steps in pressure difference ranging typically from ±10 to ±100 Pa. A characteristic leakage curve is then calculated from which the air leakage of the building envelope at a given pressure difference may be determined. A power law is the most common equation used to fit the measured data using some form of least-squares method:

[1] The Royal Institute of Technology, Division of Building Technology, S-100 44, Stockholm, Sweden.
[2] University of Alberta, Department of Mechanical Engineering, Edmonton, Alberta, Canada T6G 2G8.

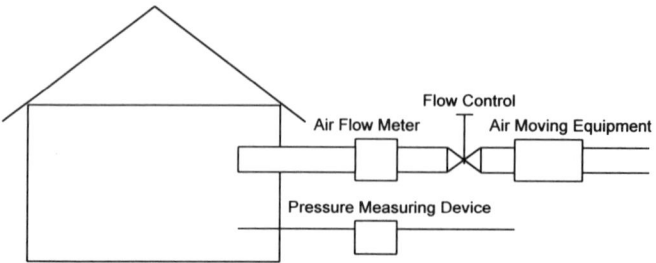

FIG. 1—*Measurement principle for the fan pressurization method.*

$$q = C\Delta p^n \tag{1}$$

where

q = airflow, m³/s
C = flow coefficient, m³/sPa^n
Δp = pressure difference, Pa, and
n = flow exponent, "−" (minus sign).

The method is fairly simple to use for detached houses compared to most tracer gas techniques and is suitable as a method for controlling building airtightness when used by local building inspectors. Many national standards exist and also a draft international standard, which are the same in principle, although they differ in detail. The most important differences are summarized in Table 1.

As can be seen from Table 1, there are different ways of interpreting and presenting the results from this method. Calculation procedures and curve representations are given in most of the standards. In Sweden, the 50 Pa average value has traditionally been the reported value [5]. The reasoning behind this was to obtain a reproducible result on building envelope air leakage that was less dependent on test conditions. For some standards, the power-law curve fit is used, while other standards represent air leakage as an equivalent open area at a reference pressure.

The purpose of this paper is to examine if differences between the standards cause differences in results that are significant for deciding the amount of air leakage. In particular the selection of pressure differences and air leakage representation at high or low pressures is examined.

The equation for equivalent leakage area, AL, given in the ASTM Method for Determining Air Leakage Rate by Fan Pressurization Test (E 779-87) is:

$$AL = C \cdot \Delta p^{n-0.5} \cdot (\rho/2)^{0.5} \qquad m^2 \tag{2}$$

where

C = flow coefficient, m³/sPa^n,
Δp = pressure difference, Pa,
n = flow exponent, "−" (minus sign), and
ρ = air density, kg/m³.

The normal reference pressure difference is 4 Pa, but other values may be used if stated. The unit for airflow measurements is m³/s. The air density is usually calculated at indoor

TABLE 1—*Overview of important differences between standards for the fan pressurization method for overall air leakage measurements (modified from Charlesworth [1] and Elmroth and Levin [2].*

Standard/Country	Differential Pressure Range[a]	Pressure Tap Location	Limiting Conditions	Equipment Accuracy: A = Airflow P = Pressure	Expression of Results
CAN/CGSB-149.10-M86 Canada [3]	15–50 only –	Averaging container with ≥4 taps	Wind speed < 5.6 m/s	A = ±5% P = ±2% Pa	Equivalent leakage area at 10 Pa
NEN 2686 Netherlands [4]	15–100 + or –	1 tap at facade	Nat. Δp < 5 Pa Wind < 6 m/s	A = ±5% P = ±5%	Flow coefficient, flow rate in m³/s at 1 and 10 Pa
SS 021551 Sweden and NS-Insta 130 Norway [5]	20–55 + and –	1 tap with T-piece 10 m from building	Wind < 6 m/s 10 m from building	A = ±6% P = ±3 Pa Overall: ±10%	Flow-ΔP graph, ACH/h at ±50 Pa (mean)
ASTM E 778-87 USA	12.5–75 + and –	1 tap, location not given	Ideal winds. < 2 m/s temp. 5–35°C	A = ±6% P = ±2.5 Pa	Flow-ΔP graph, effective leakage area (4 Pa)
ISO/DIS 9972 International [6]	10–60 + and –	1 tap, ideally at neutral plane; average of facades if differences	Nat. Δp < 3 Pa $\Delta T \cdot h$ < 200	A = ±5% P = ±5%	Flow-ΔP graph, flow coefficient and exponent, other derived quantity

[a] – = Internal underpressure; + = Internal overpressure (Pa).

conditions. Equation 2 is also given in the International Organization for Standardization (ISO)/DIS 9972 standard [6] as one way to express the results.

The Canadian General Standards Board (CGSB) 149.10-M86 standard [3] is based on the following equation for the equivalent leakage area, *ELA*, at a 10 Pa reference pressure difference:

$$ELA = 0.001\ 928\ 3 \cdot C_d \cdot \rho^{0.5} \cdot C \cdot 10^{n-0.5} \qquad m^2 \qquad (3)$$

where

C_d = coefficient of discharge, 0.60, −,
C = flow coefficient, L/sPa^n,
n = flow exponent, and
ρ = air density, kg/m^3.

The unit for airflow measurement is L/s, air density at reference indoor conditions. In the CGSB standard, the recommended regression technique, used to find C and n, is a slightly modified (weighted) least-squares method. The CGSB standard also defines leakage area differently from the ASTM and ISO standards. It assumes a sharp-edged orifice with a discharge coefficient of 0.6, compared with $C_d = 1.0$ implied in Eq 2.

Table 1 shows how the different standards prescribe pressure difference measurements. The standards call for 2 to 3 Pa or about 5% maximum error in pressure difference measurements. One major problem is finding a representative location for the outdoor pressure taps that is not strongly affected by wind speed or wind direction. In most standards, except for the Canadian and ISO standards, this has been given little attention. Moderate wind speeds can cause a significant bias in the pressure difference measurements [7].

There are also differences in methods of normalizing the results. For example, some standards include basement areas in the internal volume, while others do not.

Whether intentional ventilation openings should be open or closed will depend on the purpose of the test. Normally the method is used to assess the overall air leakage in the building envelope. In this case, intentional openings should be closed. If air infiltration calculations are to be performed, then intentional openings should, in principle, be left open. The CGSB standard prescribes how to seal intentional openings.

A related issue is whether a single measurement is capable of properly characterizing a building envelope that may experience seasonal variations. The air leakage measurements reported here were taken over a seven-month period on a wood-frame test house. Time plots of leakage area will be used to distinguish between measurement variability using different standards, and possible trends caused by structural changes. It is suspected, at least for houses with wood frame construction, that air leakage varies with season.

Description of Houses and Data Collection

Air Leakage Characteristics of Test House

The Alberta Home Heating Research Facility is made up of six permanent unoccupied test houses. They have been continuously monitored since 1980 for building envelope energy losses and air infiltration and ventilation rates. The houses have been used to test furnace efficiency, air infiltration and ventilation, envelope heat losses, moisture migration and accumulation, active and passive solar heating strategies, and radiant floor heating systems. The six houses are situated in a closely spaced line with about 2.6 m separation between their sidewalls. The inside floor area of 46.3 m^2 is about one-third to one-half the floor area of a typical single-story home. House 4, used in the present study, is sheltered on two sides by the adjacent houses in the row. A picture of the test houses is given in Fig. 2.

The test houses are built like real houses, with standard Canadian construction practices. They have full poured concrete basements and polyethylene air-vapor barriers in walls and ceilings. With the exception of the masonry house, the houses are of wood-frame construction with 41 by 92 mm studs on 406 mm centers (2-by-4 studs on 16 in. centers) with 13 mm painted drywall on the interior and 10 mm plywood exterior sheathing. The asphalt shingled roof is supported by wood roof trusses with their ends elevated 0.61 m above the ceiling by attic wall extensions. These elevated roof trusses were used to accommodate thick ceiling insulation and to provide easy access to the attic space. The door on the east side of each house has flexible weatherstrip around its outside edges.

In addition to having smaller floor area, the test modules differ from a standard house in that they have no plumbing or sewer drains, and no interior partition walls except for an entryway with an open interior doorway. The absence of interior walls and open basement stairwells promotes air mixing, and allows the house to be treated as a single air-exchange zone. The houses are heated electrically with a centrifugal fan distributing air through underfloor ducts to the main floor room. The fan in the electric heater operates continuously, recirculating 4.5 house interior volumes per hour to ensure complete mixing of air infiltration with indoor air. A standard mercury switch room thermostat located on the room side of the entryway wall maintains the indoor temperature between 21.5 and 22.5°C during the heating season. In summer, the fan continues to circulate air through the house, and room temperature is governed by ventilation and heat gains through the walls and windows. Summer indoor temperature rarely differs by more than ±5 deg C from the outdoor air.

The test house had a leakage distribution of small cracks and holes created unintentionally during construction. The major unintentional leakage sites are the partially sealed crack

FIG. 2—*The six test houses at the Alberta Home Heating Research Facility.*

between the wall and the sill plate and the top of the concrete basement wall; vapor barrier penetrations by electrical conduits and outlet boxes, flue pipes, and plumbing vents; and cracks around the frame of windows and doors. The normalized air leakage for House 4 is compared in Table 2 with air leakage data for real houses given in Ref 8. The value 1.46 cm^2/m^2 of floor area is the average of pressurization and depressurization tests. It can be seen that House 4 is very close to the Canadian average. Because House 4 was constructed with standard practice technology for wood-frame houses and that the amount of air leakage is close to Canadian average for comparatively similar houses, the air leakage paths should be of the same type as in a real wood-frame house. Therefore, the results given in this paper can be extrapolated to modern wood-frame houses built for cold climates.

This distributed unintentional air leakage was measured using a variable-speed fan and flowmeter connected to a 457-mm-diameter hole in the plywood panel that is permanently mounted over the east window of each house. The furnace flue was sealed using a motorized damper prior to each fan depressurization test.

Fan Pressurization System

An automated air leakage measurement technique was used to make continuous fan depressurization tests. Each hour the microprocessor cycled the system through a complete series of pressure and flow measurements at indoor-outdoor pressure differentials over 1 to 100 Pa. At each of 27 specified pressure differentials from 1 to 100 Pa the computer directed the measurement system through a 100 s to 110 s cycle. It first measured the indoor-outdoor pressure differential (that is, pressure offset) for 15 s with the 0.46 m axial-flow blower-door fan off and its window outlet sealed with a low-leakage motorized damper; then it opened the damper, turned on the fan (15 to 20 s delay) and measured the indoor-outdoor pressure difference and fan flow for 35 s. There was then a 35 to 40 s delay to turn off and seal the fan, after which the cycle was repeated at a new indoor-outdoor pressure differential.

The airflow through the fan was determined by measuring the pressure differential across a laminar flow element through which all of the fan flow was drawn. The flowmetering element had been calibrated using a standard American Society of Heating, Refrigerating and Air-Conditioning Engineers (ASHRAE) pitot-tube traverse in a 0.46-m-diameter duct. At the lower end of the flow range this pitot-tube traverse calibration was checked using a standard American Society of Mechanical Engineers (ASME) orifice in a 0.15-m-diameter pipe. The results of the two calibrations were in good agreement, and combined errors from all measurement effects were usually about 1%, and never more than 2%. During calibration

TABLE 2—*Results from air leakage measurements compared with House 4 expressed in AL₄ normalized by floor area (± one standard deviation).*

Description	Normalized Air Leakage[a], (cm^2/m^2)
Canadian houses	
test house No. 4	1.46
277 houses built 1961–1983	1.60 ± 1.30
91 energy efficient houses	0.36 ± 0.31
U.S. houses	
277 houses built 1961–1983	5.35 ± 3.40

[a] Data from Sherman et al. [8].

a downstream suction fan and outlet flow blockage were used to verify that fan loading had a negligible effect on flow rate measurement.

Outdoor pressure was measured using four pressure tap hoses mounted midway along each wall of the building approximately 0.5 m above the ground. The four equal-length hoses were connected to a manifold to provide a spatial average. The indoor-outdoor pressure differential was measured with a high-resolution diaphragm pressure transducer with an accuracy of 0.1 Pa (calibrated with a micrometer-micropoint manometer) and a resolution of 0.01 Pa at all pressure and flow measurements, with 15 s averages for the pressure offsets and 35 s averages for the indoor-outdoor pressure differentials and fan flow.

A test was started only if the wind was low, as determined by the average wind speed in the previous 30 min period being less than 1.5 m/s. There were 27 flow-pressure points measured in each run, with a zero offset reading between each point. The points were spaced from 1 to 100 Pa at constant multiple steps (for example, 1, 2, 4, 8, and 16) with a few points added to match the ASTM and CGSB standards.

Test Results and Testing Conditions

Each of the 105 tests in House 4 consists of 27 observations of 12 different variables, of which four were calculated from the others. Temperature differences, wind speeds, and wind directions were measured at each data point, which allow studies on both average and changing test conditions. The results are exemplified in Fig. 3, using the ASTM standard 4 Pa leakage area, AL_4, and the 50 Pa air changes per hour, ACH_{50}, according to the Swedish Standardization (SS) standard. As can be seen from the figure, a significant amount of scatter occurred in spite of the special care taken and the high accuracy of the equipment used in taking the data. A factor of two difference can be seen between the largest and smallest leakage areas. The scatter in ACH_{50} is less, as expected. Also, a trend of reduction in leakage area for the spring months can be observed.

The effect of wind speed and wind direction have been greatly reduced in this sample by performing tests only at low wind speeds, and by the use of a pressure averaging container.

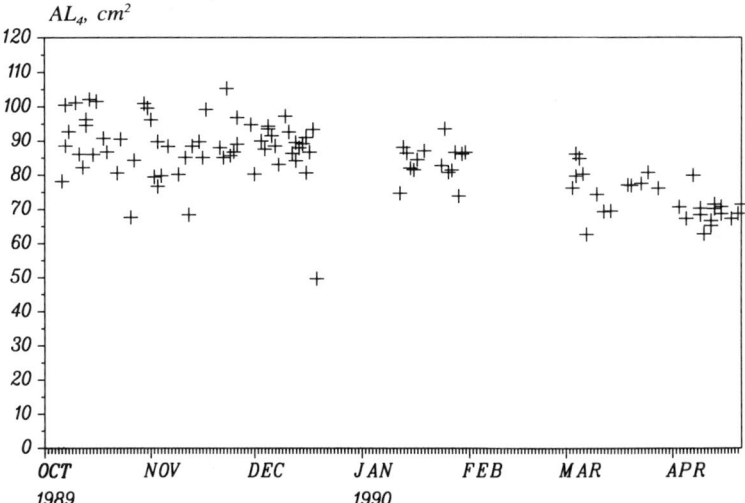

FIG. 3a—Leakage areas at 4 Pa reference pressure calculated according to the ASTM standard test method for the 105 tests in House 4.

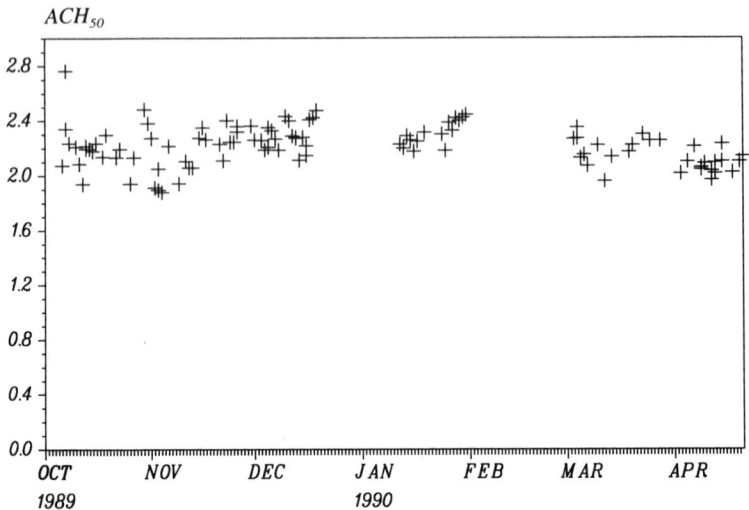

FIG. 3b—*Airflow at 50 Pa pressure difference calculated according to the SS standard test method for the 105 tests in House 4.*

Because the decision to start a test was based on the wind speed during the previous 30 min, wind speeds during the measurements sometimes exceed the target, 1.5 m/s. The average wind speed during tests ranged from 0.2 to 3.7 m/s. Figure 4 shows the average wind speeds, in monthly divisions, with standard deviation bars for the tests.

The tests included temperature differences between the inside and outside in the range of 2 to 47°C. Figure 5 gives the monthly averages and scatter for the tests.

Linear regression was used to observe correlations on the standards test result to the climatic test conditions. The strongest correlation was that the CGSB, ISO, and SS standards showed dependence on temperature difference. Determination coefficients, r^2, for the tem-

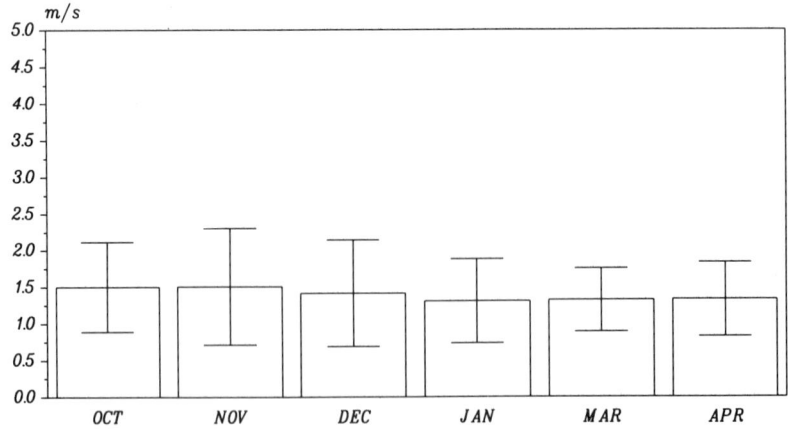

FIG. 4—*Monthly average wind speeds and standard deviations for the tests.*

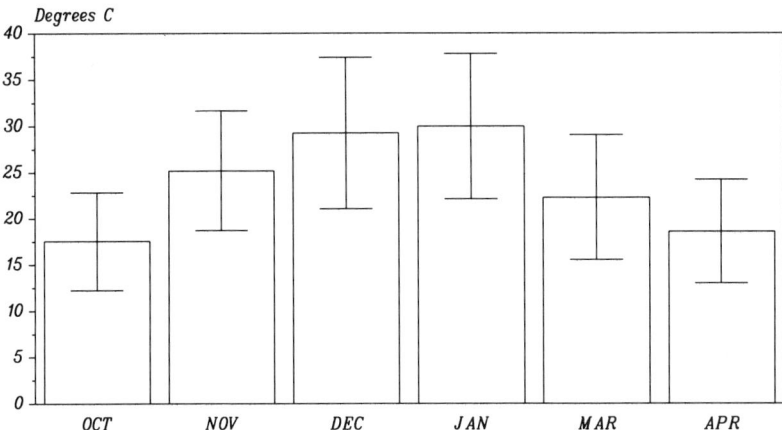

FIG. 5—*Monthly average temperature differences between inside and outside with standard deviations for the tests.*

perature difference and wind speed variables are given in Table 3. The test results, in ACH_{50}, according to the SS standard, are plotted with temperature difference in Fig. 6.

The reason for the temperature dependence is not obvious. A large temperature difference during tests influences the measurements and building envelope in a couple of ways: the correction of airflow measurement is larger, the inside air temperature changes quickly with increasing airflow, and building envelope components such as windows and doors could be warped by large temperature differences. The latter item should, however, also be seen for the ASTM test results. The correlation may have something to do with the fact that the envelope pressure difference is not measured at the neutral level, so corrections for offset pressures, that is, the pressure difference measured at zero airflow through the fan, which are not performed for the ASTM standard, vary with temperature difference. Using a non-standard method, including all data points (1 to 105 Pa) with and without offset pressure correction, reduced the r^2 for ACH_{50} versus temperature difference from 0.425 to 0.117. The recommendation that could be made from these results is that the offset pressure correction should be used only at small temperature differences.

Looking at changing test conditions, the correlations with changing temperature difference and changing wind speed were found to be very small. The results in Fig. 3 imply that the AL_4 value is more dependent than the ACH_{50} value. The low correlation with wind speed implies that the efforts taken to reduce wind influence in this data set were fairly successful.

TABLE 3—*Linear determination coefficients, r^2, of fan pressurization test results with climatic test conditions.*

	Standard			
	ASTM, AL_4	CGSB, ELA_{10}	ISO, AL_4	SS, ACH_{50}
Temperature difference	0.004	0.277	0.191	0.212
Wind speed	0.041	0.007	0.004	0.001

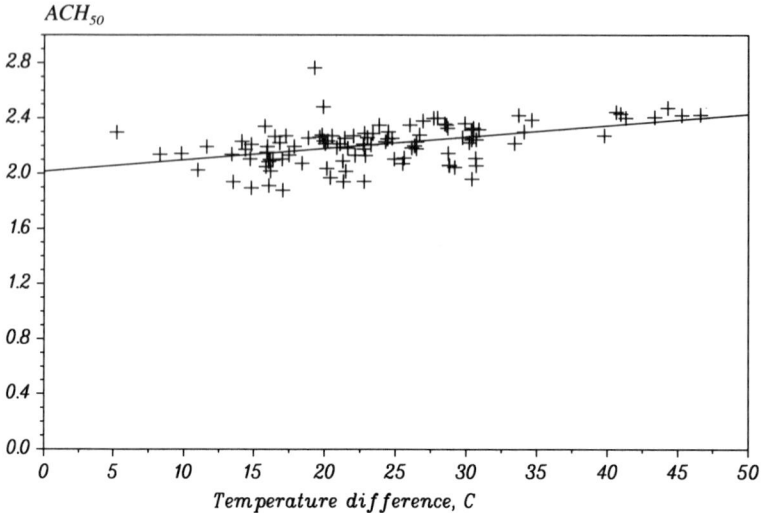

FIG. 6—*Variations in ACH₅₀ with temperature difference. A determination coefficient, r², of 0.21 was obtained in linear regression.*

Test Results Using Fan Pressurization Standards

In Table 4, leakage areas and flow from the different standards are compared. Besides the difference in pressure-flow points, the standards also differ in how the offset pressure is treated. The ASTM standard ignores the offset pressure, while the CGSB and ISO standards require pressures to be corrected for average offsets before and after testing. The SS standard adjusts in principle for offset before tests in that the pressure gage is to be zeroed at zero airflow. The automatically taken data in House 4 allow for the correction for pressure offsets at each pressure-flow point. This is normally too time-consuming for standardized field tests. As shown by Table 4, the result for the ASTM standard differs from the others, all of which are very close to each other. The CGSB leakage area has been multiplied by the discharge coefficient, $C_d = 0.6$, for comparison.

To illustrate the differences in scatter using the measurement standards, the coefficients of variation (standard deviation normalized by the mean) are plotted in Fig. 7. The 50 Pa air leakage gives the least relative error. The main difference between the ASTM and other

TABLE 4—*Average air leakage result (± one standard deviation) for all 105 tests for different standards. For each column, the same equation was used. Offset pressures were subtracted as prescribed.*

Standard	Flow exponent, n	ASTM, ISO, AL_4 cm²	CGSB, $0.6 \cdot ELA_{10}$ cm²	SS, ACH_{50} h^{-1}
ASTM	0.74 ± 0.04	83.4 ± 10.4	102.1 ± 9.6	2.11 ± 0.13
CGSB	0.67 ± 0.03	104.8 ± 9.6	120.1 ± 8.8	2.21 ± 0.15
ISO	0.67 ± 0.03	104.1 ± 9.2	119.6 ± 8.7	2.21 ± 0.14
SS	0.67 ± 0.04	104.9 ± 11.0	120.1 ± 9.7	2.21 ± 0.13

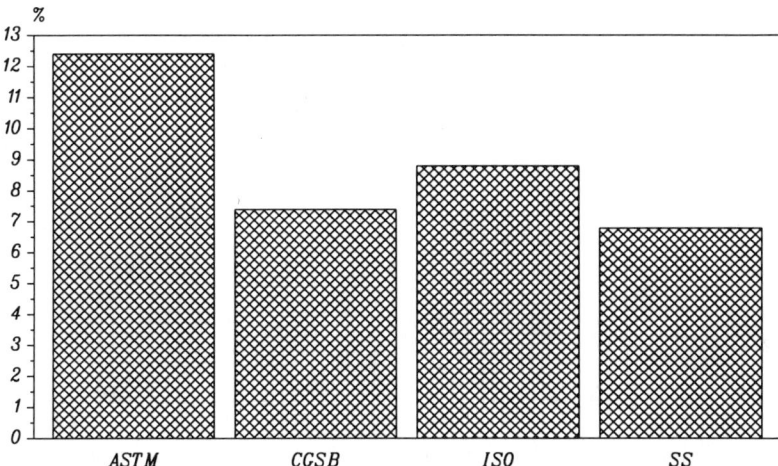

FIG. 7—*Coefficient of variation: relative errors in using the different fan pressurization method standards on the 105 tests in House 4. The relative errors are calculated as the standard deviation divided by the average result.*

standards is the correction for offset pressure. This effect can be seen comparing the scatter of calculated AL_4 according to the ASTM and ISO standards; see Table 4. The only other difference between these standards for this data set is the selection of pressure differences, which is analyzed later in the text.

Seasonal Variations in Air Leakage

Seasonal variations in air leakage that cannot be explained by measurement errors have previously been reported. A Princeton study [9] in one house showed an air leakage on the order of 25% higher in winter compared with early summer. In a Canadian study [10], fan pressurization measurements in two unoccupied houses showed a more than 20% higher airflow rate at 50 Pa in late winter compared with the fall. Dickinson and Feustel [11] reported seasonal variations in the 4 Pa effective leakage area that were high, 28% to 45% for three California alpine region houses and small, 13% to 18%, for three San Francisco Bay houses. The alpine houses in this study measured lowest air leakage in winter and highest in late summer. Moisture effects on building materials were discussed, but not quantified, in these three studies as a probable cause for the variations.

The 105 fan pressurization tests in House 4 over the seven-month period can be used to further examine the effect of seasonal variation on air leakage in a wood-frame house with a polyethylene air/vapor barrier. Ideally, a test period of more than one year would be preferred, but this was not possible in this case. To see air leakage changing with season, the results were calculated monthly. Figures 8a through 8d show the monthly average results and standard deviations for the four examined standards.

The results given in Figs. 8 show a rather large scatter for the fall months. A decrease in air leakage can be seen for the March and April data, which agree with previous findings

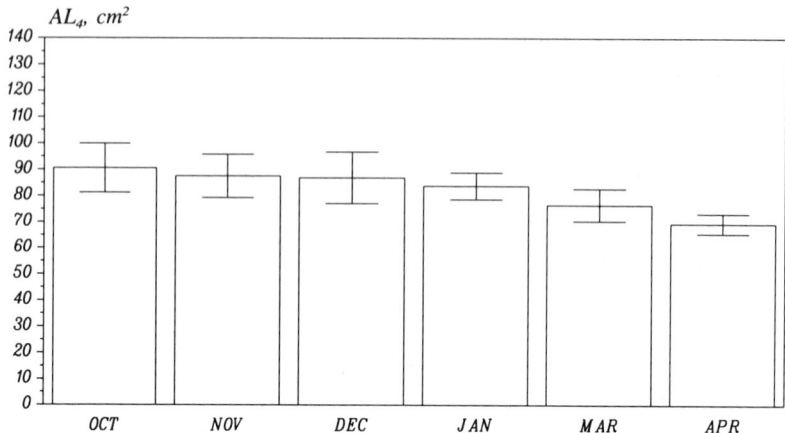

FIG. 8a—*Monthly averages and scatter (standard deviation) in* AL_4 *for the ASTM standard.*

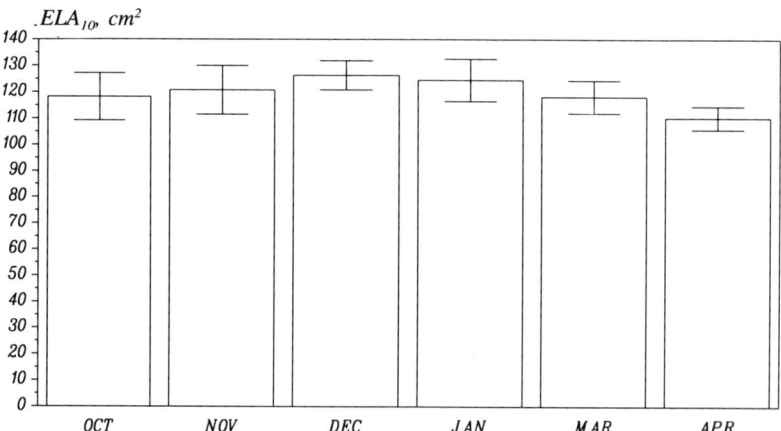

FIG. 8b—*Monthly averages and scatter (standard deviation) in* ELA_{10} *for the CGSB standard.*

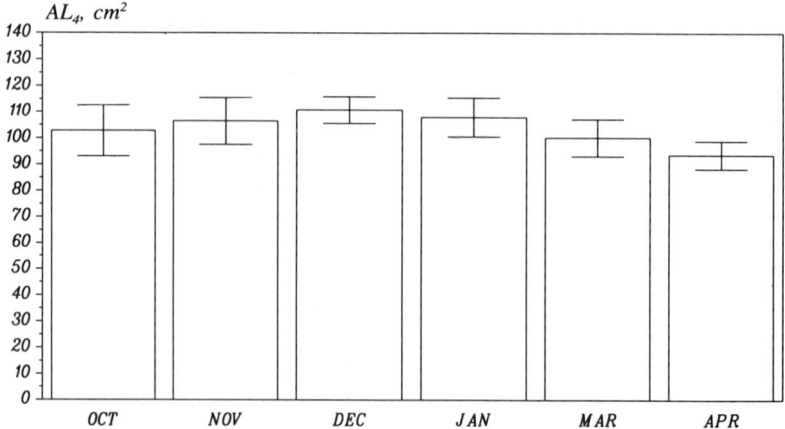

FIG. 8c—*Monthly averages and scatter (standard deviation) in* AL_4 *for the ISO standard.*

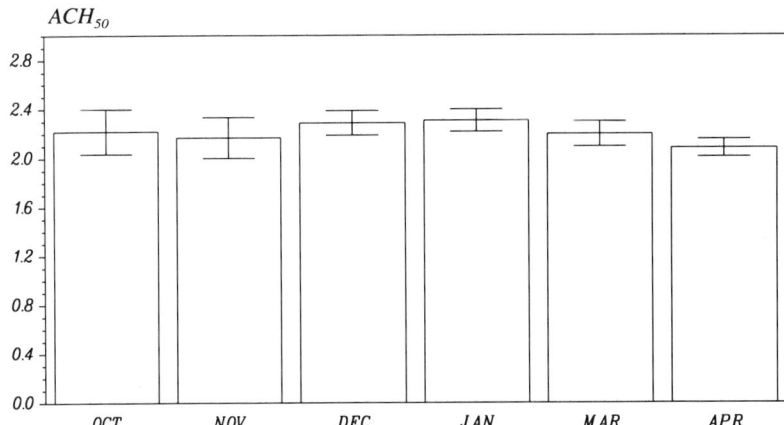

FIG. 8d—*Monthly averages and scatter (standard deviation) in* ACH_{50} *for the SS standard.*

[9,10]. The differences between many of the monthly averages for House 4 are larger than one standard deviation. It makes a little difference whether ACH_{50} or AL_4 is used as the criterion for seasonal variation.

Because the flow exponent can be an indicator of average airflow characteristics for cracks in the building envelope, changes in crack sizes with season might be seen. Figure 9 shows a monthly plot of the flow exponent, n, using the ASTM method. A significant increase of the flow exponent can be seen for the warmer and wetter spring months. Using the other standards, this effect was not visible. If these houses were occupied, moisture migration from

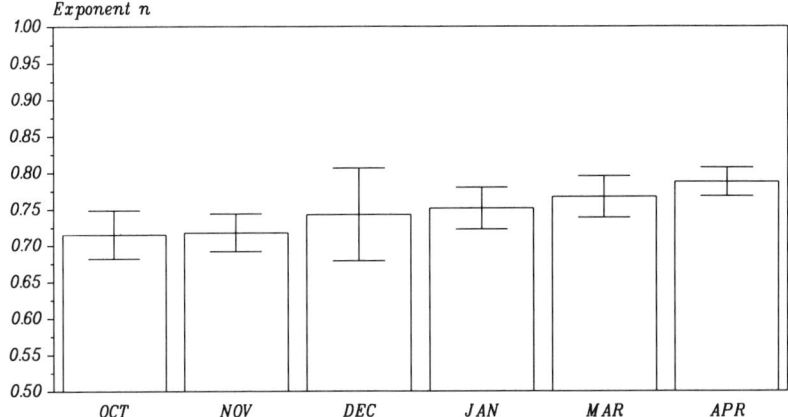

FIG. 9—*Monthly variation of the flow exponent* n *for House 4 during the test period using the ASTM standard method.*

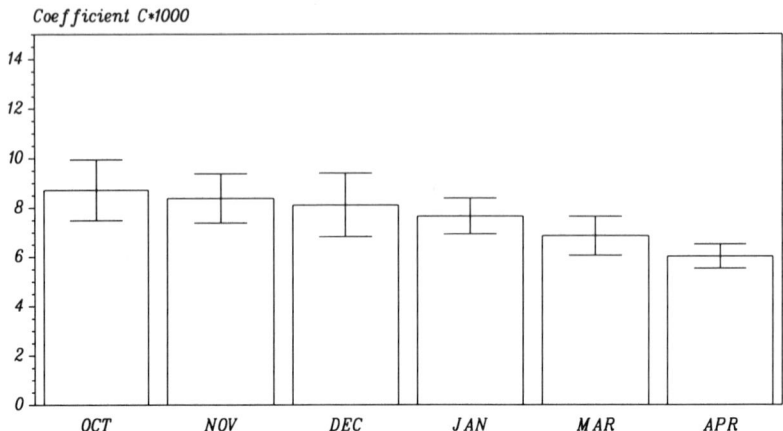

FIG. 10—*Monthly variation of the flow coefficient* C *(m^3/sPa^n) for House 4 during the test period using the ASTM standard method.*

the inside might have increased the moisture content in the wood and thus caused swelling and a subsequent reduction of envelope cracks. The houses were unoccupied and very dry inside, which is why the effect of moisture swelling might be less than in an occupied house. Also, precipitation during the tests may have been a factor. At present, there are no available corresponding moisture data for the test period. Figure 10 is a corresponding plot for the flow coefficient, C, also showing significant changes for the spring months.

To find a variable that compares the variability of the different standards, a root-mean-square (rms) coefficient of variation K_{rms}, standard deviation divided by mean, was calculated using monthly means and standard deviations. The results, given in Table 5, suggest that the errors are unevenly distributed over the test period. The ASTM standard result improves more than the others, especially for AL_4. Most of the variability over the seven-month period is caused by a systematic change in the monthly mean and not by scatter within the months. The remaining difference is in the selection of pressure points and the pressure offset treatment.

TABLE 5—*RMS coefficients in % for air leakage representation,* K_{rms}, *based on monthly averages compared with the data given in Fig. 7, showing relative errors where every measurement had equal weight (σ/m), %.*

	ASTM		CGSB		ISO		SS	
	K_{rms}	σ/m	K_{rms}	σ/m	K_{rms}	σ/m	K_{rms}	σ/m
n	4.8	5.4	4.6	4.8	4.1	4.5	5.5	6.0
AL_4	8.7	12.4	7.8	9.2	7.2	8.8	9.3	10.5
ELA_{10}	6.6	9.4	6.0	7.4	5.8	7.3	6.6	8.1
ACH_{50}	5.0	6.2	5.9	6.8	5.6	6.3	5.7	6.8

TABLE 6—*Effect of pressure difference selection using different standards and pressure intervals. Averages and standard deviations for 15 tests during April in House 4. No correction for offset pressures was performed for the standards, but for the pressure intervals, corrections for average offsets before and after each data point had to be made.*

	No of Points	Exponent, n	AL_4, cm^2	ACH$_{50}$, h^{-1}
Standards				
ASTM	6	0.79 ± 0.02	69 ± 3.8	1.97 ± 0.05
CGSB	7	0.78 ± 0.02	72 ± 3.3	2.00 ± 0.06
ISO	6	0.81 ± 0.02	67 ± 3.3	1.99 ± 0.06
SS	6	0.76 ± 0.02	74 ± 4.1	1.99 ± 0.07
Nonstandard ranges, Pa				
1–105	25	0.76 ± 0.06	81 ± 10.6	2.16 ± 0.07
1–4	5	1.18 ± 0.52	103 ± 31.1	104 ± 380
5–11	5	0.73 ± 0.03	90 ± 4.7	2.24 ± 0.12
14–26	5	0.69 ± 0.02	94 ± 4.2	2.10 ± 0.07
33–52	5	0.67 ± 0.04	98 ± 9.3	2.08 ± 0.07
58–105	5	0.70 ± 0.06	92 ± 15.9	2.08 ± 0.09

Effect of Pressure Difference Selection

The standards recommend different selections of pressure-flow points. To find out whether this selection has any influence on the air leakage result, 15 of the tests, carried out in April, were used. The rationale for selecting the April tests was that this month had the least scatter in leakage area and that the temperature differences were smaller than those in the winter months, reducing the deviations in stack pressure. The recommended pressure differences and a standard least-squares linear regression method with logarithmic transformation were used to find the coefficients C and n in the power law for the different standards. The effect of the averaging pressure container cannot be evaluated with these data because no single tap data were taken. The results are given in Table 6, where the 4 Pa leakage area and airflow at 50 Pa are used to show the combined effect of differences in C and n.

Using C and n from pressure differences recommended in the standards results thus in small differences in the 4 Pa leakage area. Note that this comparison was made so that the only difference between the standards is the selected pressure-flow points. Table 6 also gives the air leakage calculated for other nonstandard pressure intervals. Using more than the 6 or 7 pressure-flow readings of the standardized methods does not seem to reduce the scatter. With limited pressure intervals, a substantial increase in scatter was found. For low pressure readings, as can be expected, the uncertainties are large. This may be an effect of small changes in the weather conditions. Table 6 also shows that extrapolating results from low pressure differences to ACH$_{50}$ or from high pressure differences to AL_4 can cause substantial errors. The recommended envelope pressure differences for the standards are given in Table 7. The SS standard is weighted towards envelope pressure differences around 50 Pa while

TABLE 7—*Specified envelope pressure differences for the fan pressurization method standards, Pa.*

ASTM	12.5	25.0	37.5	50.0	62.5	75.0		
CGSB	15	20	25	30	35	40	45	50
ISO	10	20	30	40	50	60		
SS	20	30	40	45	50	55		

the others are evenly distributed. However, the ASTM standard covers a larger range of pressure differences and is also more biased towards higher pressure differences.

Conclusions

The test results and the analysis of the data from House 4 at the Alberta Home Heating Research Facility lead to the following conclusions:

- In spite of the four-wall outside pressure averaging, the great care taken in collecting the data, and the good accuracy of the measurement equipment, the scatter in air leakage measurement is large.
- Correction for the offset pressure difference reduced the scatter in the data. However, the correction term correlated with the temperature difference shows why this method can be recommended only for small temperature differences.
- The scatter in ACH_{50} was smaller than for AL_4, which suggests that the 50 Pa value gives more reproducible result—however less sensitive—for comparison of building envelope air leakage.
- The pressure-flow points specified in the different standards have some effect on the air leakage result since they are distributed differently across the range.
- A change in envelope air leakage that cannot be explained by scatter caused by the test conditions was found over the seven-month measuring period. The difference between monthly averages are significant for the spring months.

References

[1] Charlesworth, P., "Air Exchange Rate and Airtightness. Measurement Techniques—An Applications Guide," Air Infiltration and Ventilation Centre, University of Warwick Science Park, Coventry, U.K., 1988.
[2] Elmroth, A. and Levin, P., "Air Infiltration Control in Housing—A Guide to International Practice," Swedish Council for Building Research D2:1983, Stockholm, 1983.
[3] "Determination of the Airtightness of Building Envelopes by the Fan Depressurization Method," CAN/CGSB-149.10-M86, Canadian General Standards Board (CGSB), Ottawa, 1986.
[4] "Air Leakage of Buildings—Method of Measurement," NEN 2686, Nederlands Normalisatieinstituut, Delft, The Netherlands, 1986.
[5] "Buildings—Determination of Airtightness (Issue 2)," SS 021551, Swedish Standardization Commission (SIS), Building Standards Institution, Stockholm, 1987.
[6] "Thermal Insulation—Determination of Building Airtightness—Fan Pressurization Method," ISO /DIS 9972, Draft International Standard, International Organization for Standardization (ISO), Geneva, 1990.
[7] Modera, M. P. and Wilson, D. J., "The Effects of Wind on Residential Building Leakage Measurements," ASTM Symposium on Air Change Rate and Air Tightness in Buildings, Atlanta, GA, American Society for Testing and Materials, Philadelphia, 1989.
[8] Sherman, M. H., Kiel, D. E., and Wilson, D. J., "Variability in Residential Air Leakage," in Proceedings, ASTM Symposium on Measured Air Leakage Performance of Buildings, American Society for Testing and Materials, Philadelphia, April 1984.
[9] Persily, A. K., "Repeatability and Accuracy of Pressurization Testing," Proceedings of ASHRAE/ DOE Conference Thermal Performance of the Exterior Envelopes of Buildings II, 1982.
[10] Kim, A. K. and Shaw, C. Y., "Seasonal Variation in Airtightness of Two Detached Houses," Measured Air Leakage of Buildings, ASTM STP 904, H. R. Trechsel and P. L. Lagus, Eds., American Society of Testing and Materials, Philadelphia, 1986.
[11] Dickinson, J. B. and Feustel, H. E., "Seasonal Variation in Effective Leakage Area," LBL Report No. 19337, Lawrence Berkeley Laboratory, University of California, Berkeley, CA, 1986.
[12] Wilson, D. and Walker, I., "Passive Ventilation to Maintain Indoor Air Quality," Department of Mechanical Engineering Report No. 81, University of Alberta, Edmonton, Alberta, Canada, 1991.

Additional References

Levin, P., "Building Technology and Air Flow Control in Housing," Swedish Council for Building Research D16:1991, Stockholm, 1991.

Wilson, D. and Walker, I., "Passive Ventilation to Maintain Indoor Air Quality," Department of Mechanical Engineering Report No. 81, University of Alberta, Edmonton, Alberta, Canada, 1991.

Discussion

G. Proskiw[1] *(written discussion)*—Can your results be extrapolated to a full-scale house given that they were generated using a test hut with an ELA_{10} approximately $\frac{1}{10}$ of an energy-efficient house?

P. A. Levin (authors' closure)—The test houses at the University of Alberta are not test "huts" but are in fact built like real houses, at about half the floor area. A table has been added to the paper to address the question. It shows that, using air leakage data for real houses, House 4 is very close to the Canadian average, judging from the measurement database given in Ref *8*.

The test houses did not have as many air-barrier penetrations in the ceiling as real houses, and had no water pipes or plumbing vents, which is why we think their lower leakage area is reasonable. Despite this, House 4 is very close to the Canadian average.

The air leakage paths in House 4 should therefore be of the same type as in a real wood-frame house. We believe that the given results can be extrapolated to modern wood-frame houses built for cold climates.

---

[1] Proskiw Engineering Ltd., Winnipeg, Manitoba, Canada.

Gary Proskiw[1]

Measured Airtightness of 24 Detached Houses over Periods of up to Three Years

REFERENCE: Proskiw, G., "**Measured Airtightness of 24 Detached Houses over Periods of up to Three Years,**" *Airflow Performance of Building Envelopes, Components, and Systems, ASTM STP 1255,* Mark P. Modera and Andrew K. Persily, Eds., American Society for Testing and Materials, Philadelphia, 1995, pp. 248–265.

ABSTRACT: Regular airtightness tests were performed on 24 new houses, over periods of up to three years, to evaluate their air barrier systems and to search for evidence, air barrier degradation. Ten of the houses were constructed with polyethylene air barriers while the remaining 14 used an early version of the Airtight Drywall Approach (ADA). The 24 project houses were architecturally similar and of approximately equal size and general layout; stucco was the predominate exterior wall finish. All were exposed to similar terrain shielding.

The study found that the airtightness of the polyethylene air barrier houses remained stable over their respective monitoring periods. Although two of the ten houses demonstrated possible, albeit slight, evidence of airtightness degradation, the magnitude of these changes was small and judged not to be of practical significance. With respect to the critical issue of air barrier degradation, no evidence could be found to indicate polyethylene is unsuited for use as an air barrier material in residential construction. For example, all but one of the polyethylene houses met the airtightness requirements of the Canadian R-2000 Standard for energy efficient housing at the end of their monitoring periods.

The study found that the airtightness of the 14 ADA houses also remained stable over their monitoring periods. Although six of the 14 houses displayed possible, but slight, evidence of airtightness degradation, the magnitude of the changes was small and not of practical significance. It was concluded that no evidence could be found to indicate that the ADA system is unsuited for use in residential construction. All 14 ADA houses met the airtightness requirements of the R-2000 Program at the end of their respective monitoring periods.

KEYWORDS: residential airtightness, air leakage, blower door tests, air barriers, air barrier degradation, polyethylene air barriers, Airtight Drywall Approach

The term "airtightness" describes the building envelope's ability to resist air infiltration and exfiltration created by pressure differentials across joints, holes, and other openings in the exterior shell. Ideally, the building envelope should eliminate all air leakage since it can increase energy costs, degrade indoor air quality and comfort, produce moisture-related envelope problems, and increase the transmission of outdoor noise to the interior. In practice, air leakage cannot be eliminated, but only controlled within prescribed limits. The component, or system, responsible for providing airtightness in the building envelope is the air barrier.

The objectives of this study were to monitor the airtightness of a group of 24 new houses over a multi-year time frame, to search for evidence of air barrier degradation as would be

---

[1] Proskiw Engineering Ltd., Winnipeg, Manitoba R3H 0H1, Canada.

indicated by an increase in the measured air leakage rate, and to comment upon the suitability of the various air barrier systems for use in residential construction.

The 24 houses were constructed between 1985 and 1989 in Winnipeg, Canada by a large tract builder. With the exception of one house (#21), which was maintained as an unoccupied research structure, all were sold and occupied shortly after completion. Descriptions of the houses and their air and vapor barrier systems are provided in Table 1.

The houses were similar in size and general layout and were constructed in close physical proximity to each other within two groups on the edge of an urban environment. They were all detached bungalows with full basements and main floor areas of 60 to 85 m² (646 to 915 ft²). Electric space heating was used for all except two of the dwellings which used natural gas. Two air barrier systems were used: polyethylene and the Airtight Drywall Approach (ADA). Several versions of each system were used and in a few instances the two systems were slightly intermixed within a single structure. In such cases, the structure was classified as a "polyethylene" or "ADA" house based on the dominant system used.

Air Barriers

Air Barrier Design

The primary requirements for an effective air barrier can be summarized as (a) low permeability to airflow, (b) structural strength to withstand air pressure loads, (c) continuity to reduce leakage, (d) durability so it will last the life of the building, and (e) rigidity to provide pressure equalization behind exterior cladding. In residential construction sheet polyethylene, combined with sealed joints and penetrations, is the most commonly used material. A second system which has gained some acceptance in recent years is the ADA. The drywall, which is normally applied as the interior surface, is used as the air barrier in concert with strategically located gaskets at joints and penetrations. Vapor barrier protection is provided by either low-permeance paints or loosely applied (i.e., nonsealed) polyethylene.

At present, there is considerable debate as to which air barrier system is most appropriate for residential construction. The so-called "poly approach" is viewed as the more traditional and better-understood technique while proponents of ADA argue that the latter will have a longer life span and be better able to withstand the severe pressure forces to which air barriers can be exposed. These loads are created by the stack effect, wind action, and operation of mechanical systems. In residential construction, pressure loads due to stack effect and mechanical systems seldom exceed 10 to 20 Pa. Wind action, however, can generate pressures on an exposed building surface of over 1000 Pa for extended periods of time and perhaps 2500 Pa for short periods, i.e., a few seconds, during gusts.

Previous Studies

Persily [1] measured the airtightness of a single New Jersey house and found seasonal variations of 25% in the air change rate at 50 Pa (ac/h_{50}, see following section on monitoring program) and attributed it to changes in the moisture content of the framing members. Kim and Shaw [2] also explored this issue using two unoccupied Ottawa houses and reported seasonal variations of approximately 20%. Howell and Mayhew [3] tested six Edmonton houses over a two-year period and found the ADA houses in their sample were tighter than the conventional structures. They also found that the ADA houses became leakier with time while the conventional structures were unchanged. Buchan, Lawton, Parent Ltd. [4] reported on the airtightness levels of 90 houses over periods of 3 to 63 months and found that the

TABLE 1—Air and vapor barrier details.

HOUSE #	TYPE	AIR BARRIER					CREW	VAPOUR BARRIER		
		SEALING METHOD								
		HEADERS	CANTILEVERS	PARTITION WALLS AT CEILING	WINDOW & DOOR ROUGH OPENINGS	ELECTRICAL OUTLETS		WALLS	CEILING	BASEMENT
1-6	ADA	Closed Cell Polyethylene Gaskets	Closed Cell Polyethylene Gaskets	Gaskets	Gaskets	Poly-Pan Boxes & Gaskets	A	Paint	Paint	Paint
7,8	ADA	Closed Cell Polyethylene Gaskets	Closed Cell Polyethylene Gaskets	Gaskets	Gaskets	Poly-Pan Boxes & Gaskets	A	Paint	Paint	Paint
9,10	4 mil Polyethylene	None	None	Unsealed Polyethylene	Unsealed Polyethylene	Unsealed Polyethylene	A	Polyethylene	Polyethylene	Polyethylene
11-14	Simplified ADA	None	None	None	Ethafoam Rod Gaskets	Poly-Pan Boxes & Gaskets	B	Paint	Paint	Paint
15,16	6 mil Polyethylene	Caulking	Caulking	Sealed Polyethylene	Sealed Polyethylene	Sealed Polyethylene	B	Polyethylene	Polyethylene	Polyethylene
17,18	6 mil Polyethylene	Caulking	Caulking	Sealed Polyethylene	Sealed Polyethylene	Sealed Polyethylene	B	Polyethylene	Polyethylene	Polyethylene
19,20	ADA	Closed Cell Polyethylene & Neoprene Gaskets	Closed Cell Polyethylene & Neoprene Gaskets	Neoprene Gaskets	Ethafoam Rod Gaskets	Poly-Pan Boxes & Gaskets	B	Paint	Paint	Paint
21	Primarily 6 mil Polyethylene	Sealed Polyethylene	Sealed Polyethylene & SBPO Air Retarder	Sealed Polyethylene	Various	Sealed Polyethylene	C	Polyethylene	Polyethylene	Polyethylene
22	Primarily 6 mil Polyethylene	Sealed Polyethylene & SBPO Air Retarder	Sealed Polyethylene & SBPO Air Retarder	Saturated Urethane Open Cell Gaskets	Various	Polyethylene	D	Polyethylene	Polyethylene	Polyethylene
23	6 mil Polyethylene	Sealed Polyethylene & SBPO Air Retarder	Sealed Polyethylene & SBPO Air Retarder	Sealed Polyethylene	Various	Sealed Polyethylene	D	Polyethylene	Polyethylene	Polyethylene
24	6 mil Polyethylene	Sealed Polyethylene & SBPO Air Retarder	Sealed Polyethylene & SBPO Air Retarder	Saturated Urethane Open Cell	Various	Polyethylene	D	Polyethylene	Polyethylene	Polyethylene

airtightness only changed significantly in 9 houses. They concluded that their observations did not support the hypothesis that polyethylene degraded when used as an air barrier in residential construction. A Canada-wide study of the airtightness of approximately 200 new, conventional houses found significant variations depending on location [5].

European experiences have been somewhat different, possibly due to variations in construction practices or air barrier design or both. The Air Infiltration Centre [6] observed that changes in airtightness typically occurred in the first year after construction. They reported examples of five Swedish houses which exhibited a 70% increase in their ac/h_{50} values in the first year and thereafter maintained a constant airtightness. Three British houses were also reported to have experienced an average 83% increase in the first year. Carlsson and Kronvall [7] reported that airtightness levels remained constant in 15 Swedish timber-framed houses when tested at completion and then again after 1.5 to 4.5 years.

Monitoring Program

Testing Procedure

Airtightness tests were performed at regular intervals on the 24 project houses in accordance with the Canadian General Standards Board (CAN/CGSB) Standard 149.10-M86 Determination of the Airtightness of Building Envelopes by the Fan Depressurization Method (1986). All tests were performed with a custom-built blower door assembly which used calibrated flow nozzles to measure airflow rates and digital micromanometers to measure the nozzle pressure and indoor-to-outdoor pressure differentials. The same equipment and technicians were used for all tests.

The airtightness results are expressed using two methods: the air change rate at 50 Pa (ac/h_{50}) and the normalized leakage area at 10 Pa (NLA_{10}). The Canadian R-2000 Standard [8] for low-energy housing requires that a house's measured airtightness be less than 1.50 ac/h_{50} or that the NLA_{10} not exceed 0.7 cm^2/m^2. These values are often cited as the boundary between "tight" and "loose" construction and provide a useful benchmark for describing house performance.

Condition of Houses

Initial airtightness tests were performed shortly after completion of the houses, but prior to their occupancy. In roughly half the cases, tests were also performed prior to the application of stucco. This is noted because stucco can have a significant impact upon airtightness. During the monitoring period, regular contact was maintained with the occupants and the houses were routinely inspected to identify any physical changes which may have occurred that would have affected their airtightness. Few changes were observed and these were judged to have been typical for new houses in the first few years following construction.

Analysis Methods

Air barrier deterioration is most clearly demonstrated by a degradation in airtightness. For this study, three analysis methods were used to evaluate changes in airtightness:

1. *Variation Between the First and Last Airtightness Tests*—The absolute and percentage changes in airtightness were compared using the results of the first and last blower door tests.

2. *Variation Between the First and Last Seasonally Coincident Airtightness Tests*—This method has the advantage that the seasonal impact is eliminated since only seasonally coincident data are used.

3. *Statistical Tests*—Statistical tests can also provide useful insight. Degradation of airtightness is essentially a question of determining whether a dependency exists between airtightness and time. This can be expressed, using ac/h_{50} data as an example, with a regression equation of the form

$$ac/h_{50} = \alpha + \beta(t) \tag{1}$$

where

α = initial ac/h_{50} at the start of the monitoring period,
β = slope of the regression equation, and
t = time.

A regression equation of the form given in Eq 1 was computed to determine β, which was then compared with the zero value assumed in a null hypothesis. The significance of the difference was then evaluated by calculating the one-sided t-statistic with the appropriate number of degrees of freedom. This produced the probability that the observed difference between the sample value of β (from the regression equation) would be equal to, or greater than, the population mean (i.e., the true value). A 5% level of significance was used as a cutoff criterion, i.e., if the t-test indicated that the null hypothesis should be rejected, there was a 5% probability that the conclusion would be incorrect. Another way to quantify the variation in the dependent variable (airtightness) due to the independent variable (time) is through use of the coefficient of determination, r^2. In this application, it describes the percentage of the variation in airtightness due to the time dependency. An r^2 value of 0.60 was used to highlight possible dependencies.

These three methods were used to analyze the ac/h_{50} and NLA_{10} data for each of the project houses. No single method was regarded as providing a definitive statement on whether the airtightness was degrading with time. This assessment was made after consideration of the data, the analysis methods, and the houses themselves. The degradation analysis was restricted to data collected from the houses after the stucco had been applied.

Performance of Polyethylene Air Barrier Systems

Types

The ten houses which used the polyethylene air barrier system were constructed with four types of main wall systems:

- Standard framed walls (Houses #9, #10 and #22)
- Framed walls with exterior insulated sheathing (House #24)
- Framed walls with interior strapping (Houses #21 and #23)
- Double walls (Houses #15 to #18)

House performance is reviewed below on the basis of wall type because the walls would have been subjected to the most severe wind loadings. Airtightness results are given in Tables 2 and 3 (ac/h_{50} and NLA_{10}) and Fig. 1 (ac/h_{50} only) and the statistical analysis is summarized in Table 4. Monitoring periods typically ranged from 19 to 36 months.

TABLE 2—Polyethylene air barrier houses: ac/h_{50} results.

WALL SYSTEM & HOUSE NO.	MAR/86	JUL/86	NOV/86	FEB/87	JUL/87	NOV/87	FEB/88	MAY/88	JUN/88	JUL/88	AUG/88	NOV/88	MAR/89	JUN/89	AUG/89	DEC/89	JAN/90	MAR/90	APR/90
Standard Framed Walls																			
9	1.62	1.66	1.74	1.84	1.48	1.68	1.78				1.74	1.73	1.77						
10	1.28	1.15	1.43	1.39	1.17	1.04	1.03				1.20	1.24	1.19						
22								1.59*		0.96						0.94			1.14
Framed Walls with Exterior Insulated Sheathing																			
24											1.39				1.39	1.31		1.29	
Framed Walls with Interior Strapping																			
23								1.92*	1.43					1.32		1.27		1.34	
21 (see text)																			
Double Walls																			
15	1.47*	1.33	1.26	1.15		1.10							1.19						
16	1.26*	1.29	1.38	1.41		1.52							1.50						
17	0.55*	0.36	0.71	0.44	0.57	0.38	0.56				0.44	0.45	0.40						
18	0.49*	0.42	0.48	0.48	0.39	0.42	0.43				0.49	0.38	0.44						

CHANGE BETWEEN FIRST (POST-STUCCO) TEST AND FINAL TEST

WALL SYSTEM & HOUSE NO.	ABSOLUTE CHANGE[1] (ac/h_{50})	%[a]	NO. OF MONTHS
Standard Framed Walls			
9	0.15	9	36
10	-0.09	-7	36
22	0.19	20	21
Framed Walls with Exterior Insulated Sheathing			
24	-0.10	-7	19
Framed Walls with Interior Strapping			
23	-0.09	-6	22
21 (see text)			
Double Walls			
15	-0.14	-11	32
16	0.21	16	32
17	0.04	11	32
18	0.02	5	32

CHANGE BETWEEN FIRST (POST-STUCCO) TEST AND FINAL SEASONALLY-COINCIDENT TEST

WALL SYSTEM & HOUSE NO.	ABSOLUTE CHANGE[1] (ac/h_{50})	%[a]	NO. OF MONTHS
Standard Framed Walls			
9	0.15	9	36
10	-0.09	-7	36
22	N/A	N/A	N/A
Framed Walls with Exterior Insulated Sheathing			
24	0.00	0	12
Framed Walls with Interior Strapping			
23	-0.11	-8	12
21 (see text)			
Double Walls			
15	0.04	3	25
16	0.10	7	25
17	-0.04	-9	25
18	-0.04	-8	25

NOTES:
1. * = No Stucco
2. A -ve indicates the house became more airtight.

TABLE 3—Polyethylene air barrier houses: NLA_{10} results.

WALL SYSTEM & HOUSE NO.	MAR/86	JUL/86	NOV/86	FEB/87	JUL/87	NOV/87	FEB/88	MAY/88	JUN/88	JUL/88	AUG/88	NOV/88	MAR/89	JUN/89	AUG/89	DEC/89	JAN/90	MAR/90	APR/90
Standard Framed Walls																			
9	0.559	0.587	0.566	0.623	0.596	0.641	0.659			0.619		0.613	0.606						
10	0.588	0.418	0.642	0.805	0.404	0.441	0.392			0.468		0.517	0.644						
22								0.697*		0.372						0.361			0.445
Framed Walls with Exterior Insulated Sheathing																			
24									0.572						0.640	0.771	0.581		
Framed Walls with Interior Strapping																			
23								0.933*	0.652					0.549	0.540			0.816	
21 (see text)																			
Double Walls																			
15	0.774*	0.655	0.597	0.547		0.539				0.659		0.593							
16	0.677*	0.675	0.714	0.711		0.777						0.715							
17	0.278*	0.154	0.340	0.166	0.250	0.132	0.166			0.166		0.193	0.149						
18	0.259*	0.227	0.190	0.192	0.155	0.138	0.307			0.177		0.141	0.177						

CHANGE BETWEEN FIRST (POST-STUCCO) TEST AND FINAL TEST

WALL SYSTEM & HOUSE NO.	ABSOLUTE CHANGE[a] (NLA_{10})	%[b]	NO. OF MONTHS
Standard Framed Walls			
9	0.047	8	36
10	0.056	10	36
22	0.073	19	21
Framed Walls with Exterior Insulated Sheathing			
24	0.009	2	19
Framed Walls with Interior Strapping			
23	0.164	25	22
21 (see text)			
Double Walls			
15	-0.062	-9	32
16	0.039	6	32
17	-0.005	-3	32
18	-0.050	-22	32

CHANGE BETWEEN FIRST (POST-STUCCO) TEST AND FINAL SEASONALLY-COINCIDENT TEST

WALL SYSTEM & HOUSE NO.	ABSOLUTE CHANGE[a] (NLA_{10})	%[b]	NO. OF MONTHS
Standard Framed Walls			
9	0.047	8	36
10	0.056	10	36
22	N/A	N/A	N/A
Framed Walls with Exterior Insulated Sheathing			
24	0.068	12	12
Framed Walls with Interior Strapping			
23	-0.103	-16	12
21 (see text)			
Double Walls			
15	0.046	8	25
16	0.004	1	25
17	-0.017	-12	25
18	-0.015	-8	25

NOTES: 1. * = No Stucco
2. A -ve indicates the house became more airtight.

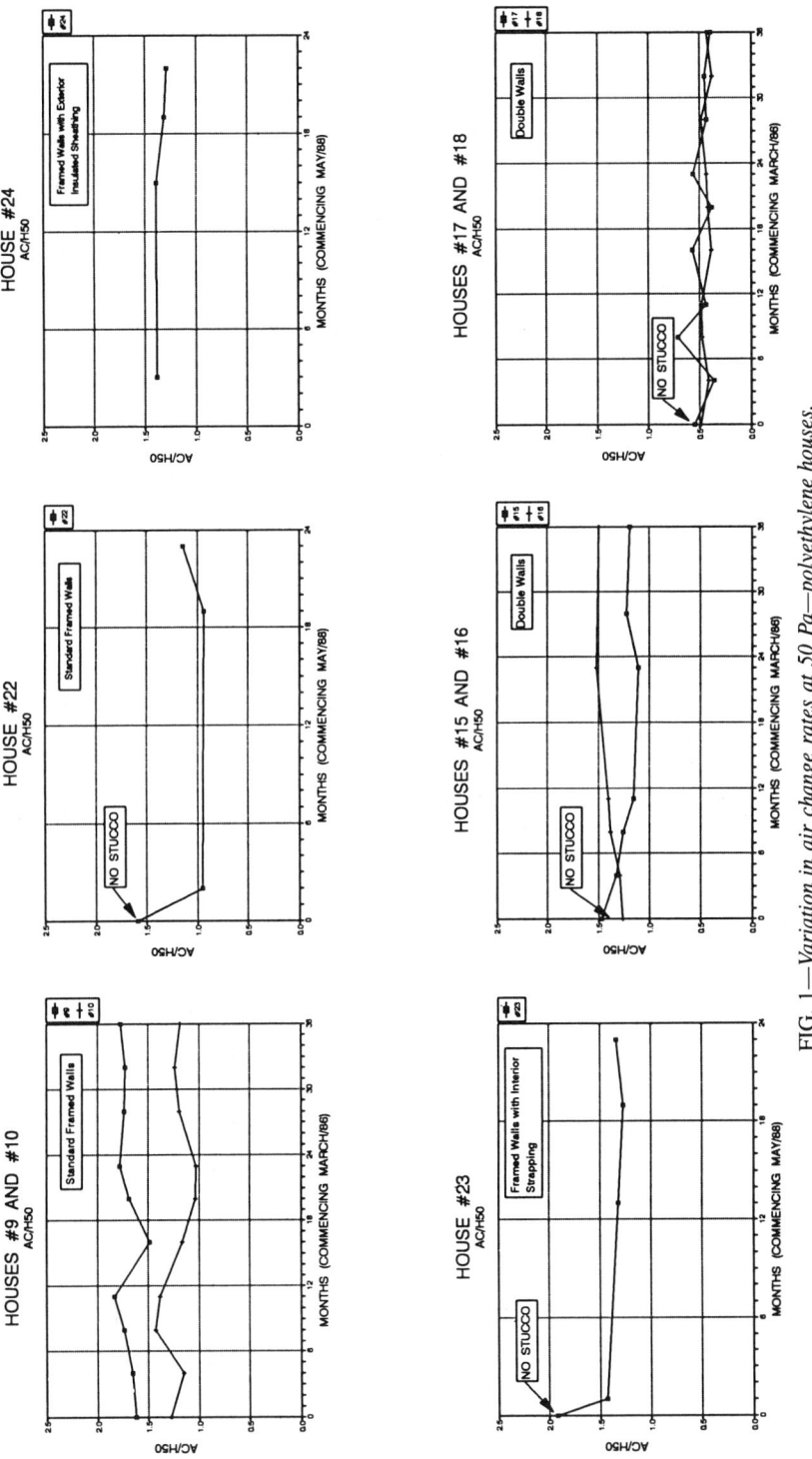

FIG. 1—*Variation in air change rates at 50 Pa—polyethylene houses.*

TABLE 4—Statistical analysis—polyethylene air barrier houses (ac/h$_{50}$ data).

WALL SYSTEM AND HOUSE NO.	α	β	n	t-STATISTIC	PROBABILITY (P)	NULL HYPOTHESIS (H$_o$)	COEFFICIENT OF DETERMINATION (r²)
Standard Framed Walls							
9	1.656	0.00273	10	-0.994	0.175	ACCEPT	0.110
10	1.280	-0.00385	10	1.091	0.154	ACCEPT	0.129
22	0.924	0.00605	3	-0.735	0.299	ACCEPT	0.351
Framed Walls with Exterior Insulated Sheathing							
24	1.401	-0.00478	4	1.770	0.110	ACCEPT	0.610
Framed Walls with Interior Strapping							
23	1.418	-0.00555	4	1.751	0.111	ACCEPT	0.605
21 (see text)							
Double Walls							
15	1.267	-0.00320	6	1.180	0.152	ACCEPT	0.258
16	1.320	0.00604	5	-2.930	0.031	REJECT	0.741
17	0.534	-0.00278	9	0.737	0.243	ACCEPT	0.072
18	0.448	-0.00063	9	0.468	0.327	ACCEPT	0.030

NLA$_{10}$ DATA

WALL SYSTEM AND HOUSE NO.	α	β	n	t-STATISTIC	PROBABILITY (P)	NULL HYPOTHESIS (H$_o$)	COEFFICIENT OF DETERMINATION (r²)
Standard Framed Walls							
9	0.580	0.00150	10	-2.007	0.040	REJECT	0.335
10	0.557	-0.00142	10	0.364	0.363	ACCEPT	0.016
22	0.360	0.00224	3	-0.657	0.315	ACCEPT	0.302
Framed Walls with Exterior Insulated Sheathing							
24	0.589	0.00444	4	-0.626	0.298	ACCEPT	0.164
Framed Walls with Interior Strapping							
23	0.587	0.00371	4	-0.408	0.361	ACCEPT	0.077
21 (see text)							
Double Walls							
15	0.603	-0.00026	6	0.129	0.452	ACCEPT	0.004
16	0.696	0.00136	5	-0.952	0.206	ACCEPT	0.232
17	0.241	-0.00173	9	0.697	0.254	ACCEPT	0.065
18	0.204	-0.00150	9	1.957	0.046	ACCEPT	0.354

NOTES:
1. See text for descriptions of α and β.
2. Number of data points = n.
3. "Probability" gives the probability that the observed difference between β and the assumed value of zero could be greater than or equal to that observed.

Standard Framed Walls

Houses #9 and #10 were typical of conventional prairie construction and used 4-mil poly-ethylene, stapled in place, directly behind the drywall with no sealing of joints or penetra-tions. House #22 used a simplified air barrier system with polyethylene on the walls, ceiling and (as the vapor barrier) in the basement with a spun-bonded polyolefin (SBPO) air retarder wrap on headers and cantilevers.

Houses #9 and #10 were found to be among the leakiest in the project, although they were still able to meet the airtightness requirements of the R-2000 program. Pre-stucco airtightness data were not available. House #22 was significantly tighter than #9 and #10, likely due to the SBPO header and cantilever wraps and the use of sealant to control leakage at joints and penetrations. The stucco was found to have a significant impact on airtightness in House #22, reducing the ac/h_{50} by 40% and the NLA_{10} by 47%.

Framed Walls with Exterior Insulated Sheathing

House #24 used a framed wall with exterior insulated sheathing (rigid glass fiber) and a simplified air barrier system with polyethylene on the main walls, ceiling and basement, and an SBPO wrap around the headers. Joints and penetrations were sealed and gaskets were applied at the tops of partition walls. Door and window penetrations were sealed using a variety of methods.

Through 19 months of monitoring, the airtightness of House #24 remained relatively constant, particularly with respect to the ac/h_{50} data.

Framed Walls with Interior Strapping

Houses #21 and #23 used a framed wall with interior strapping and a sandwiched air barrier. Exterior walls were covered with stucco (#23) and a (SBPO) air retarder and vertical vinyl siding on horizontal strapping (#21).

The 20 months of data available for House #23 showed that it became slightly more airtight over the monitoring period. Stucco produced a significant reduction in airtightness, decreasing the ac/h_{50} rate by 26% and the NLA_{10} by 30%. Although House #21 was not subjected to regular testing, some interesting anecdotal experience was gained. During one period a wind storm occurred which produced (airport-recorded) gusts of 96 km/h (60 mph). Site winds are believed to have been relatively severe since a commercial structure, located approximately 50 m (160 ft) away from House #21, had roughly 50 m² (500 ft²) of brick facade stripped off during the storm. Airtightness tests performed on House #21 approxi-mately one week prior and two days after the storm showed that the ac/h_{50} results were unchanged while the NLA_{10} increased by 13%. Although the final airtightness rate exceeded that permitted in R-2000 construction, this likely resulted from some of the nonstandard features of the house.

Double Walls

Houses #15 through #18 were built using double-wall construction with a 6-mil polyeth-ylene air barrier sandwiched between the studs and sheathing on the exterior side of the inner wall. Sealant was used extensively at major joints and penetrations including all door and window openings. Although identical envelope systems were used in the four houses, #15 and #16 contained air-to-air heat pumps with large-capacity ductwork from the indoors and outdoors. Testing subsequently found the ductwork to be a significant source of air leakage.

The two double-wall houses without heat pumps (#17 and #18) were the tightest houses in the project with airtightness levels approximately one-third that permitted by the R-2000 Standard. Leakage rates for Houses #15 and #16 were approximately 2.5 times greater than those of #17 and #18. Stucco was found to have a minimal impact on the airtightness of the four houses.

Commentary

The polyethylene air barrier systems used on the four wall types were able to achieve airtightness levels comfortably within the limits of the R-2000 Standard, with the double-wall houses displaying the lowest leakage rates of all the project houses. Possible, albeit slight, evidence of airtightness degradation was displayed by two of the ten houses. However, the observed increases in air leakage were small and all but one of the ten houses (#21) met the airtightness requirements of the R-2000 Standard at the end of their monitoring periods, including the conventional structures which were not designed to the Standard. While it obviously is not possible to project these results over the life of the houses, it was concluded that no evidence could be found to indicate that polyethylene is unsuited for use as an air barrier material in residential construction.

Based on the pre- and post-stucco airtightness data available for six of the houses, it was found that stucco had no effect on the airtightness of the four double-wall houses but did produce significant improvements in airtightness for the houses which used either standard framed walls or framed walls with interior strapping.

Performance of ADA Air Barrier Systems

Types

The Airtight Drywall Approach was used on 14 of the project houses with standard 13 mm (½ in.) drywall on the walls and 13 mm (½ in.) or 16 mm (⅝ in.) drywall on the ceilings. Sheets were fastened to framing members with drywall screws. For analysis purposes, the houses were categorized on the basis of gasket type and the presence, or absence, of an SBPO air retarder:

- Polyethylene gaskets and SBPO air retarder (Houses #1 to #6).
- Polyethylene gaskets (Houses #7 and #8).
- Simplified ADA and SBPO air retarder (Houses #11 to #14).
- Ethafoam rod, polyethylene and neoprene gaskets (Houses #19 and #20).

The 14 houses constructed with the ADA system used a variety of gasket types to seal major envelope joints and penetrations. Gasket types were mainly of the closed-cell variety, which is no longer considered to be the most suitable for ADA applications since they do not rebound as well as open-cell gaskets should the crack open with time. Monitoring periods ranged from 32 to 36 months.

It should be noted that development on the ADA system has continued since the project houses were constructed and since improved gasket types and methods have come into use. The measured airtightness results are given in Tables 5 and 6 (ac/h$_{50}$ and NLA$_{10}$) and Fig. 2 (ac/h$_{50}$ only) and the statistical analysis is summarized in Table 7.

TABLE 5—ADA houses: ac/h_{50} results.

GASKET TYPE & HOUSE NO.	MAR/86	JUL/86	NOV/86	FEB/87	JUL/87	NOV/87	FEB/88	MAY/88	JUN/88	JUL/88	AUG/88	NOV/88	MAR/89	JUN/89	AUG/89	DEC/89	MAR/90	APR/90
Polyethylene Gaskets & SBPO Air Retarder																		
1	1.67		1.48	1.57	0.98		1.48			1.20		1.65	1.45					
2		1.05	1.17	1.12		1.05	1.17					1.20	1.18					
3	1.51		1.54	1.85	1.49		1.69			1.38			1.50					
4	1.46		1.31	1.30	1.12		1.42						1.47					
5	1.12		1.26	1.10	1.14		1.05			1.20			1.03					
6	1.21		1.26	1.31	1.19		1.42						1.23					
Polyethylene Gaskets																		
7	1.17		1.52	2.20	1.34		1.44			2.06			1.42					
8			1.39	1.74						1.48			1.11					
Simplified ADA & SBPO Air Retarder																		
11	1.69*		0.96	0.88	0.88		1.01			1.06			1.07					
12	1.59*		0.96	0.98	0.88		0.98			1.26			1.25					
13	1.27*		0.83	0.76	1.04		0.94						0.89					
14	1.32*			0.96	0.99		1.16			1.10			1.32					
Polyethylene, Ethafoam Rod & Neoprene Gaskets																		
19	1.05*		0.84	0.91	0.72		1.04			0.92			1.11					
20	1.13*		0.82	0.73	1.01		0.80			0.94			0.87					

CHANGE BETWEEN FIRST (POST-STUCCO) TEST AND FINAL TEST

GASKET TYPE & HOUSE NO.	ABSOLUTE CHANGE¹ (ac/h_{50})	%²	NO. OF MONTHS
Polyethylene Gaskets & SBPO Air Retarder			
1	-0.22	-13	36
2	0.13	12	32
3	-0.01	-1	36
4	0.01	1	36
5	-0.09	-8	36
6	0.02	2	36
Polyethylene Gaskets			
7	0.25	22	36
8	-0.48	-30	36
Simplified ADA & SBPO Air Retarder			
11	0.18	20	32
12	0.13	12	32
13	0.06	6	32
14	0.18	16	32
Polyethylene, Ethafoam Rod & Neoprene Gaskets			
19	0.30	37	32
20	0.16	23	32

CHANGE BETWEEN FIRST (POST-STUCCO) TEST AND FINAL SEASONALLY-COINCIDENT TEST

GASKET TYPE & HOUSE NO.	ABSOLUTE CHANGE¹ (ac/h_{50})	%²	NO. OF MONTHS
Polyethylene Gaskets & SBPO Air Retarder			
1	-0.22	-13	36
2	0.06	5	25
3	-0.01	-1	36
4	0.01	1	36
5	-0.09	-8	36
6	0.02	2	36
Polyethylene Gaskets			
7	0.25	22	36
8	-0.48	-30	36
Simplified ADA & SBPO Air Retarder			
11	0.19	22	25
12	0.27	28	25
13	0.13	17	25
14	0.36	38	25
Polyethylene, Ethafoam Rod & Neoprene Gaskets			
19	0.20	22	25
20	0.14	19	25

NOTES:
1. * = No Stucco
2. A -ve indicates the house became more airtight.

TABLE 6—ADA houses: NLA$_{10}$ results.

GASKET TYPE & HOUSE NO.	MAR/86	JUL/86	NOV/86	FEB/87	JUL/87	NOV/87	FEB/88	MAY/88	JUN/88	JUL/88	AUG/88	NOV/88	MAR/89	JUN/89	AUG/89	DEC/89	MAR/90	APR/90
Polyethylene Gaskets & SBPO Air Retarder																		
1	0.577		0.467	0.390	0.400		0.477					0.452	0.506					
2		0.410	0.603	0.451	0.564	0.425	0.503			0.515		0.521	0.471					
3	0.513		0.517	0.762	0.437		0.656			0.546			0.525					
4	0.585		0.482	0.551	0.437		0.643						0.566					
5	0.444		0.450	0.432	0.334		0.341						0.318					
6	0.473		0.488	0.613	0.366		0.581			0.456			0.652					
Polyethylene Gaskets																		
7	0.433		0.637	0.981	0.620					0.717			0.496					
8	0.857		0.636	0.745			0.664			0.629			0.379					
Simplified ADA & SBPO Air Retarder																		
11	0.753*	0.345	0.396	0.317	0.282		0.370						0.393					
12	0.835*	0.468	0.419	0.329	0.318		0.405			0.433			0.538					
13	0.569*		0.314	0.401	0.437		0.403			0.434			0.271					
14	0.754*	0.490		0.516	0.393		0.467			0.425			0.578					
Polyethylene, Ethafoam Rod & Neoprene Gaskets																		
19	0.444*	0.232	0.320	0.347	0.279		0.402			0.381			0.444					
20	0.560*	0.298	0.287	0.208	0.444		0.299			0.391			0.343					

CHANGE BETWEEN FIRST (POST-STUCCO) TEST AND FINAL TEST

GASKET TYPE & HOUSE NO.	ABSOLUTE CHANGE[1] (NLA$_{10}$)	%[2]	NO. OF MONTHS
Polyethylene Gaskets & SBPO Air Retarder			
1	-0.072	-12	36
2	0.061	15	32
3	0.012	2	36
4	-0.019	-3	36
5	-0.126	-28	36
6	0.180	38	36
Polyethylene Gaskets			
7	0.063	14	36
8	-0.478	-56	36
Simplified ADA & SBPO Air Retarder			
11	0.048	14	32
12	0.069	15	32
13	-0.088	-25	32
14	0.088	18	32
Polyethylene, Ethafoam Rod & Neoprene Gaskets			
19	0.212	91	32
20	0.045	15	32

CHANGE BETWEEN FIRST (POST-STUCCO) TEST AND FINAL SEASONALLY-COINCIDENT TEST

GASKET TYPE & HOUSE NO.	ABSOLUTE CHANGE[1] (NLA$_{10}$)	%[2]	NO. OF MONTHS
Polyethylene Gaskets & SBPO Air Retarder			
1	-0.072	-12	36
2	0.020	4	25
3	0.012	2	36
4	-0.019	-3	36
5	-0.126	-28	36
6	0.180	38	36
Polyethylene Gaskets			
7	0.063	14	36
8	-0.478	-56	36
Simplified ADA & SBPO Air Retarder			
11	0.076	19	25
12	0.209	39	25
13	-0.130	-48	25
14	0.062	11	25
Polyethylene, Ethafoam Rod & Neoprene Gaskets			
19	0.097	22	25
20	0.134	39	25

NOTES: 1. * = No Stucco.
2. A -ve indicates the house became more airtight.

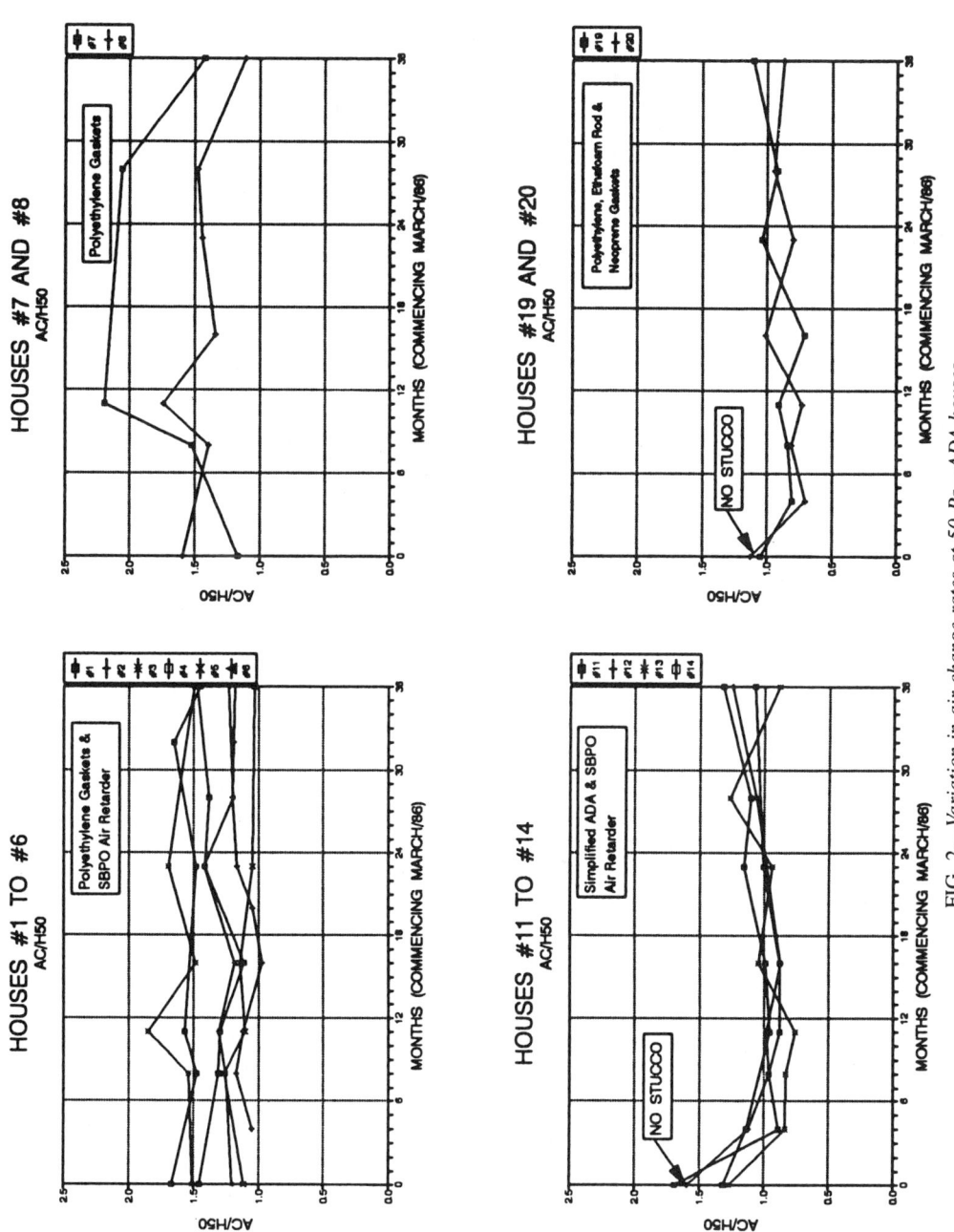

FIG. 2—*Variation in air change rates at 50 Pa—ADA houses.*

TABLE 7—Statistical analysis—ADA air barrier houses (ac/h_{50} data).

GASKET TYPE AND HOUSE NO.	α	β	n	t-STATISTIC	PROBABILITY (P)	NULL HYPOTHESIS (H_1)	COEFFICIENT OF DETERMINATION (r^2)
Polyethylene Gaskets, SBPO Air Retarder							
1	1.590	-0.00225	6	0.718	0.257	ACCEPT	0.114
2	1.047	0.00389	9	-1.672	0.069	ACCEPT	0.285
3	1.611	-0.00096	6	0.167	0.438	ACCEPT	0.007
4	1.309	0.00230	7	-0.536	0.308	ACCEPT	0.054
5	1.184	-0.00416	6	1.629	0.089	ACCEPT	0.399
6	1.248	0.00048	7	-0.162	0.439	ACCEPT	0.005
Polyethylene Gaskets							
7	1.555	0.00713	5	-0.434	0.347	ACCEPT	0.007
8	1.624	-0.01042	7	1.940	0.055	ACCEPT	0.430
Simplified ADA & SBPO Air Retarder							
11	0.857	0.00561	6	-2.861	0.023	REJECT	0.672
12	0.941	0.00504	7	-1.218	0.139	ACCEPT	0.229
13	0.805	0.00723	7	-1.283	0.125	ACCEPT	0.248
14	0.968	0.00726	6	-1.679	0.084	ACCEPT	0.414
Polyethylene, Ethafoam Rod & Neoprene Gaskets							
19	0.751	0.00856	7	-2.408	0.031	REJECT	0.537
20	0.751	0.00493	7	-1.362	0.116	ACCEPT	0.271

NLA$_{10}$ DATA

GASKET TYPE AND HOUSE NO.	α	β	n	t-STATISTIC	PROBABILITY (P)	NULL HYPOTHESIS (H_1)	COEFFICIENT OF DETERMINATION (r^2)
Polyethylene Gaskets, SBPO Air Retarder							
1	0.492	-0.00085	6	0.382	0.361	ACCEPT	0.035
2	0.459	0.00095	9	-0.430	0.340	ACCEPT	0.026
3	0.587	0.00015	6	-0.037	0.486	ACCEPT	0.000
4	0.526	0.00104	7	-0.434	0.341	ACCEPT	0.036
5	0.452	-0.00421	6	3.407	0.014	ACCEPT	0.744
6	0.464	0.00311	7	-0.926	0.199	ACCEPT	0.146
Polyethylene Gaskets							
7	0.660	-0.00041	5	0.050	0.482	ACCEPT	0.001
8	0.821	-0.00998	7	3.626	0.008	ACCEPT	0.725
Simplified ADA & SBPO Air Retarder							
11	0.330	0.00127	6	-0.699	0.262	ACCEPT	0.109
12	0.362	0.00297	7	-1.119	0.144	ACCEPT	0.200
13	0.388	-0.00076	7	0.319	0.382	ACCEPT	0.020
14	0.451	0.00139	6	-0.507	0.319	ACCEPT	0.060
Polyethylene, Ethafoam Rod & Neoprene Gaskets							
19	0.245	0.00549	7	-3.873	0.006	REJECT	0.750
20	0.275	0.00274	7	-1.012	0.179	ACCEPT	0.170

NOTES:
1. See text for descriptions of α and β.
2. Number of data points = n.
3. "Probability" gives the probability that the observed difference between β and the assumed value of zero could be greater than or equal to that observed.
4. Null hypothesis rejected if P less than or equal to 0.05 (5%) and β is positive (i.e. airtightness increasing with time).

Polyethylene Gaskets and SBPO Air Retarder

Houses #1 through #6 used an early version of the ADA system in which closed-cell polyethylene sill plate gaskets were used at the floor system/foundation intersection along with "poly-pan" boxes and foam gaskets around electrical outlets on exterior walls. An untaped SBPO air retarder was attached to the exterior insulated sheathing with the latter reversed so that the air retarder was sandwiched between the sheathing and framing.

Initial airtightness levels were found to be similar, or slightly below, that permitted by the R-2000 Standard. Pre-stucco test data were not available.

Polyethylene Gaskets

Houses #7 and #8 used the same gasket schedule as #1 to #6 but were not constructed with the SBPO air retarder on the exterior walls.

Air leakage rates in Houses #7 and #8 were slightly greater than those in #1 to #6, typically with airtightness rates close to the maximum permitted for R-2000 houses. These two houses also had the largest standard deviations of airtightness rates of the project house groups. Despite these variations, no evidence was found of a permanent change in airtightness. Pre-stucco data were not available.

Simplified ADA and SBPO Air Retarder

Houses #11 through #14 used a simplified ADA technique in which slightly less attention was paid to gasket installation, resulting thereby in somewhat higher leakage. With this technique, the houses were mechanically maintained under a negative pressure differential creating a quasi-dynamic wall effect. Gaskets—in this case ethafoam backer rod—were used around major penetrations such as doors and windows, while gaskets and covers were installed around electrical outlets. A SBPO air retarder was taped in place over the exterior insulated sheathing.

Airtightness levels measured prior to the application of stucco were found to be close to, or slightly above, those permitted by the R-2000 Standard. Once the stucco was applied, the tests were repeated and average reductions of 31% and 43% were found in the ac/h_{50} and NLA_{10} results respectively. The ac/h_{50} values increased for three of the four houses between the first and last tests and for all four between the first and last seasonally coincident tests. Similar patterns were displayed for NLA_{10} values for Houses #11 and #12.

Ethafoam Rod, Polyethylene and Neoprene Gaskets

Houses #19 and #20 used three types of gaskets: closed-celled polyethylene sill plate gaskets, primarily in the floor system/foundation area; neoprene gaskets at selected locations between the drywall and the bottom plate; and ethafoam backer rod around major penetrations. Electrical outlets on exterior walls were sealed with poly-pan gaskets and foam covers. Prior to the application of stucco, airtightness levels averaged about two-thirds that permitted by the R-2000 Standard. Average ac/h_{50} and NLA_{10} values dropped by 30% and 47%, respectively, once the stucco was applied, after which both houses become slightly leakier with time.

Commentary

Six of the 14 ADA houses displayed possible, albeit slight, evidence of airtightness degradation while the remaining eight displayed no significant change in leakage. However, the magnitude of the observed changes was small considering the air leakage rates at the end of the monitoring period were still below those permitted for R-2000 construction. Also, initial airtightness levels were less than the maximum permitted for R-2000 construction, demonstrating the capability of the ADA system to achieve low levels of envelope leakage. Stucco significantly improved the airtightness of the six houses for which pre- and post-stucco data were available. Based on these results, it was concluded that no evidence could be found to indicate that the ADA system is unsuited for use in residential construction.

Conclusions

Air Barrier Performance Relative to the R-2000 Standard—Both the polyethylene and ADA air barrier systems were able to meet the airtightness requirements of the R-2000 Standard. The tightest building envelopes were those constructed with the double-wall technique and polyethylene air barrier systems.

Stability of the Polyethylene Air Barrier System—Two of the ten stucco-covered houses constructed with polyethylene air barriers demonstrated possible, albeit slight, evidence of airtightness degradation over monitoring periods which ranged up to 36 months. However, the changes in airtightness were small and not judged to be of practical significance. All but one of the houses were able to meet the R-2000 airtightness requirements at the end of their respective monitoring periods. As a result, it was concluded that no evidence could be found to indicate polyethylene is unsuited for use as an air barrier material in residential construction.

Stability of the ADA Air Barrier System—Six of the fourteen houses constructed with (an early version of) the ADA air barrier system displayed possible, but also slight, evidence of airtightness degradation over monitoring periods which ranged up to 36 months. However, the magnitudes of the observed changes were small and not considered to be of practical significance. All 14 houses were able to meet the airtightness requirements of the R-2000 Standard at the end of their monitoring periods. Based on these results, it was concluded that no evidence could be found to indicate that the ADA system is unsuited for use in residential construction.

Impact of Stucco Upon Airtightness—Stucco produced significant improvements in airtightness for most of the polyethylene and ADA houses with the exception of the double-wall houses, which had very low initial air leakage rates and upon which stucco had little impact.

Acknowledgments

The work described in this report was conducted as part of the Flair Homes Energy Demo/CHBA Flair Mark XIV Project. Support was provided by Energy, Mines and Resources Canada under the Energy Demo Program and by Manitoba Energy and Mines under the Manitoba/Canada Conservation and Renewable Energy Demonstration Agreement (CREDA).

References

[1] Persily, A., "Repeatability and Accuracy of Pressurization Testing," *Proceedings*, ASHRAE/DOE Conference on Thermal Performance of the Exterior Envelope of the Building II, American Society of Heating, Refrigerating and Air Conditioning Engineers, 1982.

[2] Kim, A. K. and Shaw, C. Y., "Seasonal Variation in Airtightness of Two Detached Houses," *Measured Air Leakage of Buildings, ASTM STP 904,* American Society for Testing and Materials, Philadelphia, 1986.
[3] Howell, D. G. and Mayhew, W. J., "Energy Performance of Three Airtight Drywall Approach Houses," Howell Mayhew Engineering Inc., 1987.
[4] "Comparison of Airtightness Retesting Results," Buchan, Lawton, Parent Ltd., Report prepared for Energy, Mines and Resources Canada, 1988.
[5] Hamlin, T., Canada Mortgage and Housing Corporation, personal communication, 1990.
[6] "Leakage Distribution in Houses," Air Infiltration Centre, Technical Note AIC 16, Bracknell, Berkshire, United Kingdom, 1985.
[7] Carlsson, A. and Kronvall, J., "Constancy of Airtightness in Buildings," *Proceedings,* 5th AIC Conference on the Implementation and Effectiveness of Air Infiltration Standards in Buildings, Air Infiltration Centre, 1984.
[8] "R-2000 Home Program Technical Requirements," Energy, Mines and Resources Canada, 1986.

Discussion

Anton TenWolde[1] (*written discussion*)—Did you see visual degradation of ADA systems?

G. Proskiw (*author's closure*)—No visual evidence of degradation of the ADA system was observed.

---

[1] Forest Products Laboratory, Madison, WI.

Max Sherman[1] and Larry Palmiter[2]

Uncertainties in Fan Pressurization Measurements

REFERENCE: Sherman, M. and Palmiter, L., **"Uncertainties in Fan Pressurization Measurements,"** *Airflow Performance of Building Envelopes, Components, and Systems, ASTM STP 1255*, Mark P. Modera and Andrew K. Persily, Eds., American Society for Testing and Materials, Philadelphia, 1995, pp. 266–283.

ABSTRACT: The ASTM Method for Determining Air Leakage Rates by Fan Pressurization Test (E 799-87) measures the airtightness of building envelopes using fan pressurization. Uncertainty is introduced in the process from the uncertainty of the airflow and pressure measurements as well as from nonlinearities in the system to be measured. This paper analyzes the precision and bias associated with making a measurement using E 779 in typical field situations. Model specification (or modelization) errors may also contribute significantly to the overall uncertainty in the estimates of the 4 Pa leakage; the sources and sizes of these modelization errors interact with the instrumentation errors in making a fan pressurization test. Insufficient field data exist to fully include the effects of modelization and other low-pressure phenomena, but the current standard can nevertheless by improved by tightening the instrumentations and procedural specifications and by judicious choice of pressure measurement stations.

KEYWORDS: envelope leakage, air flow, infiltrator, error analysis

Nomenclature

κ = Leakage coefficient, m^3/s Pa^n
n = Flow exponent, $[-]$
$P, \Delta P$ = Outdoor-indoor pressure, Pa
Q = Airflow, m^3/s
ρ = Density of air, 1.2 kg/m^3
r = Correlation coefficient, $[-]$

Subscripts

bias = Bias error
e = Value at extrapolated pressure
modelization = Modelization error
precision = Precision error
fan = Reading from blower door
H = High pressure point
L = Low pressure point

[1] Energy Performance of Buildings Group, Energy and Environment Division, Lawrence Berkeley Laboratory, University of California.
[2] Ecotope Inc., Seattle, WA 98112.

Introduction

Whole building air leakage is an important property of residential buildings. It serves both as a quality control indicator of the airtightness of the structure and as a quantitative measure for estimating the ventilation rate for both energy use and indoor air quality purposes. Although airflow at 50 Pa is the most used measure of airtightness, the most common *quantitative* application of air leakage measurement is the effective leakage area (ELA). Various infiltration models such as the Lawrence Berkeley Laboratory (LBL) [1] infiltration model [2] use leakage data in this form. Various ASHRAE Standards such as Standard 119 for Airtightness [3] and Standard 136 on estimation of air change rates for indoor air quality purposes [4] rely on the ELA concept also.

The ASTM Method for Determining Air Leakage by Fan Pressurization Test (E 799-87) requires that a set of airflow measurements be taken over a specified pressure range from which a flow coefficient and exponent are calculated. Some infiltration models such as the AIM2 model [6] use these parameters directly. This parameterization allows extrapolation of the flow to pressures lower than that of the measurements. ELA is calculated from the flow rate at 4 Pa and can be calculated directly from the flow coefficient and exponent.

It is important to understand that the usefulness of a measured quantity is dependent upon the certainty at which we know its true value. In the E 779 procedure there are three categories of error which can increase our uncertainty: precision errors due to noise and other random errors, biases in the measurement of pressure and flow, and extrapolation errors. Although some bias and precision limits are included in the current standard, E 779 does not contain a thorough treatment of these three categories of uncertainty.

It is impossible to quantify uncertainty without first specifying the quantity for which the uncertainty is desired. Similarly, an optimal measurement protocol for measuring the airflow through the building envelope cannot be done without specifying the reference pressure. The focus of this report is on estimating the uncertainty of the flow estimation at 4 Pa (and, hence, the ELA) implicit in the ASTM procedure and on examining alternative measurement techniques for improving the estimate.

The intent herein is to evaluate different standards and protocols for measuring ELA from a general perspective, not to find the best set of procedures for a particular piece of instrumentation hardware. Although the general properties of fan pressurization equipment are considered, it is assumed that the hardware could be produced to meet the various specifications considered. For example, it is assumed that the airflow capacity of the equipment will be optimally sized to meet the error specifications. Although not being done herein, the equations developed could be used to do a detailed analysis of a specific set of instrumentation under various protocols.

This paper assumes that the reader is generally familiar with the art of making fan pressurization measurements (using a "Blower Door") and standard E 779. Both technical [7] and popular [8,9] articles are available to familiarize the reader with some of the relevant issues.

Sources of Error

Before discussing the sources of error in the experiment it is important to define certain terms related to measurement uncertainty. *Accuracy* is the ability of a particular measurement to approach some exogenously chosen reference level (usually the *true* value); it is quantified by the total *measurement error*. One's *estimate* of the measurement error is called the *measurement uncertainty*. The measurement error is made up of the precision error and the bias error. *Precision* is the property of reproducibility in a measurement; the precision error can be estimated from the standard deviation of a sample of repeated measurements. *Bias* errors are caused by systematic departures from the reference level and can be estimated from the difference between the sample average of some measurements and the reference level. In our case, the true values are not known and estimates of the measurement errors will be made by propagation of estimated errors (i.e., this is an uncertainty calculation).

The problem of estimating the ELA is not simply a measurement problem. Because the fan pressurization technique cannot directly measure the flow rate at 4 Pa, it is necessary to extrapolate the measurable behavior to determine the desired quantity. All such extrapolations require the use of a model of the physical situation and can be considered under the rubric of *inverse problem theory* [10]. Thus there are two broad sources of error: the errors associated with making the measurements themselves, and the model specification error.

Measurement error can be further subdivided. There are two independent measurements made: pressure and airflow. For each of these measurements there is the possibility of either precision errors or bias errors. Precision errors are treated as random errors that will change the value of a reading upon repeated measurements (i.e., noise). Fluctuations in pressure measurements caused by wind (speed and direction) variations can be treated as a precision error for fan pressurization measurements.

Wind is probably the most pernicious source of experimental error in field situations because it can cause both precision and bias errors. Modera and Wilson [11] have shown that increasing wind speed both increases the uncertainty of the estimate and can cause a (downward) bias in that estimate. They also consider some techniques for reducing these errors. Some of these issues are considered in following sections.

Bias errors are errors in which the reading varies in a fixed but unknown manner from the true value such as from calibration errors (i.e., systematic errors). Bias errors will not affect the reading for a given experimental situation and, therefore, cannot be estimated from the measurement. A common bias error in fan pressurization is nonlinearity associated with the pressure measurement devices. It is possible to estimate some kinds of bias error such as variations from instrument to instrument by using intercomparison procedures such as the ASTM Practice for Conducting an Interlaboratory Test Program to Determine the Precision of Test Methods (E 691-79) [12]. Murphy et al. [13] have reported on such an intercomparison.

Model specification (or modelization) errors come in many forms. The analysis in E 779 assumes that the airflow through the envelope can be characterized by a single indoor-outdoor pressure difference (i.e., *the* pressure is measured by a single pressure location), that the flow through the envelope is equal to the flow through the fan (i.e. *the* flow), and which is described as a simple power-law function in the pressure:

$$Q = \kappa \Delta P^n \tag{1}$$

For brevity we shall drop the "Δ" from in front of the pressure, remembering that the pressure, P, indicates a pressure difference.

The ELA can be found from the flow at the extrapolated pressure of 4 Pa:

$$\text{ELA} = \sqrt{\frac{\rho}{2P_e}}\, Q_e = \kappa \sqrt{\frac{\rho}{2}}\, P_e^{(n-1/2)} \tag{2}$$

In an actual E 779 test this value is calculated separately for pressurization and depressurization and then averaged. The procedure reduces certain kinds of uncertainties and potentially creates additional model specification errors. The present report, however, does not analyze either of these effects.

Since the ELA is proportional to the flow at 4 Pa, we shall henceforth deal with that extrapolated flow and determine its uncertainty with the understanding that it can then be converted into ELA.

We shall treat the three types of errors[3] separately and then after Coleman and Steele [14] based on ASME [15] measurement uncertainty principles, we shall combine the independently calculated precision, bias, and modelization errors together:

$$\delta Q_e = \sqrt{\delta^2 Q_{precision} + \delta^2 Q_{bias} + \delta^2 Q_{modelization}} \tag{3}$$

We shall examine each of these three terms separately, but first we must describe the sources of error from the instrumentation itself.

Pressure Measurement

In Eq 1 the pressure is treated as the independent variable. In most common forms of regression analysis, errors in the independent variables are ignored; unfortunately we cannot ignore them here.

Weather-induced effects can cause significant errors in the measurement of the pressure, but in their absence there can still be biases in the measurement due to nonlinearities in the gage or calibration errors. The most commonly used pressure measurement device, the magnetically coupled mechanical gage, is known to have a significant nonlinear response as well as hysteresis and sticking problems. Normally, however, this systematic error is within the specifications of the standard (i.e., within 2.5 Pa), but such a systematic error can have profound effects. The size of these biases are likely to vary over the range of the measurement. The electronic pressure transducers sometimes used do not suffer from these problems and can be calibrated to a much higher level of intrinsic accuracy.

Steady wind and temperature difference can cause the pressure drops across different parts of the envelope to vary, thus causing the model assumption of a single representative pressure to be violated. If the "preferred test conditions" are met, then this violation should be less than about 4 Pa. Furthermore, the standard requires that this violation be held to less than 10% of the test pressure. It is doubtful, however, if this requirement is ever verified in the field. Slow changes in these steady weather conditions can manifest themselves as zero drift in the pressure measurement (at zero airflow); otherwise this error is independent of the size of the measured pressure.

The most noticeable error in the pressure measurement is the fluctuation of the reading caused by variations in the wind speed. According to the standard, these fluctuations must be included in the 2.5 Pa and 10% of reading limits also. It is difficult (and probably rarely accomplished) to meet the 10% specification at the lowest pressure stations.[4] Some research-

---

[3] We use the notation δQ to indicate the expectation value of the accuracy (i.e., the uncertainty) with an optional subscript indicating the source of the error. The uncertainties in Eq 3 and the uncertainties that will make them up (e.g., Eq 9) should be interpreted as the 95% confidence levels (about two standard deviations). (The equations are, however, also valid if the terms are all treated as standard deviations.) The notation $\delta^2 Q$ refers to the square of this confidence level.

[4] Stations refer to measurements made at specified nominal values of the house pressure. For example, E 779 currently requires that airflow be measured at six pressure stations: 12.5, 25, 37.5, 50, 62.5, and 75 Pa.

ers are accomplishing it, however, by using an electronic transducer and time-averaging the signal. Regardless of the size of this error, it can be treated as independent of the measured pressure.

Airflow Measurement

Method E 779 requires that the equipment used for measuring the airflow be accurate to within 6% of average value. The standard is mute on the interpretation of this statement, but for the purposes of this report we assume the requirement is to be accurate to 6% of the reading. The most common type of blower door uses the pressure drop across a calibrated orifice plate or nozzle assembly to estimate the flow through the fan. That pressure measurement is usually made with an instrument of design similar to the one for measuring the "house" pressure above, but at a higher pressure range.

For such a door the errors caused by the weather-related effects are normally not significant, because of the higher operating range. Pressure fluctuations induced by the fan as well as the nonlinearity and hysteresis effects can still cause error in the "fan" pressure measurement. Additionally, the properties of the orifice plates used with the door have uncertainties associated with them.

Other types of blower doors are occasionally used (e.g., RPM-calibrated doors); also, different manufacturers calibrate their doors differently. Whether it be an orifice door or an RPM door the percentage error is always higher at the lower end of a "range" (e.g., one of the orifice plates for an orifice door). In order to avoid considering each of the possible options independently, we shall make the reasonable assumption that the uncertainty in the measured flow is a constant value, independent of the pressure for any given experimental design.

One-Point Analysis

If our objective is to determine the flow rate through the envelope at a specified pressure, the most straightforward approach would be to directly measure it at that pressure. We can construct the measurement uncertainty for such a case as follows:

$$\frac{\delta Q_e}{Q_e} = \sqrt{\frac{\delta^2 Q_{\text{precision}} + \delta^2 Q_{\text{bias}}}{Q_{\text{fan}}^2} + n^2 \frac{\delta^2 P_{\text{precision}} + \delta^2 P_{\text{bias}}}{P_e^2}} \qquad (4)$$

The second term is computed from Eq 1 and is due to the fact that uncertainty in the measured pressure implies an additional uncertainty in the desired flow because we may be measuring at the wrong pressure. Thus we include the subscript e on the desired value.

For such an experimental design it would not be difficult to assure that the total error (i.e., both precision and bias) in the flow could be held to 5%. Similarly, we could assume that the bias in the pressure measurement could be held to 5%, but that the precision error would be about 2.5 Pa. If we apply these limits for a high pressure (e.g., 50 Pa), we get an uncertainty of about 7%.

If we apply these limits at a low pressure (e.g., 4 Pa), we get over a 40% uncertainty. This uncertainty does not include any modelization errors associated with the assumption of a single representative pressure difference. An alternative approach would be to measure the airflow at 50 Pa and extrapolate to 4 Pa by guessing an exponent of $\frac{2}{3}$; the overall uncertainty, however, is not improved because of the large uncertainty of the exponent. (See *Analysis of Modelization Errors* in the next section for more detail).

If our objective is to measure the 50 Pa airflow, a one-point measurement variant would be sufficient. As our objective is to measure the 4 Pa airflow, we need a better experimental design than a single-point measurement (at least with the current generation of instrumentation). A second measurement point would allow an improved estimate of the exponent and, therefore, of the extrapolated flow. Since Eq 1 conventionally represents the leakage process, such a two-point measurement is the minimal requirement for characterization and is worthy of a more detailed examination.

Two-Point Analysis

In this section we analyze the situation assuming that the airflow is measured at two pressures. These two pressures are chosen to be representative of an applicable field experiment. That is, the high-pressure point is in the neighborhood of 50 to 75 Pa and the low-pressure point would typically be about a factor of six smaller. Furthermore, we wish to consider the case in which we analyze the data to obtain an extrapolation at a yet lower pressure value (e.g., 4 Pa).

The experimental design is that the airflows (Q_H, Q_L) are measured at two pressure stations (P_H, P_L), respectively. From these two measurements we can uniquely determine the parameters of Eq 1:

$$n = \frac{\log(Q_H/Q_L)}{\log(P_H/P_L)} \tag{5}$$

(where "log" is the natural logarithm)

$$\kappa = \frac{Q_H}{P_H^n} = \frac{Q_L}{P_L^n} \tag{6}$$

The extrapolated flow can then be found by applying Eq 1 at the desired pressure. Using Eq 1 as the defining relation, we can determine the dependence of the extrapolated flow on the measurements for the power-law model:

$$\frac{dQ_e}{dQ_H} = \frac{Q_e}{Q_H} \times \frac{\log(P_e/P_L)}{\log(P_H/P_L)} \qquad \frac{dQ_e}{dP_H} = -\frac{nQ_e}{P_H} \times \frac{\log(P_e/P_L)}{\log(P_H/P_L)}$$

$$\frac{dQ_e}{dQ_L} = \frac{Q_e}{Q_L} \times \frac{\log(P_H/P_e)}{\log(P_H/P_L)} \qquad \frac{dQ_e}{dP_L} = -\frac{nQ_e}{P_L} \times \frac{\log(P_H/P_e)}{\log(P_H/P_L)} \tag{7}$$

From these expressions we can calculate the uncertainties in the extrapolated flow due to the different sources of error.

Analysis of Precision Errors

Here assume that there are no bias or model specification errors and analyze only for uncertainties due to random fluctuations in the measurements around their true values. For such precision errors there is no bias and, therefore, no first-order error terms; also, it is reasonable to assume that the fluctuations in the measurements are uncorrelated, in which case all of the cross terms disappear.

The uncertainty (due to precision errors alone) in the estimated flow is thus described by the familiar quadrature formula:

$$\delta Q_e = \sqrt{\left(\frac{dQ_e}{dQ_H}\right)^2 \delta^2 Q_H + \left(\frac{dQ_e}{dQ_L}\right)^2 \delta^2 Q_L + \left(\frac{dQ_e}{dP_H}\right)^2 \delta^2 P_H + \left(\frac{dQ_e}{dP_L}\right)^2 \delta^2 P_L} \qquad (8)$$

Examination of these terms for our prototypical situation (i.e., extrapolation point below the low-pressure station which is a factor of six below the high-pressure station, plus pressure and flow uncertainties of constant size) indicates that the uncertainty in the extrapolated flow is dominated by the low-pressure point. Thus we can approximate the uncertainty by the following expression:

$$\frac{\delta Q_e}{Q_e} = \frac{\log(P_H/P_e)}{\log(P_H/P_L)} \sqrt{\left(\frac{\delta^2 Q_{precision}}{Q_L^2} + n^2 \frac{\delta^2 P_{precision}}{P_L^2}\right)} \qquad (9)$$

Evaluating this expression for our prototypical values and using the uncertainty specifications gives a precision error in the extrapolated flow (and hence the ELA of about 27%). This uncertainty is dependent upon the pressure stations. For example, increasing the lower-pressure station to about 18 Pa can reduce this uncertainty to about 20%.

Analysis of Bias Errors

We now examine the effect when bias errors alone are at work. A bias error is caused when the instrument reports a difference from the true value; this difference may be a function of the size of the measurement, but it does not fluctuate. An example would be an instrument that always read 10% high. If one knew that fact it could be corrected for. Therefore, one does not know, in general, the size of the potential biases that are left after any corrections.

Although bias errors are fixed, their size (and even sign) is unknown. One might infer that the same formulas used for precision errors would apply, but that is rarely the case. Bias errors are usually highly correlated, especially for the same quantity. For the errors we consider herein, we can assume that the two pairs of measurements are completely correlated, but that the biases in the flow are uncorrelated with the pressure. In this case the bias error can estimated as

$$\delta Q_e = \sqrt{\left(\frac{dQ_e}{dQ_H}\delta Q_H + \frac{dQ_e}{dQ_L}\delta Q_L\right)^2 + \left(\frac{dQ_e}{dP_H}\delta P_H + \frac{dQ_e}{dP_L}\delta P_L\right)^2} \qquad (10)$$

For most experimental designs used, reasonable calibration of the fan would imply that the bias errors associated with the fan flow would be much smaller than the bias errors associated with the pressure measurement. If we ignore the bias errors from the flow, then the uncertainty becomes

$$\frac{\delta Q_e}{Q_e} = \left(\frac{-n}{\log(P_H/P_L)}\right)\left(\log(P_H/P_e)\frac{\delta P_L}{P_L} + \log(P_e/P_L)\frac{\delta P_H}{P_H}\right) \qquad (11)$$

Using this formula we can see that if the pressure were consistently 10% high, for example, the extrapolated flow would be about 7% low. If the pressure gage were nonlinear such that it read 10% high at the low reading and 10% low at the high reading, then for our prototypical pressure stations the extrapolated flow would be about 15% low at an extrapolated pressure of 4 Pa.

For some types of bias errors (e.g., zero drift) the bias is a fixed size (e.g., 1 Pa), in which case only the low-pressure station significantly contributes towards the bias in the extrapolation. If we assume such a zero error for our prototypical case we get a 9% error in the extrapolated flow. Zero drifts can be caused by the instrumentation itself, but can also be caused by reasonably steady weather conditions that shift the reading away from the average pressure drop across the envelope.

The net error from most of these sources of bias can be approximated by just using the bias caused by the low-pressure point:

$$\frac{\delta Q_e}{Q_e} = -n \frac{\log(P_H/P_e)}{\log(P_H/P_L)} \frac{\delta P_{bias}}{P_L} \tag{12}$$

Analysis of Modelization Errors

There are a virtually unlimited sources of potential model specification errors. In any experiment there are many implicit assumptions that could be incorrect. For example, one normally assumes that the density of air passing through the fan does not change during the course of the test; it could and it would make difference to the estimate, but we will not consider it.

We also assume that all of the leaks experience the same driving pressure. If the leakage were linear, violation of this assumption would not be a problem (assuming the pressure measurement represented an unbiased average). The fact that the flow exponent is not unity means that there will be an error associated with this assumption failure, which will be larger for the low-pressure station. The size of this effect varies over the range of house pressures typically used. The effect can be quite large when the fan-induced pressure is not enough to dominate the weather-induced pressure everywhere on the envelope. If, however, we exclude such cases from consideration, this model specification error acts like a bias in the pressure and is normally less than about 10% of the weather-induced pressures. If we restrict the pressure range to be greater than about twice the weather-induced pressure, we can neglect this effect.

Another key model specification error that we address is the assumption that the flow can be extrapolated using a power-law formulation. It is always suspect to extrapolate data beyond the measurement limits, unless there is strong exogenous evidence (e.g., a physical law) to suggest that the functional form of the curve is valid.

The functional form of the fan pressurization curve (i.e., Eq 1) is principally justified by empirical evidence from measurements. The first author [16] has shown that from first principles the relationship can look like a power-law over any narrow pressure range, but that over a wider pressure range it must deviate. In fact the deviation can go in either direction (i.e., increasing or decreasing exponent), depending on the distribution and type of the leaks.

If we assume that the exponent can vary as a function of pressure[5] and that there will be a deviation in the average exponent between the two pairs of pressure stations of δn, then there will be a model specification error in the extrapolation of the approximate size

---

[5] The coefficient must also vary as a function of pressure to keep the airflow continuous, but we do not need to consider this effect.

$$\frac{\delta Q_e}{Q_e} = \log(P_e/P_L)\delta n \qquad (13)$$

We have examined specific, low-pressure sets of our own data and unpublished data from Walker and Wilson [6] in an attempt to quantify the change in exponent. Ostensibly large (e.g., $\delta n = 0.1$) variations can be found. If, however, the nonlinearity errors and instrumentation errors are taken into account, the only conclusion that can be drawn is that the variation is not consistently larger than 0.1. For the purposes of the following subsection on error minimization we will assume half that value with the understanding that such an estimate is quite crude.

Error Minimization

If we combine (in quadrature) these three sources of error we can estimate the total uncertainty of the measurement assuming that the precision and bias errors are dominated by the low-pressure station:

$$\frac{\delta Q_e}{Q_e} = \sqrt{\left(\frac{\log(P_H/P_e)}{\log(P_H/P_L)}\right)^2 \left(\frac{\delta^2 Q}{Q_L^2} + n^2\frac{\delta^2 P}{P_L^2}\right) + (\log(P_e/P_L)\delta n)^2} \qquad (14)$$

where we have combined the precision and bias errors for each measurement type. If we evaluate this expression for our prototypical situation the combined error is 36%.

Given that we have an expression that allows us to estimate the combined error, we can use it to optimize the experimental design. The conditions we can control in the experimental design are the high- and low-pressure stations and to some extent the errors in the instruments. All else being assumed fixed, we can find a lower-pressure station that minimizes the error by using the previous equation. Table 1 summarizes the results of using Eq 14. For each of these calculations we have assumed that $n = \frac{2}{3}$, $\delta n = 0.05$, $P_e = 4$ Pa.

Analysis of Uncertainties in the Parameters n and κ

It is sometimes desirable to characterize the data by the two parameters of Eq 1. One can calculate the uncertainties in these parameters from the data, but must also remember that the parameters may be (and are, in fact) highly correlated. The dependence of these parameters on the measured data

TABLE 1—*Total errors in extrapolation using two-point method.*

P_H, Pa	$\delta Q/Q_L$	δP, Pa	P_L, Pa	$\Delta Q_e/Q_e$	$\Delta Q_e/Q_e$ ($\delta n = 0$)
75	10%	4	14	29%	29%
75	10%	1	9	20%	20%
75	5%	4	19	25%	23%
75	5%	1	12	16%	14%
50	10%	4	12	33%	32%
50	10%	1	8	21%	21%
50	5%	4	15	29%	28%
50	5%	1	11	17%	16%

$$\begin{bmatrix} dn \\ d\kappa \end{bmatrix} = (\log(P_H/P_L))^{-1} \cdot \begin{bmatrix} 1 & -1 \\ -\log P_L & \log P_H \end{bmatrix} \cdot \begin{bmatrix} \dfrac{dQ_H}{Q_H} - n\dfrac{dP_H}{P_H} \\ \dfrac{dQ_L}{Q_L} - n\dfrac{dP_L}{P_L} \end{bmatrix} \tag{15}$$

can be used to estimate the uncertainties in the parameters from the errors in the measured data.

As can be inferred from this formulation, the two parameters are highly correlated and do not, therefore, have an independently measurable value. We can demonstrate this fact by calculating the uncertainty in the two parameters assuming that the errors in the measurements are uncorrelated fluctuations (see earlier *Analysis of Precision Errors*). In such a case the uncertainties are as follows:

$$\delta n = \frac{1}{\log(P_H/P_L)} \sqrt{\left(\frac{\delta^2 Q_L}{Q_L^2} + n^2 \frac{\delta^2 P_L}{P_L^2}\right) + \left(\frac{\delta^2 Q_H}{Q_H^2} + n^2 \frac{\delta^2 P_H}{P_H^2}\right)}$$

$$\frac{\delta\kappa}{\kappa} = \frac{1}{\log(P_H/P_L)} \sqrt{\log^2 P_H \left(\frac{\delta^2 Q_L}{Q_L^2} + n^2 \frac{\delta^2 P_L}{P_L^2}\right) + \log^2 P_L \left(\frac{\delta^2 Q_H}{Q_H^2} + n^2 \frac{\delta^2 P_H}{P_H^2}\right)} \tag{16}$$

We can express the correlation with the normalized covariance between the two parameters (i.e., the cross-correlation coefficient), which is close to -1:

$$r_{\kappa,n} = -1 + \varepsilon$$

$$\varepsilon \approx \frac{1}{2}\left(1 - \frac{\log P_L}{\log P_H}\right)^2 \frac{\dfrac{\delta^2 Q_H}{Q_H^2} + n^2 \dfrac{\delta^2 P_H}{P_H^2}}{\dfrac{\delta^2 Q_L}{Q_L^2} + n^2 \dfrac{\delta^2 P_L}{P_L^2}} \tag{17}$$

For our prototypical case, ε is about 3%, indicating that the uncertainties in the two parameters are about 97% (negatively) correlated. In order to use them to calculate other uncertainties, we would have to keep track of the cross terms.

If one is using the uncertainty in the two parameters to estimate the accuracy of an extrapolation, the covariance must be taken into account or the estimate will be too large. For example, κ is equal to the flow rate at an extrapolated pressure of 1 Pa. The uncertainty in that flow is given correctly by the lower half of Eq 16 without regard to the uncertainty of the exponent.

Regression Analysis

Method E 779 requires multipoint pressure sampling and a regression analysis to determine parameters κ, n of Eq 1. A regression analysis has several advantages over a two-point approach: it allows an estimation of the uncertainty (due to precision errors) from the data and, by virtue of making multiple measurements, it has the potential for reducing the random errors. A general example of this is that when making multiple measurements of a single variable the standard deviation can be estimated and the error of the mean is reduced as more measurements are made.

In this section we examine the usefulness and uncertainties associated with the regression technique using standard approaches. The reader is directed to the literature [17,18] for a discussion of regression and least-squares analysis. Regression analyses explicitly exclude effects arising from bias or modelization errors, but include precision errors. Below we analyze a few regression designs excluding the bias and model specification errors.

Numerical Analysis

For the analyses that follow we assume that conventional blower door equipment is used, but that equipment minimally meets the specifications of each analysis. Thus, the specifications of the hardware will be different for each analysis and cannot be used to determine the optimal protocol for a given set of equipment. Rather, this section allows us to compare different standards and different potential hardware designs.

The current standard requires that measurements be made every 12.5 Pa up to 75 Pa and then analyzed with an (unweighted) linear least-squares approach (using the logarithms of the variables). Since conventional blower door equipment uses orifice-type flow measurement, the 6% specification on airflow measurement accuracy will be interpreted to mean that at the lowest pressure station the airflow uncertainty will be 6%; such an assumption leads to a higher level of accuracy at higher pressure stations. We also assume that the exponent of the leaks is 0.65 (compared with 0.5 for the orifice plate) and that the absolute accuracy of the (house) pressure is 2.5 Pa. We have done an error analysis of this approach using synthetic data with measurement errors otherwise similar to our prototypical case and found the following for the *E 779 analysis*:

$$\delta n = 0.07$$

$$\delta \kappa = 29\%$$

$$\varepsilon = 0.1\%$$

$$\frac{\delta Q_e}{Q_e} = 19\% \qquad (18)$$

By using an unweighted regression of the logarithms of the data, this type of analysis implicitly assumes that the uncertainty (due to precision errors) is a constant fraction of the measured value. As discussed earlier, this is not a good assumption for the pressure measurement. The net effect of this incorrect assumption is that the low point has an unduly large influence on the result of the regression. We can correct this error by weighting each point according to its estimated accuracy. The weights are derived from the total uncertainty of each measured point and can be derived from the One-Point Analysis section (the weight is the inverse square of Eq 4). Doing so downweights the low points appropriately. We call this the *weighted E 779 analysis*:

$$\delta n = 0.04$$

$$\delta \kappa = 18\%$$

$$\varepsilon = 0.3\%$$

$$\frac{\delta Q_e}{Q_e} = 12\% \qquad (19)$$

It is apparent that weighting the points reduces the precision error of the estimated flow. One must also be aware that by downweighting the low-pressure points there could be an increase in the extrapolation error associated with model specification. It is, however, quite difficult to estimate the impact of this model specification error without a better model of how the exponent might vary with pressure.

The Canadian General Standards Board (CGSB) (1986) has a standard analogous to ASTM E 779 in which only depressurization is done and for which the pressures range from 15 to 45 Pa in 5-Pa steps; the accuracy requirement on the flow is 5% rather than 6%. We can perform a similar analyses to determine the uncertainties using the *CGSB analysis*:

$$\delta n = 0.08, \ 0.06$$

$$\delta \kappa = 27\%, \ 22\%$$

$$\varepsilon = 0.2\%, \ 0.3\%$$

$$\frac{\delta \ Q_e}{Q_e} = 17\%, \ 14\% \tag{20}$$

where the first number is for an unweighted regression and the second for the weighted regression. The CGSB weighted regression technique is based on standard statistical methods [19] to compensate for differential errors in the flow measurement only. As can be seen from the two-point analysis the uncertainty in the final answer due to the flow error is small compared to that caused by the pressure error. Our analysis uses the regression weights (as stated earlier) to include both flow and pressure error, but did use the CGSB pressure stations and instrumentation assumptions.

Since all the points in the previous analyses are a fixed pressure spacing apart, they actually are more heavily weighted towards the high-pressure side for the logarithmic analysis. An alternative approach is one in which one attempts to overcome this problem by logarithmically spacing the points (at 10, 14.3, 20.5, 29.3, 41.9, and 60 Pa) and by weighting the points using an estimate of the measurement uncertainty. Using synthetic data for the same assumptions as above for the E 779 analysis, we find the following for the *log-spaced analysis*:

$$\delta n = 0.08, \ 0.05$$

$$\delta \kappa = 28\%, \ 19\%$$

$$\varepsilon = 0.3\%, \ 0.6\%$$

$$\frac{\delta \ Q_e}{Q_e} = 17\%, \ 12\% \tag{21}$$

It has been suggested that a reasonable compromise between these two approaches might be to use equally spaced points from 10 to 60 Pa, but to include the appropriate weighting in the analysis. For such an experimental design our analysis yields

$$\delta n = 0.05$$

$$\delta\kappa = 21\%$$

$$\varepsilon = 0.4\%$$

$$\frac{\delta\,Q_e}{Q_e} = 13\% \tag{22}$$

Because the percentage error (i.e., the weights) change substantially over the range of measurement, it is usually superior to use a weighted regression when analyzing fan pressurization data. As these examples demonstrate, the precision error is reduced because the points contributing more to the uncertainty are weighted less. Although not apparent from the results, the same can be true for some types of bias errors.

As discussed in the Two-Point Analysis section, bias and model specification errors can come in many guises. Because bias errors can be highly correlated between measurements, the multipoint advantages of a regression may not reduce their impact. Although, in general, the regression analysis will give larger bias errors than an analogous two-point analysis, we will make the same approximations and use the same expression, Eq 12, to estimate the bias errors. Similarly, we can assume that modelization errors can be approximated by Eq 13. Accordingly, for each of the regression analyses above, terms must be added to the uncertainties to account for the bias and model specification errors.

Improving the Test Method

The previous sections have indicated the sources and ranges of the errors that affect the test method for determining the extrapolated flow at 4 Pa. This section discusses ways of improving the method to decrease the uncertainty.

Reducing Precision Errors

The use of multiple measurements, as in a regression analysis, can reduce the uncertainty of the result, due to precision errors. This uncertainty can also be reduced by increasing the precision of the measurement itself. Such precision increases can come from either improved instrumentation or a reduction or limitation on the exogenous noise sources through improved experimental design.

Weather and specifically wind play in important role in the precision of the house pressure measurement, especially at the low end of the measurement range. Wind pressures around buildings are quite turbulent and it is reasonable to anticipate that the variations of these pressures will be a substantial fraction of their mean value. Limiting the mean wind conditions of the test is one way to reduce precision errors, but it also reduces the applicability and, therefore, usefulness of the measurement procedure.

Noise can also be reduced by making time-averaged measurements for each pressure station. This can be accomplished by making multiple measurements and averaging as part of the analysis or by having the instrumentation average as part of making the measurement. For example, six measurements of a single pressure can reduce the precision error from 2.5 to 1 Pa. Although it is normally the pressure measurement that contributes the most to the uncertainty, the increase in precision is equally applicable to the flow measurements.

Multipoint Versus Two-Point Testing

Given that we can use the various techniques described above to improve the precision of the measurements, multipoint testing (i.e., regression analysis) is, theoretically, inferior to two-point testing for extrapolating to lower pressures.[6] It is, however, a more robust method because it can bring to light systematic errors not anticipated in the design such as instrumentation failures, changes in the experimental setup (e.g., changes in leakage behavior caused by the pressure), and unusual variations in the parameters.

Optimal design would then include a small number of pressure stations to be (optionally) used in a regression analysis.[7] The precision of the measurement at each pressure station should then be used to weight the regression. It is important to recognize, however, that the uncertainty in the extrapolation will still be heavily dependent on the properties of the lowest pressure point(s).

Location of Low-Pressure Station

The optimal selection of the low-pressure station is a tradeoff between measurement and model specification errors. The precision and (especially) bias considerations of the pressure measurement would suggest that a higher value of the pressure station would provide better estimates. The model specification error due to a nonconstant exponent would suggest a lower value. Table 1 demonstrates these tradeoffs for a particular set of values.

In addition to these considerations it is important that other types of systematic errors be avoided. For example, if the low-pressure station were too low, the model specification error associated with the single pressure difference assumption would have to be considered.

Location of High-Pressure Station

As Table 1 indicates, the uncertainty for that set of assumptions is not highly dependent on the location of the high-pressure station for fixed low-pressure and extrapolated pressure values. There are, however, two other concerns that one should consider in making this selection: the range of measurement and compatibility with other uses.

The range of pressure measurements can be expressed by the ratio of the highest pressure station to the lowest. All else being equal, the larger this ratio is the better is the determination of the exponent and, hence, the extrapolated flow. A large ratio, however, implies that the measurement equipment needs a large dynamic range. Since increasing the dynamic range of an instrument usually increases its error, it is important not to make the ratio too big. Values of the ratio in the 3 to 5 range should work well.

Many practitioners use blower doors to measure the flow at 50 Pa as a construction quality indicator. As discussed in the One-Point Analysis section, the accuracy and precision demands for this purpose are much less than for our purposes. Therefore, it might be useful to have one of the pressure points be at 50 Pa.

Other ASTM test methods, such as the Test Method for Rate of Air Leakage Through Exterior Windows, Curtain Walls, and Doors (ASTM E 283-84) and the Method for Field Measurement of Air Leakage Through Installed Exterior Windows and Doors (ASTM 783-

---

[6] It can be shown the least uncertainty can be achieved when all measurements are concentrated at the end points rather than distributed over the range.

[7] The extra points between the endpoints can cause the extrapolation to get worse if there are certain types of systematic errors in operation; thus the term "optionally."

84) [5], suggest that 75 Pa should be used for the leakage testing of components. Although this value is not required, but only suggested, in these standards similar considerations apply as above.

Reducing Bias and Modelization Errors

Bias errors can be reduced by appropriate choice of instrumentation and experimental design. Instruments should be chosen whose intrinsic bias (caused by such things as nonlinearity and hysteresis) is smaller than the bias due to calibration error and the necessary precision of the measurement. Once the instrument bias is sufficiently reduced, the bias in the pressure measurement is primarily due to weather conditions. To reduce the bias effect of the weather, a few changes in the experimental technique can be made.

The wind spectrum has many frequency components. The highest ones will behave as noise and be part of the precision error. Lower frequencies (i.e., those on the order of the time of the measurements) will appear as drift in the reading made with no flow through the fan (i.e., the effective or relative zero). Therefore, it is important to measure the effective zero before and after each measurement. The pairs of zero measurements can be used both to correct the measurement and to estimate the size of the zero drift bias. Physical pressure averaging devices such as those used with the CGSB [20] standard can help minimize this bias error as well as precision errors.

A steady wind causes a modelization error because it induces differential pressures on the faces of the envelope. These pressures must be kept significantly smaller than any of the pressure stations to avoid introducing another error. As long as the peak variation from building face to building face is less than half the pressure station, the effect can be ignored, especially if the both pressurization and depressurization tests are made. The *preferred conditions* of the current ASTM Standard E 779-87 should normally insure this.

An Improved Measurement Protocol

In this section we propose an improved measurement procedure based on the analyses above. We include instrumentation specifications, experimental procedures, and analysis methods to be used. These recommendations combine both theoretical arguments with practical considerations and thus may not be unique.

Instrumentation Specifications

The flow measuring device must be unbiased to within 2%; that is, all biases including calibration errors and nonlinearities must be no more than 2% of the reading. The intrinsic precision error of the instrument should be no more than 5% of reading; that is, any internally generated (electronic or mechanical) fluctuations may be no more than 5% of the reading.

The pressure measuring device must be unbiased to within 5% of reading, and also to within 1 Pa of the reading, whichever is more restrictive. Similarly, the intrinsic precision error must be no more than 5% of reading or 1 Pa.

Procedure

Because of the need for precision higher than is usually afforded by a single measurement, provision must be made to estimate the mean of both the pressure and the airflow rate at a pressure station and to estimate the error of that mean.

The airflows are to be measured at four pressure stations: 10, 15, 25, 50 Pa.

At each pressure station sufficient data must be taken to assure that the error the mean is no more than 2% of the mean airflow reading and 5% of the mean pressure reading.

Before and after each pressure station data set the zero pressure must be measured to a precision of 5% of the pressure station. The pressure station reading should be corrected by the average of these two values. The corrected mean value of the pressure must be within 10% of the desired station pressure. The difference between the two zero measurements must be within 5% of the station pressure.

Analysis

The data are to be analyzed using a weighted regression technique. We have used synthetic data with the specifications of this section to determine the expected (precision) uncertainties for both the unweighted and weighted techniques:

$$\delta n = 0.02, 0.01$$

$$\delta \kappa = 7\%, 5\%$$

$$\varepsilon = 0.4\%, 0.7\%$$

$$\frac{\delta Q_e}{Q_e} = 4\%, 3\% \tag{23}$$

The uncertainty of the final result has been significantly reduced (compared with the E 779 Standard method, Eq 18), primarily because the measurement uncertainties have more stringent specifications. The advantages of weighting the regression still exist, but are not as apparent due to the improved measurement accuracy.

Discussion and Conclusions

Because we are interested in extrapolating an apparent power-law curve below the lower limit of the measurements, an analysis of uncertainties is dominated by the low-end behavior. The precision, bias and model specification errors must all be minimized.

The precision errors are mostly caused by the environmental variations during the test. For any given set of circumstances the precision error can be minimized by averaging time series measurements at a single pressure station. The number of measurements required to meet a particular precision target will depend on the environmental conditions and other sources of noise. Increasing the number of points in the regression analysis is another way of reducing the overall precision error.

Bias errors cannot usually be reduced with multiple measurement. Rather, the specifications and calibration of the instrumentation and the experimental protocols must be optimized. For example, in order to extrapolate to 4 Pa it is important to keep the pressure bias below 1 Pa. Because it may be difficult to keep the environmentally induced biases below such a level, it is important to make sure that the instrumentation itself has a bias much smaller than that. Thus, instrumentation exhibiting significant zero drift, nonlinearity, or hysteresis must be avoided.

Modelization errors principally affect measurements or extrapolations made near the range of naturally occurring pressures. If measurements are made at too low a pressure—even if the measurements are quite accurate—there may be an error due to the nonlinearities of the leaks. The procedure recommended above, for example, could be used to a wind speed of

3 m/s and a temperature difference of 30 K without inducing that type of modelization error. Similarly, if the measurements are made at too high a pressure there may be a significant uncertainty due to the departure from a true power law.

There does not currently exist a body of sufficiently well-characterized data that would allow an empirical determination of the size of these modelization errors. Typical blower door technology is unable to make the necessary measurements, but a thorough understanding of the low-pressure behavior of the leakage function is necessary to improve the accuracy of the technique over the limits indicated herein. Such an understanding can be gained only by making detailed, high-accuracy measurements across many building types; such measurements may require or lead to the development of novel leakage measurement techniques.

Within the confines of our current knowledge and fan pressurization technology, uncertainty can be optimized by appropriate choices for the pressure stations. The most striking conclusion of this study is that the accuracy of current fan pressurization tests is limited by the uncertainties associated with measuring the lowest pressures. Uncertainty associated with the extrapolation would suggest that measurements be made at low pressures, while the need to minimize percentage uncertainties suggests the measurements be made at high pressures. Equipment limitations suggest the range of pressures should be small, while statistical arguments suggest that a large pressure range might be better. Rules of thumb suggest that the low-pressure point should be in the range of 10 to 20 Pa, the high-pressure point in the 40 to 60 Pa range, and the range should be between a factor of three and five. These rules represent tradeoffs that involve some judgment based on typical equipment specifications and noise sources; other ranges may also prove useful in some circumstances.

There are many factors we did not address in how Standard E 779 is or should be constructed. These factors include issues relating to the preparation of the envelope (e.g., the sealing of vents), whether pressurization or depressurization or both should be carried out, the effects of internal resistance to airflow, the changes in temperature and related effects caused by density changes, and errors caused by specific blower door designs. While all of these issues are potentially important and must be addressed, they are beyond the scope of this report.

Acknowledgment

This work was sponsored by the Residential Ventilation Consortium operated by the Electric Power Research Institute and also supported by the Assistant Secretary for Conservation and Renewable Energy, Office of Building Technology of the U.S. Department of Energy.

References

[1] Sherman, M. H. and Grimsrud, D. T., "Infiltration-Pressurization Correlation: Simplified Physical Modeling," *ASHRAE Transactions,* Vol. 86, No. 2, 1980, pp. 778–807; Lawrence Berkeley Laboratory Report LBL-10163.

[2] *ASHRAE Handbook of Fundamentals,* Chapter 23, American Society of Heating, Refrigerating and Air Conditioning Engineers, 1989.

[3] ASHRAE Standard 119, Air Leakage Performance for Detached Single-Family Residential Buildings, American Society of Heating, Refrigerating and Air Conditioning Engineers, 1988.

[4] ASHRAE Standard 136 (proposed), A Method of Determining Air Change Rates in Detached Dwellings, American Society of Heating, Refrigerating and Air Conditioning Engineers, 1992.

[5] *ASTM Book of Standards,* Vol. 04.07, American Society for Testing and Materials, Philadelphia, 1991.

[6] Walker, I. S. and Wilson, D. J., "AIM 2: The Alberta Air Infiltration Model," University of Alberta, Department of Mechanical Engineering, Report No. 71, Edmonton, Alberta, Canada.

[7] *Air Change Rate and Airtightness in Buildings, ASTM STP 1067,* M. H. Sherman, Ed., American Society for Testing and Materials, Philadelphia, 1990.

[8] *Energy Auditor and Retrofitter,* Vol. 3, No. 5, A. K. Meier, Ed., Energy Auditor & Retrofitter, Inc., Berkeley, CA, 1986.

[9] *Energy Design Update,* Vol. 4, No. 4, N. Nissan, Ed., Cutter Information Corp., Arlington, MA, 1985.

[10] Tarantola, A., *Inverse Problem Theory,* Elsevier, Amsterdam, 1987.

[11] Modera, M. P. and Wilson, D. J., "The Effects of Wind on Residential Building Leakage Measurements," in *Air Change Rate and Airtightness in Buildings, ASTM STP 1067,* M. H. Sherman, Ed., Philadelphia, 1990.

[12] *ASTM Standards on Precision and Bias for Various Applications,* 2nd ed., American Society for Testing and Materials, Philadelphia, 1985.

[13] Murphy, W. E., Colliver, D. G., and Piercy, L. R., "Repeatability and Reproducibility of Fan Pressurization Devices in Measuring Building Air Leakage," *ASHRAE Transactions,* Vol. 97, No. 2, American Society of Heating, Refrigerating and Air Conditioning Engineers, 1991.

[14] Coleman, H. W. and Steele, W. G., *Experimentation and Uncertainty Analysis for Engineers,* Chapter 2, Wiley, New York, 1986.

[15] *Supplement on Instruments and Apparatus: Part 1, Measurement Uncertainty,* ASME PTC 19.1-1985, American Society of Mechanical Engineers, 1990.

[16] Sherman, M. H., "A Power Law Formulation of Laminar Flow in Short Pipes," *Journal of Fluids Engineering,* Vol. 114, No. 4, 1992, pp. 601–605.

[17] Deming, W. E., *Statistical Adjustment of Data,* Dover, New York, 1964.

[18] Weisberg, S., *Applied Linear Regression,* Second ed., Wiley, New York.

[19] Jones, L. E., "Linearizing Weight Factors for Least-Squares Fitting," *Journal of the Hydraulics Division, Proceedings,* American Society of Civil Engineers, 1971, pp. 665–667.

[20] Determination of the Airtightness of Building Envelopes by Fan Depressurization Method, CGSB Standard 149, Canadian General Standards Board, 1986.

Discussion

David Saum[1] (*written discussion*)—Could the measurement analysis options be eliminated on existing real data sets (e.g., D. Wilson's data)?

M. Shermann (*author's closure*)—No. Although the analysis approach can be tailored to the properties of the measurements, the basic principles still apply.

---

[1] Geomet Technologies, Germantown, MD 20874.

Mark P. Modera[1]

Field Comparison of Alternative Techniques for Measuring Air Distribution System Leakage

REFERENCE: Modera, M. P., **"Field Comparison of Alternative Techniques for Measuring Air Distribution System Leakage,"** *Airflow Performance of Building Envelopes, Components, and Systems, ASTM STP 1255,* Mark P. Modera and Andrew K. Persily, Eds., American Society for Testing and Materials, Philadelphia, 1995, pp. 284–298.

ABSTRACT: ASTM has recently standardized a methodology for measuring the leakage of residential air distribution systems to unconditioned zones [ASTM Test Method for Determining the External Air Leakage of Air Distribution Systems by Fan Pressurization (E 1554-93)]. The standard includes two alternative leakage measurement techniques. One technique requires only a blower door, whereas the second technique requires a flow-capture hood as well as a blower door. This paper reports on the results of field measurements in 30 houses using both measurement techniques, and analyzes the relative strengths and weaknesses of the two techniques. The repeatability of each of the techniques, as well as the comparability of the results from the two techniques, is examined. A key issue addressed is the importance of duct pressure measurements in each of the two techniques. Analyses show that the leakage measured with the blower-door-only technique would have been negatively biased by 30% to 50% if the duct pressure had not been incorporated into the measurements and analyses as specified in the standard. Similarly, it is shown that supply leakage measurements with the flow-capture-hood technique would have been negatively biased by 33% if the envelope pressure differential had been used instead of the duct pressure differential.

KEYWORDS: ducts, leakage, airtightness, fan pressurization, measurement, infiltration

Over the past several years there has been a tremendous increase in the number of air leakage measurements on residential air distribution systems [*1–5*]. These measurements have focused mainly on determining the leakage between the duct system and outdoors (including unconditioned spaces such as attics or crawl spaces). As the groups performing these measurements were accustomed to using fan pressurization (that is, blower doors) to measure envelope leakage, the first technique to be commonly used involved blower doors. This technique is based upon subtracting the results from two blower-door tests of building envelope leakage, one with the duct system registers open to the house, the second with the registers sealed. The second technique to come into common use involved the use of a direct measurement of duct leakage flow in combination with the blower door [*1*]. The main impetus for developing the second technique was that the changes in blower-door flows between the sealed and unsealed configurations is generally small compared with the total blower flow, and therefore the duct leakage results can be dramatically affected by the uncertainties

---

[1] Energy and Environment Division, Lawrence Berkeley Laboratory, 1 Cyclotron Road, Berkeley, CA 94720.

in blower-door measurements. Measuring the duct-leakage flow directly with a flow-capture hood can significantly reduce this flow measurement uncertainty.

The American Society for Testing and Materials (ASTM) recently completed the standardization process for the blower-door-only and the combination blower-door/capture-hood measurement techniques [ASTM Test Method for Determining the External Air Leakage of Air Distribution Systems by Fan Pressurization (E 1554-93)]. This paper reports on the results of field measurements in 30 houses using both measurement techniques, and analyzes the relative strengths and weaknesses of the two techniques, as well as the importance of direct duct-pressure measurements for each.

Field Procedure and Apparatus

The results reported in this paper are based upon a field study of 30 houses in California, each of which was submitted to a two-day set of diagnostic measurements designed to characterize the performance of the duct system [3]. The diagnostic measurements performed included leakage measurements following both ASTM measurement methods, measurements of duct pressures and register flows during normal system operation (auxiliary measurements in ASTM Standard E 1554), measurements of pressure differentials across internal doorways with the fan on and the doors closed, and measurements of building air exchange rates with and without the distribution-system fan in operation.

The leakage measurements were performed using a blower door that required angular velocity (rpm) and pressure differential measurements to calculate the flow through the fan, a commercially available flow capture hood, electronic pressure transducers, and a computerized data acquisition system. All measurements were automatically time-block averaged within the computer, including the angular velocity of the fan. This involved simultaneously measuring the blower-door pressure differential, the envelope pressure differential, the blower-door angular velocity at approximately 1 Hz, and automatically recording ten-second average data [6]. The program employed would digitally filter and display on the screen the pressure differential across the building envelope (which utilized a four-wall pressure averaging probe), the pressure difference across the fan, and the angular velocity of the fan. When the operator was satisfied with the pressure differential across the envelope, hitting any key would initiate a time-block average of those three values, and would initiate a series of time-block averages of various other pressure differentials of interest. The additional pressure measurements were made with two transducers. One port of each of the transducers was connected to a rotary-valve pressure-port multiplexer. For one transducer the second port was connected to the main living zone, and for the other transducer its second port was connected to the four-wall pressure averaging probe. The additional pressures measured included the pressures in the supply and return plenums, in the return duct near the register, in the longest and shortest supply ducts near their registers, in the attic, and in the crawl space. The computer would control the multiplexer so as to obtain ten-second averages for each pressure relative to the main living zone and the four-wall outside average. The pressure differentials across duct leaks reported in this paper are referenced to the zone in which the ducts are located. As an example, the pressure differential across return-plenum leaks to the attic would be computed as

$$\Delta P_{leak} = (P_{attic} - P_{4\text{-wall}}) - (P_{retplen} - P_{4\text{-wall}})$$

Although a standard-size flow-capture hood was used to measure register flows and duct leakage flows, particular attention was paid to data precision and accuracy so as to extend its measurement range down to low flows (that is, 10 m³/h). To accomplish this, all flows were determined from the measured pressure differential across the pressure sensing probes provided by the manufacturer, which was read and time-block averaged by the computer. All flow-capture pressure differentials were measured simultaneously with two different pressure transducers built by different manufacturers. To minimize the impact of zero drift in the pressure transducers, the zeroes of the transducers were automatically measured prior to each set of flow measurements (register flows or duct leakage flows).

Blower-Door-Only Protocol (Method A)

The basic protocol in ASTM Standard E1554 for blower-door-only measurements consists of three distinct fan pressurization tests whose results are compared to obtain the leakage of the supply and return ducts separately. The first test is a fan pressurization test with all of the duct registers open to the house. This test measures the leakage of the building envelope plus the entire duct system. For the second fan pressurization test, all return registers are sealed, the return side is sealed from the supply side at the fan, and the supply registers are kept open to the house. This test is intended to measure the leakage of the envelope plus the supply-duct leakage. For the third test, both the supply and return registers are sealed, and the seal between the supply and return sides at the fan is removed. The third test is intended to measure the leakage of the envelope only. The standard specifies that the pressures in the return plenum of the duct system be monitored during the second and third pressurization tests to account for any leakage from the house to the ducts during these sealed-duct tests.

The blower-door-only measurement protocol utilized in the field study differed somewhat from the protocol specified in ASTM Standard E1554. The major differences were that in the field study the fan seal was kept in place during the third pressurization test (that is, when both the supply and return registers are sealed), and the return air filter was not removed in the field study, contrary to the E1554 protocol. Because the fan seal was not removed in the third fan pressurization test, the supply-plenum pressure was not always equal to the return-plenum pressure during the third fan pressurization test. As a result, the supply-plenum pressure was substituted for the return-plenum pressure in the correction factor for the total duct leakage calculations (see Table 1). The impact of not removing the filter during the field tests is not expected to be large for the Method A tests, as the filter should only impact the unsealed-duct tests, and would show up as a pressure drop between the ducts and the house (see discussion under later subsection "Comparison of Method A and Method B"). The impacts of these variations from the ASTM E1554 protocol were generally found to be small for the measurements presented.

Blower-Door/Capture-Hood Protocol (Method B)

The blower-door/capture-hood measurement protocol in ASTM E1554 basically consists of two fan-pressurization tests that use a flow capture hood to directly measure the leakage flow through the ducts. One test measures the leakage of the supply ducts, while the second test measures the leakage of the return ducts. In both cases, all registers except for one are sealed on the duct section (that is, supply or return) to be measured, the supply side is sealed from the return side at the fan, and the flow-capture hood is installed on the unsealed register. The unsealed register is chosen to be the one with the least flow resistance to the plenum, which is the register with the largest flow under normal operating conditions, and is usually the closest to the plenum. Standard E1554 allows the supply and return tests to be performed simultaneously if two flow-capture hoods are utilized. E1554 specifies that the pressure differential across the duct system be separately monitored for use as the pressure reference in the leakage calculations. This duct pressure reference is measured at two locations on the return side [the return plenum and farthest (sealed) return register], two locations on the supply side [the supply plenum and the farthest (sealed) supply register], and is referred to the zone in which the majority of the duct system is located (for example, the attic or the crawl space).

The protocol used in the field study was essentially identical to that specified in the standard, the only differences being that: (1) the return filter was not removed in field tests (contrary to the ASTM E1554 requirement), and (2) in a few houses the return registers were kept sealed during the supply-leakage measurements. The return filter could create nonuniform pressures in the return duct during leakage testing (see discussion alluded to earlier), and the sealed return grilles could potentially increase any leakage across the fan seal during the supply-leakage test.

Data Analysis

The data analysis protocol for the flow-capture method (Method B) is relatively straight-forward; however for the blower-door-only technique (Method A), the data analysis protocol specified in ASTM Standard E1554, and that used to analyze the field data collected, requires some explanation.

The aspect of the E 1554 data analysis for Method A that stands out is the correction factor applied to the subtractions of consecutive blower-door test results. This correction factor accounts for the bias in subtractions of sealed-duct test results from unsealed-test results, which stems from leakage between the house and the theoretically sealed duct. More precisely, even if the duct system registers are perfectly sealed, if there are leaks in sections of ductwork that pass through conditioned spaces, air will flow through those leaks from the house into the duct system, and then from the duct system to unconditioned spaces through the duct leaks to those spaces. Thus, the flow through the external duct leaks is not zero, even when the registers are sealed. The correction is based on treating the house/duct system as a series combination of leaks. More specifically, the measured flow through the blower door fan while the duct system remains unsealed (first fan pressurization test) can be expressed as:

$$Q_{FAN_{unseal}}(\Delta P) = Q_{ENV}(\Delta P) + Q_{duct}(\Delta P_{duct_{unseal}})$$ (1)

while the fan flow when the ducts are sealed (second or third test) can be expressed as

$$Q_{FAN_{seal}}(\Delta P) = Q_{ENV}(\Delta P) + Q_{duct}(\Delta P_{duct_{seal}})$$ (2)

Subtracting Eq 2 from Eq 1 yields

$$\Delta Q_{FAN}(\Delta P) = Q_{duct}(\Delta P_{duct_{unseal}}) - Q_{duct}(\Delta P_{duct_{seal}})$$ (3)

which, assuming that $\Delta P_{duct\ unseal}$ is equal to ΔP (see cited discussion), and that the flow through the duct leaks is described by a power law, can be reexpressed as

$$\Delta Q_{\text{FAN}}(\Delta P) = Q_{\text{duct}}(\Delta P) \left(1 - \left(\frac{\Delta P_{\text{duct}_{\text{seal}}}}{\Delta P} \right)^{n_{\text{duct}}} \right) \tag{4}$$

Solving Eq 4 for $Q_{\text{duct}}(\Delta P)$ yields

$$Q_{\text{duct}}(\Delta P) = \Delta Q_{\text{FAN}}(\Delta P) \left(\frac{1}{1 - \left(\frac{\Delta P_{\text{duct}_{\text{seal}}}}{\Delta P} \right)^{n_{\text{duct}}}} \right) \tag{5}$$

The term in parentheses in Eq 5 is the correction term incorporated into ASTM E1554. It is used to correct the return duct leakage obtained by subtracting results of the second fan pressurization test (that is, sealed return grilles and sealed fan) from the first fan pressurization test results (that is, all grilles open). It is also used to correct the total duct leakage obtained by subtracting the results of the third fan pressurization test (that is, sealed return and supply grilles) from the first fan pressurization test results (all grilles open). In both cases, $\Delta P_{\text{duct seal}}$ is the average pressure differential across the duct leaks (between the duct section and its surroundings) during the test in which that duct section is theoretically sealed from the house.

Measurement Results

The measurements performed in this study were designed to examine the consistency of the results obtained with Method A and Method B, to examine the comparability of the results obtained with the two techniques, and to examine the sources of bias or uncertainties in the two techniques that affect both consistency and comparability. In addition, as mentioned above, E1554 requires the measurement of several pressures that have not consistently been measured as part of duct leakage testing. In particular, the pressures at the supply plenum, the furthest supply register, a return grille, and the return plenum are required to be measured for both the blower-door subtraction method (Method A) and the flow-capture/blower-door method (Method B). To provide some evidence of the importance of making these additional pressure measurements, the ratios of these pressures to the indoor-outdoor pressure difference, and their ultimate impact on the leakage results, are also examined. Finally, the results of some of the auxiliary measurements required by the standard are presented in order to provide an expected range of results and point out the potential utility of these measurements.

Method A Results

The supply-side and return-side duct leakage measured by the blower-door subtraction method are summarized in Figs. 1 through 3, in which the leakage areas at 4 Pa and the leakage flows at 25 Pa are compared between pressurization and depressurization for supply and return leakage, and the total leakage flow at 25 Pa is compared between pressurization and depressurization. Examination of these figures brings out two points: (1) there is reasonably good agreement between pressurization and depressurization results, with no apparent bias, and (2) the scatter between pressurization and depressurization results is significantly smaller for the flows at 25 Pa compared with the leakage areas at 4 Pa. It is worth noting that pressurization results need not necessarily agree with depressurization results, as one can easily imagine that some duct leaks could act like valves, changing their effective

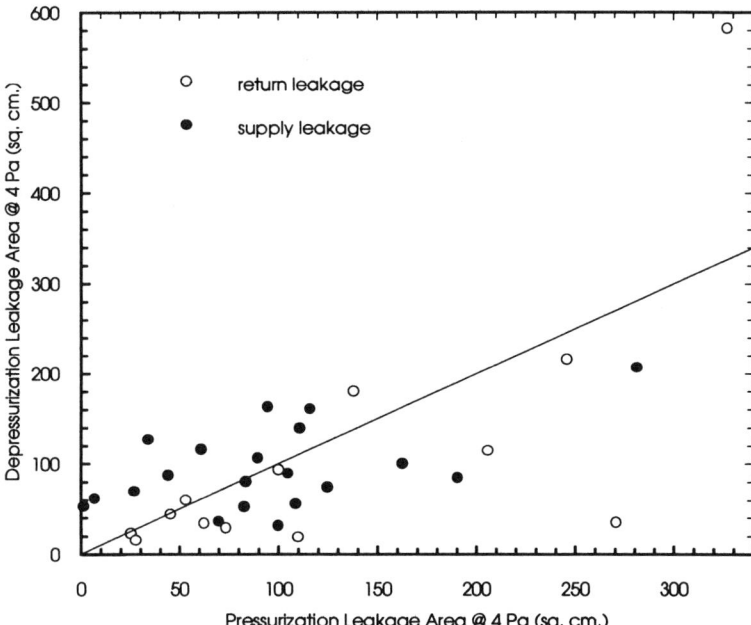

FIG. 1—*Comparison of pressurization and depressurization results for return duct and supply duct leakage areas at 4 Pa obtained with Method A* (*blower-door subtraction*).

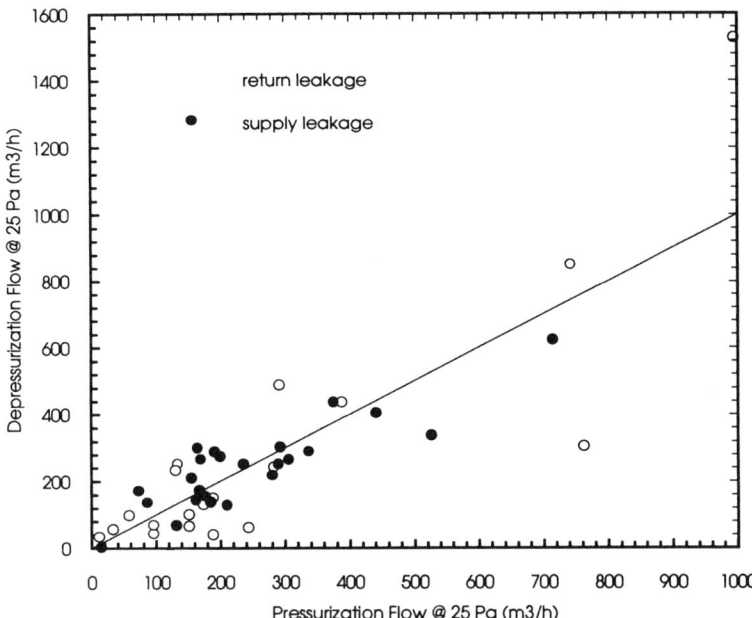

FIG. 2—*Comparison of pressurization and depressurization results for return duct and supply duct leakage flows at 25 Pa obtained with Method A* (*blower-door subtraction*).

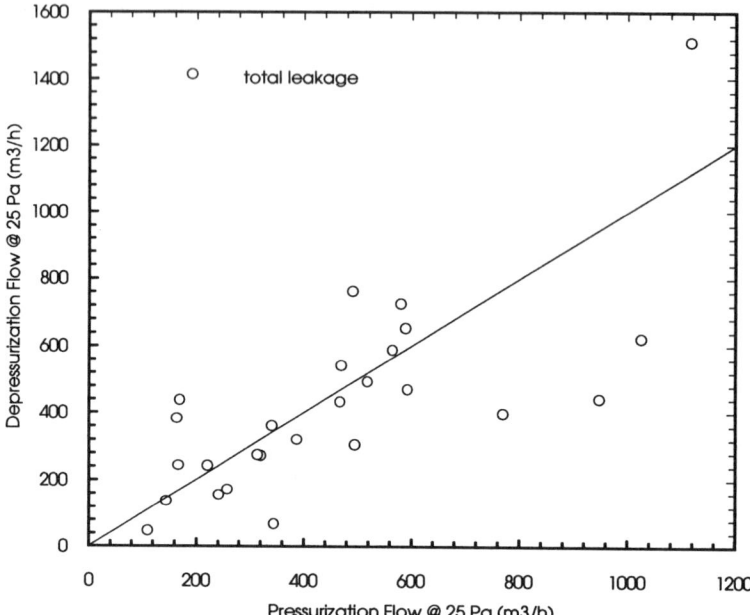

FIG. 3—*Comparison of pressurization and depressurization results for total duct leakage flows at 25 Pa obtained with Method A (blower-door subtraction).*

leakage area due to flexing under pressurization or depressurization. The good agreement between pressurization and depressurization results suggests that this may in fact be a small effect.

The houses for which subtraction results yielded negative leakage areas are not included in Figs. 1 through 3. For the effective leakage area at 4 Pa, 15 out of 56 measurements yielded negative return leakage estimates based upon subtraction results, and 10 out of 56 subtractions yielded negative supply leakage areas. For the 25 Pa leakage flows, the comparable results were 9 negative return results and 5 negative supply results. It should be noted that the supply leakage results were based upon subtracting zero from the total duct leakage results whenever the return leakage was calculated to be negative. This treatment of the data assures that there will not be any cases in which the supply leakage could be computed to be larger than the total leakage of the duct system.

The better performance of the measurement technique at 25 Pa is not surprising, as those results are less sensitive to the uncertainty in the measured flow exponent because they do not depend upon extrapolation outside the measurement range. In general, these negative results are somewhat discouraging, as they imply null measurement results 22% of the time for leakage areas, and 13% of the time for the leakage flow at 25 Pa. Based upon the premise that Method A performs better for leakier duct systems in tight houses due to the larger percentage uncertainties associated with taking small differences of large numbers, the distribution of negative results was checked relative to the ratios of duct leakage to envelope leakage. The results were that two thirds of the negative results were in the lowest 50% of duct-to-envelope leakage ratios, both for the effective leakage area and 25-Pa flow results. As would be expected based on their larger magnitude, the total duct leakage results [based upon subtractions of the results of the third pressurization test (all registers sealed) from the first test (all registers open)] only turn out to be negative in 4 instances out of 56 leakage area calculations (7%) and in 2 out of 56 calculations of 25-Pa leakage flow (4%). A comparison of pressurization and depressurization results for total duct leakage is shown in Fig. 3.

To examine the importance of measuring duct pressures during sealed-duct conditions, the magnitude of the pressures observed in the ducts and the corrections based on Eq 5 are summarized in Table 1.

The results in Table 1 are quite dramatic, basically indicating that return duct leakage would be underestimated by almost 50% [100 (1 − 1/(1.89)) = 47%] on average if the correction factor is not applied to the data, and that the equivalent underprediction for the total leakage would be almost 30% on average. The maximum underpredictions are as much as 76% for the return leakage, and 64% for the supply leakage. These results indicate that the measurement of duct pressures during sealed-duct conditions is a critical component of the standard. It is worth noting that the field measurements performed differ from those specified in ASTM E1554 in that the fan seal was not removed for the third pressurization test (that is, when both supply and return grilles are sealed) in the field test. It is principally for this reason that the return pressures are not equal to the supply pressures in the third fan pressurization test, during which both the supply and return grilles were sealed (the supply-plenum pressures were used for the corrections). From the results in Table 1, one can also infer that the fan seal was not perfect, as indicated by the change in return-side duct pressures between the second and third tests. Any leakage at the fan seal is exacerbated by the fact that the pressure difference across that seal can change sign between the two tests.

Method B Results

The supply-side and return-side duct leakage measured by the flow-capture-hood method are summarized in Figs. 4 and 5, in which the leakage areas at 4 Pa and the leakage flows at 25 Pa are compared between pressurization and depressurization. Examination of these figures brings out two points: (1) the agreement between pressurization and depressurization results is even better than that observed with the blower-door subtraction method, and (2) once again, the scatter between pressurization and depressurization results is significantly smaller for the flows at 25 Pa compared with the leakage areas at 4 Pa. Once again, it is

TABLE 1—*Ratios of duct pressure differentials to envelope pressure differentials during sealed-duct conditions, and associated correction factors for pressurization tests (Method A) (28 houses).*

Pressure Ratio (Test)	Mean Ratio	Maximum Ratio	Minimum Ratio	Mean Correction Factor (median)
Return grille/outdoor to indoor/outdoor (sealed return)	0.27	0.66	0.01	1.89 (1.85)
Return grille/outdoor to indoor/outdoor (sealed supply and return)	0.17	0.57	−0.01	···
Supply plenum/outdoor to indoor/outdoor (sealed supply and return)	0.14	0.52	−0.02	1.41 (1.30)

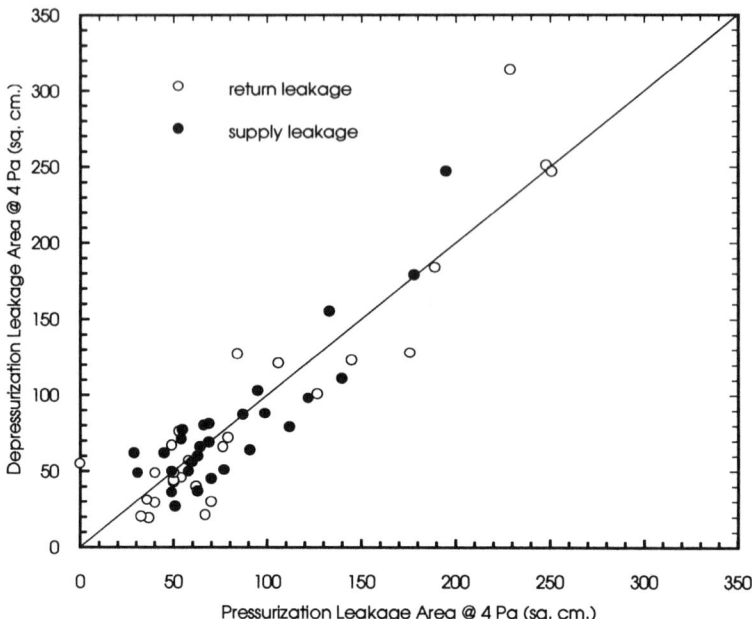

FIG. 4—*Comparison of pressurization and depressurization results for return duct and supply duc leakage areas at 4 Pa obtained with Method B (flow-capture hood).*

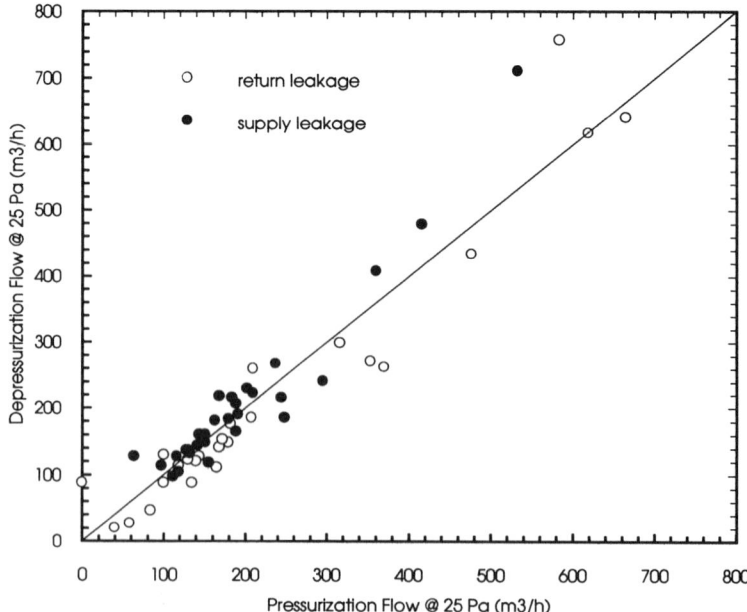

FIG. 5—*Comparison of pressurization and depressurization results for return duct and supply duct leakage flows at 25 Pa obtained with Method B (flow-capture hood).*

worth noting that due to valve-like leak behavior, pressurization results need not necessarily agree with depressurization results, so the good agreement between pressurization and de-pressurization results suggests that valve-like behavior may be negligible.

For the flow-capture-hood measurements (Method B), the measured pressure difference across the ducts is used directly in the flow/pressure regressions used to characterize the duct leakage. In Table 2, the return-duct pressure differentials in the return leakage tests, and the supply duct pressure differentials in the supply leakage tests, are compared with the envelope pressure differentials measured simultaneously during these tests.

The results in Table 2 clearly demonstrate the importance of direct measurements of duct pressures for determining duct leakage characteristics with Method B. The pressure differ-ential is generally not dramatic for the return ducts; however, in at least one case the pressure differential across the ducts is only two-thirds that across the envelope. The effect is clearly more pronounced for supply duct measurements, which is due to the larger resistance of supply registers and ducts to airflow. More specifically, there appears to be a large pressure drop across the supply grille to which the capture hood is connected, after which there is a non-negligible pressure drop through that supply duct, and a negligible pressure drop after the supply plenum. This is not surprising, as the resistance of the supply grille is large compared to the duct resistance, which in turn is large compared to the resistance of all the remaining ducts in parallel. To understand the impact of not measuring duct pressure, the results in Table 2 indicate that the calculated supply duct leakage would be 33% low on average if the house pressure were substituted for the measured duct pressure, and that the underprediction could be as much as 76%.

Comparison of Method A and Method B

The duct leakage flows at 25 Pa determined with Method A and Method B are compared in Figs. 6 and 7, as well as in Table 3. Figure 6 is a scatter plot of Method A results against Method B results for supply and return leakage flows at 25 Pa, and Fig. 7 is a similar plot for total duct leakage flows at 25 Pa. What is clear in both figures is that the agreement between the two techniques is much worse than the agreement between pressurization and depressurization results for either of the two techniques. The negative results obtained with Method A are included in these figures, as they make it clear that the frequency of null results (that is, negative duct leakage values) is much smaller for the total duct leakage compared to that for supply or return leakage only. These results are disappointing but not that surprising when one considers that the best precision to be expected on a regular basis

TABLE 2—*Ratios of duct pressure differentials to envelope pressure differentials for capture-hood tests of duct leakage (Method B).*

Pressure Differential Ratio	Mean	Minimum	Maximum
Return grille/outdoor to indoor/outdoor (28 houses)	0.94	0.63	1.0
Return plenum/outdoor to indoor/outdoor (25 houses)	0.91	0.65	1.0
Near supply grille/outdoor to indoor/outdoor (30 houses)	0.65	0.32	1.0
Supply plenum/outdoor to indoor/outdoor (30 houses)	0.54	0.11	1.0
Far supply grille/outdoor to indoor/outdoor (30 houses)	0.53	0.11	0.99

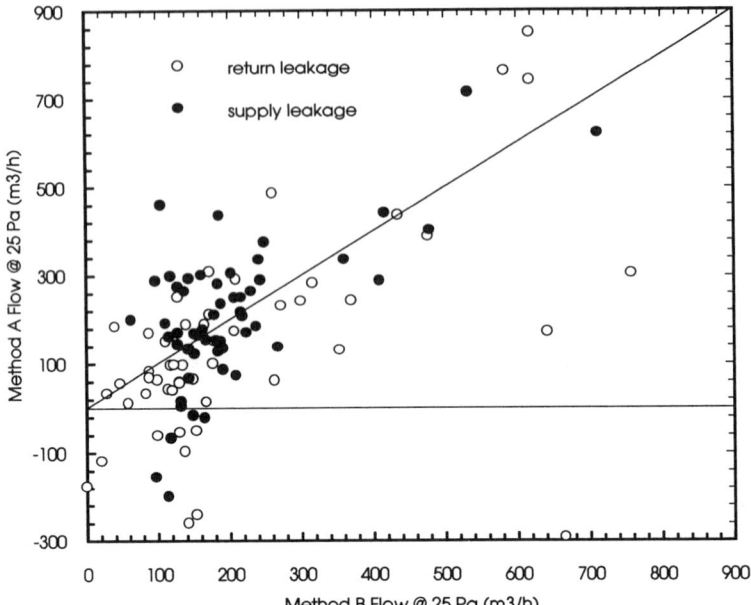

FIG. 6—*Comparison of return duct and supply duct leakage flows at 25 Pa obtained with Method A (blower-door subtraction) and Method B (flow-capture hood).*

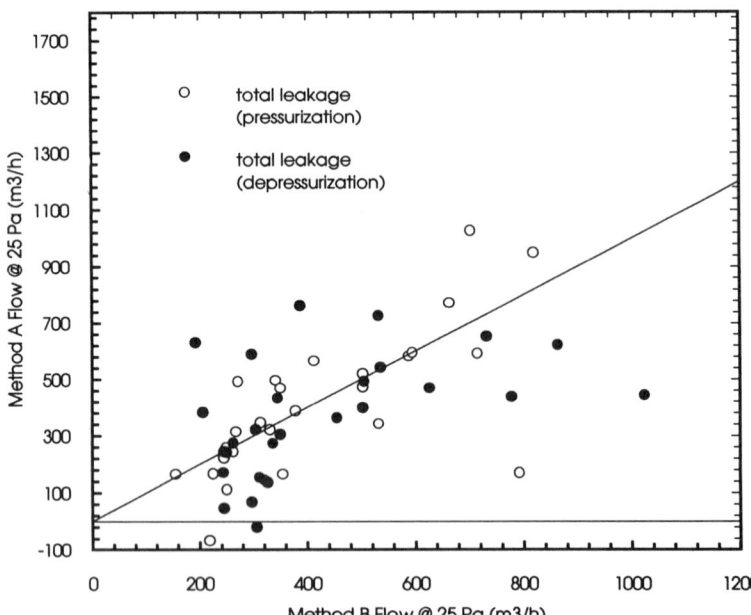

FIG. 7—*Comparison of total duct leakage flows at 25 Pa obtained with Method A (blower-door subtraction) and Method B (flow-capture hood).*

TABLE 3—*Summary of total duct leakage flows at 25 Pa obtained with Method A and Method B (28 houses).*

Data Set	Mean Duct Leakage, m³/h	Median Duct Leakage, m³/h	Minimum Duct Leakage, m³/h	Maximum Duct Leakage, m³/h
Total leakage with Method A (pressurization)	400	345	−66	1025
Total leakage with Method B (pressurization)	421	351	155	820
Total leakage with Method A (depressurization)	376	382	−21	761
Total leakage with Method B (depressurization)	425	337	193	1025

for fan pressurization tests would be 3% for the leakage area at 4 Pa [6]. For Method A, this uncertainty translates to an average uncertainty of 27% for total duct leakage and 53% for either supply or return leakage, based upon the observed average fraction of envelope leakage in the ducts of 16% (split about evenly between supply and return).

For comparison purposes, although a similarly rigorous uncertainty analysis has not been performed for Method B, the major influence on the uncertainty of that technique is the extrapolation to 4 Pa. The basic sources of uncertainty—the uncertainty in the pressure measurement in the duct/house and the uncertainty in the flow measurement with the flow-capture hood—are both significantly smaller in magnitude than the pressure and flow uncertainties associated with Method A. An analysis of the uniformity of the pressures across the duct leaks for the supply and return side measurements indicated that there is a bias in the supply-side results due to the pressure drop between the duct fitted with the flow-capture hood and the supply plenum (see discussion below), but that the pressures in the remainder of the supply side and the pressures in the return side during the return leakage test are quite uniform. The ratio of the far supply register pressure to the supply-side plenum pressure had a mean value of 0.99 and a standard deviation of 7%, and the ratio of the return register pressure to the return plenum pressure had a mean value of 1.00 and a standard deviation of 8%. Based upon these results, and an estimated uncertainty of 5% in the flow measurements, the uncertainty in the Method B results should be on the order of 10%.

Table 3 presents a simple statistical summary of the data in Fig. 7.

The results in Fig. 3 suggest that despite the large scatter between the Method A and Method B results in Figs. 6 and 7, there does not seem to be a significant bias between the two measurement techniques. On the other hand, it is worth noting that several sources of bias in the two techniques were identified from the data taken in this field study. Specifically, the assumption in Method A that the pressure in the ducts is the same as the house pressure when the registers are not sealed was checked. On average, the pressure differentials across the duct leaks were 3% lower than the pressure differentials across the building shell, although in one case the difference was as much as 15%. Propagating this result through the data analysis procedure indicates that the Method A results are biased high by 2% to 3% on average.

The Method B supply-side data suggest that those results are biased somewhat high due to nonuniformities in duct pressures during the leakage tests. More specifically, the pressure differential across the leaks in the supply duct that is fitted with the flow-capture hood is on average 11% higher than the average value used for the remainder of the duct system. Assuming that that duct represents 10% of the duct system, and that the flow exponent is 0.65, the positive bias in the overall leakage result would be 0.7%. The Method B return-side data indicate that the duct pressure was relatively uniform for the return-side leakage tests. Excluding two houses which had pressure drops of 24% and 38% between the return register and plenum, the mean ratio of the return-grille pressure to the return-plenum pressure was 1.00, and the average of these two pressures is used for the analysis. One possible explanation for the two extreme results is dirty return filters, which are specified to be removed as part of ASTM E1554.

Another potential source of bias in Method B is a negative bias due to leaks in the fan seal. Such leaks would tend to negatively bias both the supply and return leakage, and therefore the total duct leakage. This underprediction results from the fact that it is implicitly assumed that all of the flow through the duct leaks is passing through the flow-capture hood. However, if the fan seal is leaking, it is leaking in parallel with the flow-capture hood, thereby causing an underestimation of the flow through the duct leaks. It should also be noted that any leakage between the house and the duct system has the same effect as fan-seal leakage on Method B results. The results from Method A in Table 1 made it clear that, even with the care taken in the field test to seal the fan as well as possible, the fan seal can be expected to leak. The ramification of this observation is that duct leakage results will tend to be underestimated with Method B. The size of this effect is dependent on the pressure difference across the fan seal and the duct-to-house leaks. An analysis of the auxiliary data collected indicates that for the return leakage tests the return-plenum pressure was on average 9% lower than the supply plenum pressure (that is, the ratio of the differential between the return and its surroundings to the supply surroundings differential was 0.91 on average), suggesting that leakage across the fan seal should be small for these tests. On the other hand, for the majority of the supply tests (18 tests with complete data in which the return grilles were not blocked), the supply plenum pressure was 31% lower than the return plenum pressure on average, suggesting that there could be significant leakage across the fan seal, thereby causing a negative bias. For the houses in which the return grilles were kept sealed during the supply-leakage tests (7 tests), the return plenum pressures were sometimes much lower than the supply plenum pressures, in which case the results would be positively biased by any fan-seal leakage. Thus, for the sample reported on in this paper, underprediction due to fan-seal leaks could matter for the supply-side leakage, but is unlikely to have had a significant impact on the return-side measurements.

Auxiliary Measurement Results

ASTM Standard E1554 also specifies that duct-system pressures be measured under normal distribution-fan operation. The results of these measurements in the field study are summarized in Table 4.

The results in Table 4 bring to light several points: (1) The pressures across duct system leaks during normal fan operation are significantly higher than those across leaks in building shells; (2) the pressure differentials across return leaks are typically higher than those across supply leaks; and (3) there is a large variability in operating pressures across duct leaks. The third observation is most pertinent to ASTM E1554, as it points out the importance of performing the auxiliary duct pressure measurements if one would like to be able to estimate the leakage rate of a given duct system during normal operation.

TABLE 4—*Pressure differences between ducts and their surroundings (for example, attic, crawl space, garage) during normal system operation.*

Location	Mean Value, Pa	Standard Deviation, Pa	Minimum, Pa	Maximum, Pa
Supply plenum	46	28	9	138
Supply duct average	29	17	7	83
Return plenum	−88	43	−14	−181
Return duct average	−57	31	−5	−126

Supply duct average = 2 × plenum + (near register) + (far register)/4.
Return duct average = (plenum + grille)/2.

Discussion

Based upon the data collected and analyzed, two potential variations of Method B merit some discussion. Both of these variations address the potential negative bias due to fan-seal leakage. The first variation would be to measure total duct leakage and return-duct leakage, rather than supply leakage and return leakage. The second variation, which could be combined with the first, would be to actively control the pressure in the ducts so as to make it equal to the house pressure.

The first variation is based upon the results in Table 2, which point out that the pressure drop through the return grille was less than 10% of the total pressure drop, whereas the pressure drop through the supply grille and the short adjoining duct was almost 50% of the pressure drop across the envelope. The potential modification to the standard would be to measure return leakage with the fan sealed and then to measure total duct leakage with the fan unsealed and the flow-capture hood installed on the return grille. Under these circumstances, the pressure is likely to be far more uniform throughout the duct system, and should be fairly close to the pressure in the house (as observed for the return side). With such a protocol, the supply leakage would be obtained by subtracting the measured return leakage from the measured total leakage. It should be noted, however, that the results in Table 2 stem from the type of duct-system construction observed in California. That is, large central return grilles and ducts are typically used in sunbelt houses instead of multiple smaller return grilles, which are more prevalent in some regions of the country.

The second variation on Method B would be to use a fan in conjunction with the flow capture hood to assure that the average pressure inside the duct system is equal to the pressure inside the house. This modification should dramatically reduce the fan-seal leakage and house-to-duct leakage problems discussed above by making the pressure difference across those extraneous leaks close to zero. The best situation might be to combine the fan-assist with the switch to measuring flows only on the return side, in which case it seems that a uniform pressure equal to that in the house could be achieved for the entire duct system (unless the duct system is very leaky).

Conclusions

Several conclusions can be drawn based upon the results presented in this paper. First and foremost is that internal duct-pressure measurements, and appropriate analysis procedures based upon those measurements, are necessary to provide reliable unbiased estimates of duct leakage to unconditioned spaces. This was demonstrated to be the case both for Method A and Method B. Another important conclusion is that although no significant bias was observed between the two methods, Method B seemed to be superior, both in its better internal consistency between pressurization and depressurization results, and in that it does not provide negative duct leakage results. In addition, although Method A has more scatter, neither technique showed any significant bias between pressurization and depressurization results, suggesting that valve-type leakage was not significant. Finally, and not surprisingly, the data presented also made it clear that the reproducibility and the degree of physical confidence in the results were significantly better for the leakage flow at 25 Pa versus the effective leakage area at 4 Pa.

Acknowledgments

The research reported here was funded in part by the California Institute for Energy Efficiency (CIEE), a research unit of the University of California. Publication of research results does not imply CIEE endorsement of or agreement with these findings, nor that of any CIEE sponsor. This work was also supported by the Assistant Secretary for Conservation and Renewable Energy, Office of Building Technologies, of the U.S. Department of Energy under Contract No. DE-AC03-76SF00098.

References

[1] Robison,.D. H. and Lambert, L. A., "Field Investigation of Residential Infiltration and Heating Duct Leakage," *ASHRAE Transactions,* Vol. 96, Part 2, 1989.

[2] Cummings, J. B., Tooley, J. J. Jr., and Dunsmore, R., "Impacts of Duct Leakage on Infiltration Rates, Space Conditioning Energy Use, and Peak Electrical Demand in Florida Homes," in *Proceedings,* ACEEE Summer Study, Pacific Grove, CA, American Council for an Energy Efficient Economy, Aug. 1990.

[3] Modera, M. P., "Characterizing the Performance of Residential Air Distribution Systems," *Energy and Buildings,* Vol. 20, No. 1, 1993, pp. 65–75. (Also, Lawrence Berkeley Laboratory Report LBL-32532, 1991).

[4] Kinert, R. C., Engel, D. C., Proctor, J. P., and Pernick, R. K., "The PG&E Model Energy Communities Program: Offsetting Localized T&D Expenditures with Targeted DSM," in *Proceedings,* ACEEE Summer Study, Pacific Grove, CA, American Council for an Energy Efficient Economy, Aug. 1992.

[5] Palmiter, L. and Bond, T., "Impact of Mechanical Systems on Ventilation and Infiltration in Homes," in *Proceedings,* ACEEE Summer Study, Pacific Grove, CA, American Council for an Energy Efficient Economy, Aug. 1992.

[6] Modera, M. P. and Wilson, D. J., "The Effects of Wind on Residential Building Leakage Measurements," *Air Change Rate and Air Tightness in Buildings, ASTM STP 1067,* M. H. Sherman, Ed., American Society for Testing and Materials, Philadelphia, 1990, pp. 132–145. (Also, Lawrence Berkeley Laboratory Report LBL-24195).

Author Index

Subject Index

A